Animals, Health, and Society

CRC One Health One Welfare

Learning from Disease in Pets: A 'One Health' Model for Discovery
Edited by Rebecca A. Krimins

Animals, Health, and Society: Health Promotion, Harm Reduction, and Health Equity in a One Health World
Edited by Craig Stephen

Animals, Health, and Society

Health Promotion, Harm Reduction, and Health Equity in a One Health World

Edited by

Craig Stephen

CRC Press
Taylor & Francis Group
Boca Raton London New York

CRC Press is an imprint of the
Taylor & Francis Group, an **informa** business

First edition published 2021
by CRC Press
6000 Broken Sound Parkway NW, Suite 300, Boca Raton, FL 33487-2742
and by CRC Press
2 Park Square, Milton Park, Abingdon, Oxon, OX14 4RN

Library of Congress Cataloging-in-Publication Data

Names: Stephen, Craig, editor.
Title: Animals, health, and society: health promotion, harm reduction, and health equity in a One Health world/edited by Craig Stephen.
Description: First edition. | Boca Raton: Taylor & Francis, 2021. | Series: CRC One Health one welfare | Includes bibliographical references and index. | Summary: "This timely book reframes the historic narrative of people, animals, and nature as risks to each other, to one where we think about health as a shared capacity. This new narrative promotes the positive contributions made to health across species and generations and addresses growing calls to shift from a reactive to proactive approach in One Health. Editor Craig Stephen takes the reader on a tour of the situations wherein we can all, regardless of our job description, work across species, sectors, and generations to motivate action. Perspectives and methods from a variety of fields and experts are shared and adapted to promote collaborative understanding of and action on determinants of health at the animal-society interface. Case studies demonstrate that the principles and practices presented are feasible, empowering people to make choices that concurrently benefit the health of animals, societies, and ecosystems. The first book to adapt and explain health promotion, harm reduction, and health equity issues in a One Health context, and in terms of animal health, this is necessary reading for students of and practitioners working in planetary health, conservation, EcoHealth, public health, health promotion, veterinary medicine, and animal welfare"-- Provided by publisher.
Identifiers: LCCN 2020032525 (print) | LCCN 2020032526 (ebook) | ISBN 9780367336226 (paperback) | ISBN 9780367642600 (hardback) | ISBN 9780429320873 (ebook)
Subjects: LCSH: Veterinary public health. | Animal welfare. | Environmental health.
Classification: LCC SF740.A57 2021 (print) | LCC SF740 (ebook) | DDC 636.08/32—dc23
LC record available at https://lccn.loc.gov/2020032525LC ebook record available at https://lccn.loc.gov/2020032526

ISBN: 978-0-367-64260-0 (hbk)
ISBN: 978-0-367-33622-6 (pbk)
ISBN: 978-0-429-32087-3 (ebk)

Typeset in Times LT Std
by KnowledgeWorks Global Ltd.

Contents

Foreword

David Waltner-Toews

In the midst of the wreckage wrought by decades of reckless economic development and pandemics viewed as enemy armies, Craig Stephen and his co-authors show us how to creatively reorganize ourselves for a more convivial future.

For many decades, those working in public health, environmental management, and veterinary public health have been warning the rest of us about the collapse of ecosystems, the extinction of species, climate catastrophes, and the coming, apparently inevitable, waves of diseases of people, animals, and plants that will sweep the globe. The literature and practitioner communities and platforms are replete with jeremiads, slogans, and calls to think and work across disciplinary boundaries, to rise above our expertise. In 2020, many have been wringing their hands, too deeply despairing to even whisper a "told you so."

Animals, Health, and Society: Health Promotion, Harm Reduction, and Health Equity in a One Health World is feminist narrative therapy for our dysfunctional human family, acknowledging our troubled past, but now saying: okay, we've survived, so where do we go from here? In this book, Craig Stephen and his colleagues grab us firmly by the shoulders, turn us around, and shift our gaze from the disheartening chaos around us, and energize us to take on the tasks of building a more convivial multi-species future.

With just the right mix of theory and practice, and drawing on tools and practices from health promotion, harm reduction, health equity, and Indigenous traditions, these authors would have us consider not just human health, but the fate of urban rats and invasive monkeys, unloveable lampreys, plastic waste, and climate change.

As Dr. Stephen says in the Preface, "This book strives to broaden the continuum of care we offer in One Health and allied fields by encouraging health-focused approaches that emphasize building capacities, skills, and resources needed to cope with the Anthropocene in advance of harms."

Animals, Health, and Society: Health Promotion, Harm Reduction, and Health Equity in a One Health World is the perfect book for this moment in time – one of those rare moments when we have an opportunity to collectively and collaboratively reframe destruction into creative destruction and reorganization. Where some see only ends, these authors would have us consider beginnings. This book should be required reading for any aspiring physicians, veterinarians, natural resource managers, and public health workers.

Preface

The purpose of this book is to encourage us to borrow, adapt, and modify perspectives, methods, and tools across different health sectors to empower people to make choices that concurrently benefit the health of animals, societies, and ecosystems.

This collection of chapters provided by a diverse group of authors hopes to inspire people to see win-win-win solutions when confronting health challenges in a One Health world. Decisions usually have winners and losers. Too often in the past and in the present, those with power and with a voice win, while others like other species and future generations lose. The global forces shaping societies and ecosystems no longer allow us the luxury of having an unbalanced tally sheet of winners and losers. Concurrently, promoting and protecting the health of ourselves, each other, and our fellow travellers on Earth now must be a priority. Reacting to harms as they emerge is no longer a viable or sustainable option in the Anthropocene. The harms are coming too fast, are spread too far over space and time, and are compounding, preventing human and animal communities to keep up and continue to thrive in a time of unprecedented change. Many books on animal health, One Health, and ecosystems health have taken a deficits approach, focusing on adverse outcomes like diseases and hazards like pathogens or pollutants. This book strives to broaden the continuum of care we offer in One Health and allied fields by encouraging health-focused approaches that emphasize building capacities, skills, and resources needed to cope with the Anthropocene in advance of harms.

This book is organized in three parts.

PART 1 – MAKING THE CASE FOR HEALTH AND RECIPROCAL CARE

Chapters 1–4 make the case for an expanded focus for health promotion and harm reduction on reciprocal care wherein we work collaboratively to create the circumstances that allow us to make decisions that encourage care of ourselves, our societies, and Nature so that we move towards win-win-win actions. These chapters introduce the ideas of health, health promotion, and reciprocal care from an interspecies and intergenerational perspective.

PART 2 – CORE CONCEPTS

Chapters 5–13 introduce tools, skills, perspectives, and approaches that can help expand ideas of harm reduction and health promotion beyond their usual sphere of influence in human health into a wider world of reciprocal care across species and generations. These chapters introduce foundational ideas that can help researchers, practitioners, and managers design, implement and advocate for projects, policies, and programs to comprehensively promote resilience by concurrently and collaboratively tending to the determinants of health and resilience, for each other, our communities, animals, and our shared environments.

PART 3 – ACTIONS AND APPLICATIONS

Chapters 14–23 offer case studies of the application of the core concepts and foundational ideas presented in earlier chapters. They demonstrate how these ideas can be helpful in practice and show their relevance across species, context, and generations. By showing the feasibility and utility of a health promotion and harm reduction in a One Health world, these chapters aim to inspire people to creatively adapt and share ideas, successes, and techniques across the spectrum of public health, conservation, and animal health to generate shared benefits that promote and sustain health across species and time.

Acknowledgements

This book would not have been possible without the willingness of each contributing author to share their experience, expertise, and ideas. Each has helped push the boundaries of One Health and expand the tools and opportunities we can use to make the world a healthy place for all species.

The encouragement, support, and conversations with Joy Wade were indispensable and irreplaceable parts of this book and my life.

Editor

Craig Stephen, DVM, PhD, has been working at the interface of human, animal, and environmental health for over 25 years. Most recently, he was Professor at the Western College of Veterinary Medicine at the University of Saskatchewan and Clinical Professor at the School of Population and Public Health at the University of British Columbia. He has served in a number of adjunct and affiliate academic positions, including Clinical Professor at the Ross University School of Veterinary Medicine. Dr. Stephen is a veterinarian and epidemiologist who has held a variety of One Health leadership positions, including being the Founding President and Director of the Centre for Coastal Health, the Scientific Director of the Animal Determinants of Emerging Diseases Research Network, the Scientific Director of the British Columbia Occupational and Environmental Health Network, a Canada Research Chair in Integrating Human and Animal Health, and the CEO of the Canadian Wildlife Health Cooperative. His work has focused on applying population health principles to a variety of human, wildlife, fish, and domestic animal populations dealing with issues from food sustainability and poverty reduction to conservation to emerging environmental threats.

Contributors

Evan Adams
Office of the Chief Medical Officer
First Nations Health Authority
British Columbia, Canada

Lindsay Beck
Office of the Chief Medical Officer
First Nations Health Authority
British Columbia, Canada

Daniele Behn-Smith
Office of the Provincial Health Officer
Ministry of Health
British Columbia, Canada

John Berezowski
Institute of Veterinary Public Health
University of Bern
Bern, Switzerland

Ryan K. Brook
College of Agriculture and Bioresources
University of Saskatchewan
Saskatchewan, Canada

David Butler-Jones
Health Gnosis Inc.
Ontario, Canada

Luis Pedro Carmo
Institute of Veterinary Public Health
University of Bern
Bern, Switzerland

Carolee Carson
Laboratory for Foodborne Zoonoses
Public Health Agency of Canada
Ontario, Canada

Luis Cruz-Martinez
Biomedical Sciences
Ross University School of Veterinary Medicine
St. Kitts and Nevis

Pierre-Yves Daoust
Department of Pathology and Microbiology
Atlantic Veterinary College
University of Prince Edward Island
Prince Edward Island, Canada

Colleen Duncan
Department of Microbiology Immunology, and Pathology
College of Veterinary Medicine
Colorado State University
Fort Collins, Colorado

Tricia Fry
School of Veterinary Medicine
University of Wisconsin–Madison
Madison, Wisconsin

Christa Gallagher
Biomedical Sciences
Ross University School of Veterinary Medicine
St. Kitts and Nevis

Maya Gislason
Faculty of Health Sciences
Simon Fraser University
British Columbia, Canada

Bonnie Henry
Office of the Provincial Health Officer
Ministry of Health
British Columbia, Canada

Luis-Pablo Hervé-Claude
Departamento de Ciencias Clínicas
Facultad de Ciencias Veterinarias y Pecuarias
Universidad de Chile
Santiago, Chile

Chelsea Gardner Himsworth
School of Population and Public Health
University of British Columbia
and
British Columbia Ministry of Agriculture
British Columbia, Canada

Dawn Hoogeveen
Department of Geography
University of Northern British Columbia
British Columbia, Canada

Harmony Johnson
Harmony Johnson Consulting
British Columbia, Canada

Perry Kendall
Public Health Consultant
British Columbia, Canada

Irene Lambraki
School of Public Health and Health Systems
University of Waterloo
Ontario, Canada

Shannon E. Majowicz
School of Public Health and Health Systems
University of Waterloo
Ontario, Canada

Margot Parkes
School of Health Sciences
University of Northern British Columbia
British Columbia, Canada

E. Jane Parmley
Department of Population Medicine
Ontario Veterinary College
University of Guelph
Ontario, Canada

Pushpakumara D.B. Nihal
Department of Wildlife Conservation
Wildlife Health Management Centre
Konduwatuwana, Sri Lanka

Colin Robertson
Department of Geography and Environmental Studies
Wilfrid Laurier University
Ontario, Canada

Todd K. Shury
Parks Canada
Saskatchewan, Canada

Krista Stelkia
Faculty of Health Sciences
Simon Fraser University
British Columbia, Canada

David Stephen
Max Plank Institute for Quantum Optics
Munich, Germany

Craig Stephen
School of Population and Public Health
University of British Columbia
and
Ross University School of Veterinary Medicine
St. Kitts and Nevis

Joy Wade
Fundy Aquaservices Inc.
British Columbia, Canada

Part 1

Making the Case for Health and Reciprocal Care

1 The Call to Action

Craig Stephen

CONTENTS

The guiding premise of this book is that health is the product of our relationships with each other and the world around us. This is not new. Hippocrates wrote over 2,000 years ago about the web of social and environmental connections that influence health and well-being. Long-held aboriginal perspectives of health and wellness are founded on the interdependence of the land, culture, self, and community (Hill, 2009). Aldo Leopold, one of the most widely read Western environmentalists of the 1900s, saw the health of the land connected to the health of human and non-human communities. The unprecedented rate and global reach of social and ecological changes occurring today are reinvigorating a worldview of connectedness wherein health is the outcome of the interplay of animals, environments, and societies.

We are living in the Anthropocene which is the name for the current geological age when people have the dominant influence on climate and the environment. The Anthropocene is characterized by the Great Acceleration which refers to the rapid and sometimes exponential growth of human impacts, starting in the 1950s characterized by increased carbon pollution, consumption, human population growth, habitat loss, species extinction, and more. The exponential growth of the human population along with ever-increasing global movements of people, goods, and biota are creating landscapes that generate new and sometimes unforeseen health risks shared by people, animals, and environments. In 2017, the United Nations Department of Economic and Social Affairs projected a human population of 8.6 billion in 2030, 9.8 billion in 2050, and 11.2 billion in 2100 (UN, 2017). This anticipated rate of population growth, along with a growing middle class, will put extraordinary strains on the ecological services provided by our biotic and abiotic environments, which in turn will diminish the resources needed for healthy and sustainable biodiversity. The authors of the 2014 5th Intergovernmental Panel on Climate Change assessment report were very confident that global warming will lead to very high risks of severe, widespread, and irreversible impacts on individuals, ecosystems, and communities by the end of the 21st century. The United Nations Intergovernmental Science-Policy Platform on Biodiversity and Ecosystem Services concluded in 2018 that we were losing species 1,000 times faster than the natural rate of extinction. The 2016 Living Planet Index linked these losses to human-induced factors: habitat degradation, invasive species, climate change, pollution, unsustainable freshwater use, and species overexploitation. Accelerating environmental and human behavioural changes have been creating new ecological niches that drive microbial evolution and epidemiological shifts that have fostered the emergence of new infectious diseases, like the COVID-19 pandemic of 2020. Persistent poverty, food insecurity, and urbanization remain major challenges to achieving sustainable development. Many international initiatives are rising to address these shared challenges through multisectoral action plans and strategies (Table 1.1). Success in these plans will require people capable of working across boundaries and able to inspire collaborative actions in the animal, environment, social, and health sectors.

This book asks: What actions can we take at the animal-society interface to foster reciprocal care of the health of ourselves our communities and the life with which we share Earth? The socio-ecological model of health is instrumental in answering this question. The origins of the socio-ecological model can be found in the Lalonde report (1974), "A New Perspective on the Health of Canadians," which concluded that lifestyle and the environment contributed to our health as much or more than health care and biology. It was followed in 1986 by the Ottawa Charter for Health Promotion which repositioned health as a positive aspiration to pursue rather than the result of negative consequences avoided. It extended the reach of health management beyond the health care system to the upstream social and ecological factors that provide the opportunities and capacities to live a healthy life. The prevailing public health perspective for understanding what makes a population healthy or not is based on the interactions and contributions of abiotic, biotic, and social elements that determine health outcomes. The population health perspective focuses on the state and interactions among the many contributing factors that influence health rather than measuring health as a physiological state. A similar conception of health began appearing in the animal health literature in the 2010s (e.g. Wittrock et al., 2019; Hanisch et al., 2012) (Figure 1.1).

Population welfare, whether human or otherwise, is coherence between a species' adapted capacities and expectations and the realities of its current social and biophysical environments (Stephen and Wade, 2018). Health management is being reframed as a collaborative enterprise that continually creates and improves physical and social environments that provide the raw material for people and non-human communities to mutually support each other's health.

The socio-ecological model of health sees health as a series of interconnected, co-dependent, and interacting factors – in other words, as a system (Diez Roux, 2011). This creates challenges to those wishing to study health. The fact that ecosystems are ever-changing and human systems undergo

TABLE 1.1
Examples of International Agreements Dependent on Co-Management, Cooperation, and Reciprocal Care of Social, Animal, and Environmental Health

Agreement	Purpose	Focus
Global Health Security Agenda (2014)	Help create a world safe and secure from infectious disease threats	Multisectoral "action packages" to build capacity to prevent, detect, and respond to infectious diseases and thereby contain threats at their source
Convention on Biological Diversity. Strategic Plan for Biodiversity (2011–2020)	Conserve and sustain the fair and equitable use of biological diversity	Strategies to promote living in harmony with nature by (i) initiating action to address the underlying causes of biodiversity loss; (ii) ensuring that biodiversity concerns are mainstreamed throughout government and society; (iii) acting to decrease the direct pressures on biodiversity; (iv) protecting access to ecosystem services, especially for the poor who most directly depend on them
United Nations Sustainable Development Goals (2015)	Peace and prosperity for people and the planet, now and into the future	End poverty and other deprivations by working together with strategies to improve health and education, reduce inequality, and spur economic growth – all while tackling climate change and working to preserve oceans and forests
United Nations Framework Convention on Climate Change – 2016 Paris Agreement	Preventing dangerous anthropogenic interference with the Earth's climate system by strengthening the global response in the context of sustainable development and poverty reduction	Peaking and reducing greenhouse gas emissions, protecting carbon sinks found in biodiversity, strengthening resilience, and reducing vulnerability to climate change, with a view to contributing to sustainable development
United Nations Conference on Environment and Development – Agenda 21 (1992)	Rethink economic development and find ways to halt the destruction of irreplaceable natural resources and pollution of the planet	Rights and responsibilities of nations in environmental protection and sustainable development. Balancing natural resource use with preserving the environment and ensuring natural resources for future generations

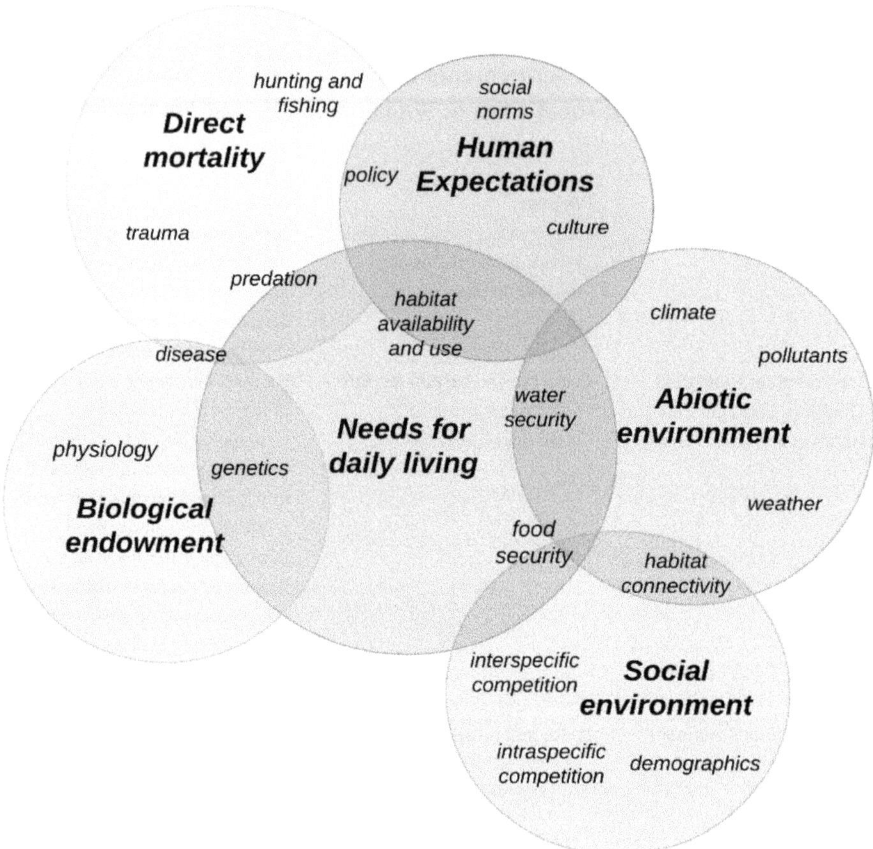

FIGURE 1.1 Illustrative determinants of health model for fish and wildlife. Categories of determinants are in the circles, with illustrative contributing determinants within each circle. The size, shape, interaction, and contributions of each determinant will vary with species, ecosystem, and social context. (Adapted from Wittrock et al., 2019.)

ongoing transformations makes socio-ecological systems inherently unknowable, unpredictable, and not well suited to research seeking a mechanistic truth in a reductionist fashion. Health gains made in the 19th and 20th centuries came largely from advances in knowledge on individual drivers of death and diseases. Discrete disciplines emerged and grew to further our understanding of health and diseases by analysing their distinct components. Humanity can be proud of the increases in longevity and quality of life experienced by an ever-increasing proportion of people; of advances in animal health, welfare, and productivity that increase access to safe and nutritious food; and in reductions in poverty and suffering that has come with improving living standards. Improvements in human well-being have unfortunately and too often come at the expense of wild and domesticated biodiversity and environments. The Millennium Ecosystem

Assessment (2005) showed how human changes to ecosystems have resulted in a substantial and largely irreversible loss in the diversity of life on Earth, which in turn is diminishing the essential ecological services upon which we depend. Biodiversity and ecosystems have responded in kind. New emerging infections, lost income opportunities due to pests and invasive species, and the climate's responses to carbon pollution are examples of how nature has reacted by inadvertently producing new threats to humanity, biodiversity, and living systems. Our responses to these emerging threats are too often too late. The focus on clinical care in human health and on eliminating hazards from animals and environments after they emerge put us in a constant state of reaction and recovery rather than investing in the protection of the assets we need to stay healthy. It is increasingly being recognized that focusing health care services on reducing diseases and prolonging life through clinically oriented services is insufficient to protect well-being and ensure health of subsequent generations and all species.

The good news is that there is a way forward. The solutions to modern health problems require us to embrace health as a product of interacting and interdependent animal, human, and environmental systems. Governments, businesses, and communities are recognizing the advantages of integrated, proactive approaches that focus on protecting the social and environmental services that lay at the heart of healthy people, animals, and systems. Several new forms of collaboration emerged towards the end of the 20th century and began nudging researchers and practitioners back towards seeing health as a positive socio-ecological phenomenon that should be viewed as a whole and not reduced to its parts. Whether termed EcoHealth, One Health, Planetary Health, or something else, the last two decades saw a rapid expansion in the number of people and programs dedicated to promoting health equity across populations, species, and generations.

SYSTEMS THINKING AND HEALTH

Systems thinking began in the 19th century but flourished in the later part of the 20th century in fields as varied as biology, physics, psychology, management, and computer science. All were looking for new ways to find a common understanding about complex issues. Rather than reducing an issue to its parts, systems thinking explores interconnections and relationships between the parts to better anticipate the system's behaviour and work more efficiently and proactively towards shared goals. Systems thinking is used to better understand how a system works, including how new phenomena can emerge that are more than the sum of their parts. Systems thinking encourages us to expand the boundaries of what is being studied as well as the set of people involved in generating that knowledge.

The call for systems thinking in health sciences came from the frustration that evidence derived from highly controlled trials was artificial and failed to represent the large numbers and variety of variables and circumstances that influence efforts to modify population health outcomes (Green, 2006). Past failure of single discipline programs to inspire actions against emerging infectious diseases, for example, fostered the need for multisectoral, systems-based methods (Burns and Stephen, 2015). The inadequacy of wildlife health management decisions to protect wildlife health has been linked to failures to appreciate the unpredictability of complex systems or to incorporate ethical and cultural dimensions of decisions (Stephen, 2016). Conservation biologists turned to systems thinking to inspire cross-scale efforts that include the many perspectives and stakeholders needed to support cross-cultural conservation efforts that lead to effective, acceptable, and adaptive co-management (Berkes, 2004). Government agencies responsible for Pacific salmon management in Canada, for example, have been encourage to wed social sciences with cumulative impact assessments to generate the necessary research findings to manage salmon health as a product of socio-ecological interactions rather than investing exclusively in detecting hazards that cannot practically be managed in free-ranging marine species (Stephen, 2016). The inadequacy of fragmented and mechanistic science leading to sustainability pointed sustainable development researchers towards systems thinking (Hjorth and Bagheri, 2006). Mechanistic and fragmented public policy has been identified as an impediment to effective food security programs (Ashe and Sonnino, 2013). New systems research and policy approaches are needed at the animal-health-society nexus: approaches that recognize that context matters and that there is not a single truth out there that can be seen, understood, and controlled in a rational manner.

THE PROBLEM OF TOO MANY PROBLEMS

Health, as it is experienced by people and animals in their daily lives, is influenced by numerous interacting factors. Food sustainability, emerging diseases, habitat alteration, pollution, climate change, social conflict, and other determinants of health happen concurrently rather than in isolation. Many health challenges faced by people intersect those challenging animals and vice versa. It seems reasonable, therefore, to propose that an interdisciplinary approach to a single problem is insufficient to protect health in a sustainable way. Population health research and management needs to evolve from an interdisciplinary approach to single problems to one that is "interprobleminary" – an approach that examines the interactions and implications of multiple problems occurring simultaneously in a place or population.

In a world of concurrent problems, unique solutions for each problem are neither feasible nor effective (Fried et al., 2012). The actions needed to keep populations healthy are found in interacting determinants of health in a shared social and physical space. If we remain fixated on discovering the "scientific laws of health" to predict the next problem rather than building resilience against the next inevitable unknown problem, we will get farther behind. The goal of an interprobleminary approach is not to build more and more complex models to predict the next hazard, but rather to build robustness against negative events that occur and be able to exploit positive ones. Achieving success will require researchers and practitioners to act as information brokers who can connect specialized pools of knowledge. The socio-ecological approach asks us to think of problems in terms of the places and spaces shared by people and animals in which everyday health occurs and focus on the conditions that influence their capacity to cope with the multiple interacting problems. The looming challenges of climate change, depletion of ecological services, and exponentially growing human population suggest that achieving animal, environmental, and human health by separate science, policies, and actions will be impossible.

ANIMALS, HEALTH, AND SOCIETY

Ours is an animal society. Most of the world's poor still depend on subsistence agriculture and fisheries to meet their daily nutritional and income needs. Animal protein fuels the diets of the growing middle class and wealthy. Animals play a major role in aboriginal cultures, income in rural and remote communities, our mental health, and scientific innovations. Conversely, there are few animals that can escape mankind's influence. Our dietary preferences, natural resource management, and attitudes to animal welfare and rights greatly influence how animals can access the resources they need to be healthy. Anthropogenic factors weigh heavily in the list of major threats to fish and wildlife. The decisions we make for housing and husbandry of our pets and agricultural animals determine their fate. Which species we protect is dictated by sociopolitical considerations, resource limitations, and social values. Popularity too often plays an illogical but heavily weighted role in how we decide to protect animal health (Kirkwood and Sainsbury, 1996). Conservationists unavoidably grapple with conflicting social and economic issues that impede actions to secure critical resources that meet the evolved needs and social expectations for wildlife.

The contributions of animals to our well-being can be illustrated by examining the implications of sick wildlife. Wildlife diseases can affect individuals (e.g. influence survival or fitness), economies (e.g. impact agricultural trade), communities (e.g. increase social conflict), psychologies (e.g. create fear of natural places), ecosystems (e.g. altering functions as species decline), and politics (e.g. create policy priorities and spending conflicts) (Stephen et al., 2018). Wildlife acts as sentinels of public health threats (Kuiken et al., 2005) and supports people's mental health and sense of community (Berto, 2014). Aboriginal communities derive an enormous cultural meaning from wildlife, and many rural and remote communities rely on wildlife for sustenance and food security (Stephen and Duncan, 2017). Fishing, hunting, guiding, and tourism, both recreationally and commercially, are huge industries upon which many people rely for income (POC, 2015). Wildlife provides ecosystem services with health and economic implications, such as in the case of bats whose massive consumption of insect pests saves the agriculture and forestry industries billions of dollars in pesticides (Boyles et al., 2011), in turn, reducing workers and communities exposure to toxic pesticides. The inseparable links between these varying harms and benefits and the array of factors that determine and modify these links, coupled with tremendous

BOX 1.1 AN EXAMPLE OF SOCIO-ECOLOGICAL THINKING IN HEALTH ACTION

CASE STUDY: VACCINE EFFECTIVENESS THROUGH A HEALTH IN SOCIO-ECOLOGICAL SYSTEMS (HSES) LENS

Zinsstag et al. (2011) use the term HSES to describe the perspective of animal and human health as quantitative and qualitative products of socio-ecological systems in which there is no distinction between the social and ecological. The authors illustrated the utility of this concept by applying it to the design and assessment of a vaccine program under real-world conditions. A vaccine's effectiveness can be influenced by biological efficacy of the vaccine (which can be influenced by how well the diseases ecology in the field matches conditions under which the vaccine was developed), the immune status of the vaccine recipient (which can be influenced by host nutritional status, genetics, or concurrent diseases), the availability and affordability of the vaccine, personal beliefs in vaccination, equitable access to the health care system delivering the vaccine (which in turn can be influenced by gender, race, or species), and the accuracy of diagnoses leading to vaccination (which is a product of the equality and accessibility of the health care system). An interdisciplinary, interprobleminary systems view is needed to see the critical boundaries and connections influencing vaccine program effectiveness.

challenges in eliminating the drivers of harm, suggest multi-level intervention, ranging from influencing policy to targeted biomedical interventions, are needed.

Despite the many co-dependencies between human and animal health, most policy and scholarship acts as if human and animal determinants of health operated independently. What, where, and how we choose to look at a problem shapes our opportunities to see critical connections that influence options for action (e.g. Box 1.1). Our ability to find a common ground and connections between disciplines, species, and stakeholders can often be the factor that determines if we can find common ground for collaborative action that spans interests (Best et al., 2003). An uncritical acceptance of boundaries can marginalize one interest (e.g. wildlife conservation) to benefit another (e.g. zoonotic disease control).

IS THERE A RIGHT WAY TO DO THIS?

Socio-ecological systems are too complex to ever be fully known or understood (Checkland, 2005). It is futile to try to know all the facets of a health-producing system. What then is the goal of socio-ecological systems research or action at the animal, health, and society interface?

One Health, health promotion, global health, conservation medicine, EcoHealth, and related fields can be classified as multisectoral, socio-ecological, systems-based, collaborative (MSSC) approaches. This group of related approaches is linked by a goal of integrating knowledge from multiple sectors, species, and/or disciplines in a collaborative fashion to restore, protect, or improve health. Despite their growing use and profile, they have relatively limited evidence to use to select the types of problems for which they may be most useful. There remain significant challenges of bridging different disciplines and perspectives. Some of these challenges are linked to collaboration, while others are concerned with how to integrate, weigh, and assess different types of information to develop a holistic view of health. And yet the profile of MSSC approaches, their acceptance, and application continue to grow.

MSSC work is of mixed pedigree. It relies on multiple disciplines and methods with many underlying theories rather than being its own disciplines with its own theoretical foundation. Unlike physics or chemistry, there are no general laws that determine how various determinants of health combine to create healthy circumstances, populations, or individuals. The relative weight of each subdiscipline within an MSSC project varies based on the problem, the funder, and the team. This results in a variety of conceptual understandings and applications, which in turn makes it almost impossible to generate enough evidence that is comparable and transferable between studies and settings. Thus, it is hard to find definitions of the scope of practice for each MSSC approach that are universally accepted among their practitioners.

Each type of MSSC approach has a different "brand" (Table 1.2). The problems they work on and their approaches to finding improvements vary. Less complex systems with well-defined and confined problems that can be easily identified can often be solved by using a more reductionist approach. Usually a few people are involved, and there is agreement on what defines the problem and how to improve the situation. The goal for action for less complex problems tends to favour identifying the main problem and the best solution (Naaldenberg et al., 2009). As more people, species, parts, interactions, and connections increase, the origin of problems gets harder to identify and there can be multiple entry points to find a solution. The values and perspectives of the increasing number of actors in these more complicated and complex problems start to vary, leading to different opinions on the nature of the problem and its solution. There is rarely one main problem in more complex systems, nor is there one easy fix. The goal in these situations is not to find the single solution but rather to find collective actions that can make progressive improvements towards a more sustainable and equitably healthy system.

The proposed value of MSSC work is twofold. Firstly, simultaneous enquiry about different aspects of a problem may better reflect how that problem is manifest in a "real-world" setting and thus recommendations arising from investigations may be better suited to identify the implications of actions in response to the problem. Secondly, by engaging people whose actions influence (or can be influenced by) the problem, the time between discovery and knowledge mobilization

TABLE 1.2
Examples of Multisectoral, Socio-Ecological, Systems-Based Collaborative (MSSC) Approaches Working at the Animal, Human, and Environmental Health Interface

Approach	Description
Veterinary Public Health	Applies veterinary science to protect and improve the physical, mental, and social well-being of people
One Health	Promotes multiple sectors to work together to design and implement programs, policies, legislation, and research to achieve optimal health outcomes recognizing the interconnection between people, animals, and their shared environment
Conservation Medicine	Combines health and ecology to understand the ecological context of health and remediate ecological health problems. It recognizes the role of health in biodiversity conservation and the functioning of ecosystems as well as the effects of changing species diversity or abundance on disease maintenance and transmission
Environmental Health	Protects against environmental factors that may adversely impact health or the ecological balances essential to long-term human health and environmental quality
Health Promotion	Enables people to increase control over, and to improve, their health by focusing beyond individual behaviour towards a wide range of social and environmental interventions
Ecosystem Health	Examines and manages the relationships between biophysical changes and the social structure and economic sustainability of human communities in terms of their influence on ecosystem vigour, organization, and resilience
EcoHealth	Aims for sustainable human and animal health and well-being through healthier ecosystems by connecting ideas of environmental and social determinants of health with those of ecology and systems thinking in an action-research framework
Global Health	Promotes worldwide improvement of health for all and protection against global threats that disregard national borders through collaborative transnational research and action
Planetary Health	Aims for the highest attainable standard of human health, well-being, and equity worldwide through attention to the human systems that shape the future of civilization and the state of the natural systems on which it depends

Many of these approaches lack definitive definitions. The descriptions herein provide a sense of the focus of each approach.

may shorten. Although there is no clear consensus on what defines an MSSC approach, their pragmatic, problem-solving focus may be the lens through which to identify when MSSC methods can be useful.

Pragmatic research is action-oriented and serves to inspire change under real-world conditions (Stephen et al., 2016). The goal of pragmatic research is to nominate effective interventions under realistic conditions rather than discover mechanisms of causation. It is more concerned with clarifying meaning by tracing out practical consequences than on insisting on finding the antecedent factors that explain why a phenomenon occurred (Glasgow, 2013). An MSSC practitioner considers if her/his approach to a problem can produce evidence deemed credible, believable, and evocative by the targeted knowledge users. Pragmatism's commitment to uncertainty acknowledges that causal relationships are transient and hard to identify in a changing world, allowing researchers to be flexible and open to emerging phenomena (Feilzer, 2010). Pragmatic researchers focus on the problem rather than the methods and use whichever methods allow them to understand and act to improve the problem.

MSSC approaches share some features (Burns and Stephen, 2015). They see health as a cumulative effect of interactions between individuals with systems. They are applied in the setting where results will be used under conditions experienced by the populations of concern. They have an action-oriented ethic. MSSC approaches strive to inspire effective, ethical, and sustainable actions to address problems important to people in the affected settings. Paying attention to settings, circumstances, and context is, therefore, paramount. MSSC practitioners must negotiate scientific uncertainty and conflicting values, both of which can impede action. There is often a strong focus on accelerating the integration of research with policy and practices to promote prompt problem-solving action. Sometimes they need to discover how new insights and innovations can break the status quo. Finally, they need to see health as a team sport. MSSC practitioners and researchers need to be adept at directing and working with teams towards shared goals. They serve better as the coach than the star of the team.

WHY THIS BOOK?

A new generation of problem solvers is needed to confront 21st-century health challenges shared by animals and societies. The purpose of this book is to show how working together, in concert with the non-human world, we can reconnect our health and animal health with the health of the world around us and by doing so transform our ailing socio-ecological systems into healthy ones. Innovative and disruptive approaches are needed to address shared health threats such as climate change, urbanization, and pollution, because business as usual is insufficient to inspire the necessary actions to protect the health of one species without risking the health of another. This book introduces and adapts systems thinking, health promotion, health equity, and harm reduction concepts for an audience with the interests and positions to align research, policy, and practice to protect and sustain the co-dependence of animals, health, and society.

Protecting health at the human-animal environment nexus is not a straightforward technical process. It requires complicated social processes that allow a variety of information, ideas, and decisions to come together. Disruption to the status quo should lead to innovations and actions that incrementally return us to a sustainable, healthy coexistence with our fellow creatures and environments. Disruption can come from rethinking the roles for health researchers and practitioners as collaborative problem solvers who comfortably work across boundaries to develop a new and shared sense of meaning among disparate groups. The following chapters help us rethink our roles by reframing the historic narrative of people, animals, and nature as opposing forces and risks to each other to thinking about health as reciprocal care wherein we promote the positive contributions we make to each other's health by sustaining shared determinants and drivers of health. The issues and examples highlighted in this book provide a tour of the landscape wherein the specialized generalists can contribute. These are people who, regardless of their job description, can work across species, sectors, and generations to motivate action to protect the healthy co-dependence of animals and societies.

Principles, perspectives, and methods are shared and adapted in subsequent chapters to promote collaborative understanding of and action on upstream determinants of health at the animal-society interface. The case studies found in later chapters help the reader gain confidence that the presented principles and practices are feasible and useful. You will find that there are few new theories or methods presented throughout this book. People familiar with health promotion, sustainable agriculture, One Health, EcoHealth, and other MSSC approaches will find familiar concepts in the following chapters. The authors have set out to adapt and apply those familiar concepts in unfamiliar settings and in doing so discover unexamined ways to act together to mitigate harms from, or build resilience to, shared challenges confronting animals, health, and society.

REFERENCES

Ashe, Leah M., and Roberta Sonnino. "At the Crossroads: New Paradigms of Food Security, Public Health Nutrition and School Food." *Public Health Nutrition*, 16, no. 6 (2013): 1020–1027.

Berkes, Fikret. "Rethinking Community-Based Conservation." *Conservation Biology*, 18, no. 3 (2004): 621–630.

Berto, Rita. "The Role of Nature in Coping with Psycho-Physiological Stress: A Literature Review on Restorativeness." *Behavioral Sciences*, 4, no. 4 (2014): 394–409.

Best, Allan, Gregg Moor, Bev Holmes, Pamela I. Clark, Ted Bruce, Scott Leischow, Kaye Buchholz, et al. "Health Promotion Dissemination and Systems Thinking: Towards an Integrative Model." *American Journal of Health Behavior*, 27, no. 1 (2003): S206–S216.

Boyles, Justin G., Paul M. Cryan, Gary F. McCracken, and Thomas H. Kunz. "Economic Importance of Bats in Agriculture." *Science*, 332, no. 6025 (2011): 41–42.

Burns, Theresa E., and Craig Stephen. "Finding a Place for Systems-Based, Collaborative Research in Emerging Disease Research in Asia." *EcoHealth*, 12, no. 4 (2015): 672–684.

Checkland, Peter. *Systems Thinking, Systems Practice* (Chichester: John Wiley & Sons, Inc., 2005), 330.

Diez Roux, Ana V. "Complex Systems Thinking and Current Impasses in Health Disparities Research." *American Journal of Public Health*, 101, no. 9 (2011): 1627–1634.

Feilzer, Martina Y.. "Doing Mixed Methods Research Pragmatically: Implications for the Rediscovery of Pragmatism as a Research Paradigm." *Journal of Mixed Methods Research*, 4, no. 1 (2010): 6–16.

Fried, Linda P., Peter Piot, Julio J. Frenk, Antoine Flahault, and Richard Parker. "Global Public Health Leadership for the Twenty-First Century: Towards Improved Health of All Populations." *Global Public Health*, 7, no. suppl.1 (2012): S5–S15.

Glasgow, Russell E. "What Does It Mean to Be Pragmatic? Pragmatic Methods, Measures, and Models to Facilitate Research Translation." *Health Education and Behaviour*, 40, no. 3 (2013): 257–265.

Green, Lawrence W. "Public Health Asks of Systems Science: To Advance Our Evidence-Based Practice, Can You Help Us Get More Practice-Based Evidence?" *American Journal of Public Health*, 96, no. 3 (2006): 406–409.

Hanisch, Shauna L., Shawn J. Riley, and Michael P. Nelson. "Promoting Wildlife Health or Fighting Wildlife Disease: Insights from History, Philosophy, and Science." *Wildlife Society Bulletin*, 36, no. 3 (2012): 477–482.

Hill, Dawn Martin. "Traditional Medicine and Restoration of Wellness Strategies." *Increasing the Safety and Wellness of Women Impacted by Violence* (2009): 62.

Hjorth, Peder, and Ali Bagheri. "Navigating Towards Sustainable Development: A System Dynamics Approach." *Futures*, 38, no. 1 (2006): 74–92.

Kirkwood, James K., and Anthony W. Sainsbury. "Ethics of Interventions for the Welfare of Free-Living Wild Animals." *Animal Welfare*, 5 (1996): 235–244.

Kuiken, Thijs, Frederick A. Leighton, Ron A.M. Fouchier, James W. LeDuc, Joseph S. Malik Peiris, Alejandro Schudel, Klaus Stöhr et al. "Pathogen Surveillance in Animals." *Science*, 309, no. 5741 (2005): 1680–1681.

Naaldenberg, Jenneken, Lenneke Vaandrager, Maria Koelen, Anne-Marie Wagemakers, Hans Saan, and Kees de Hoog. "Elaborating on Systems Thinking in Health Promotion Practice." *Global Health Promotion*, 16, no. 1 (2009): 39–47.

POC. "Licenced Hunting and Trapping in Canada." Parliament of Canada, Government of Canada, 2015. http://www.parl.gc.ca/HousePublications/Publication.aspx?DocId=8045718&Mode=1&Parl=41&Ses=2&Language=E (accessed March 2020).

Stephen, Craig, "Using One Health to Benefit Animals: The Case of Declining Sockeye Salmon," in *One Health Case Studies: Addressing Complex Problems in a Changing World*, eds. Susan C. Cork, David C. Hall, and Karen A. Lijebjelke (Sheffield: 5M Publishing, 2016), 160–175.

Stephen, Craig, Theresa Burns, and Ana Riviere-Cinnamond. "Pragmatism (or Realism) in Research: Is There an EcoHealth Scope of Practice?" *EcoHealth*, 13, no. 2 (2016): 230–233.

Stephen, Craig, and Colleen Duncan. "Can Wildlife Surveillance Contribute to Public Health Preparedness for Climate Change? A Canadian Perspective." *Climatic Change*, 141, no. 2 (2017): 259–271.

Stephen, Craig, and Joy Wade. "Wildlife Population Welfare as Coherence between Adapted Capacities and Environmental Realities: A Case Study of Threatened Lamprey on Vancouver Island." *Frontiers in Veterinary Science*, 5 (2018): 227.

Stephen, Craig, Julie Wittrock, and Joy Wade. "Using a Harm Reduction Approach in an Environmental Case Study of Fish and Wildlife Health." *EcoHealth*, 15, no. 1 (2018): 4–7.

United Nations. “World Population Prospects: The 2017 Revision, Key Findings and Advance Tables.” Working Paper No. ESA/P/WP/248 (2017): 46.

Wittrock, Julie, Colleen Duncan, and Craig Stephen. “A Determinants of Health Conceptual Model for Fish and Wildlife Health.” *Journal of Wildlife Diseases*, 55, no. 2 (2019): 285–297.

Zinsstag, Jakob, Esther Schelling, David Waltner-Toews, and Marcel Tanner. “From “One Medicine” to “One Health” and Systemic Approaches to Health and Well-Being.” *Preventive Veterinary Medicine*, 101, no. 3–4 (2011): 148–156.

2 Whose Health?

Craig Stephen

CONTENTS

THE CHALLENGE OF DEFINING HEALTH

In 2009, the government of Canada embarked on one of its largest and most expensive public inquiries. The Commission of Inquiry into the Decline of Fraser River Sockeye Salmon aimed to explain why this iconic species returned to its natal river in surprisingly and alarming low numbers in one year. The final report of the Commission used the word health 400 times. Never once did it describe how to recognize a healthy salmon or a healthy population. Optimal health seems to be an almost universal goal. Healthy individuals, healthy communities, and healthy ecosystems feature prominently in government policies, community plans, resource management guidelines, and personal goals. But it is rare for such aspirations to clearly prescribe how to recognize health. We are much more adept at recognizing when health is absent or tracking the conditions and contributors to health. This chapter introduces some of the ways health is conceived for people, animals, or ecosystems and discusses the need to understand these differences before embarking on an interspecies health program.

The definition of health is the subject of long-standing debate in the philosophy of medicine. The dividing line between health and diseases and the combinations of capacities, attributes, and attitudes that define a healthy state have varied over time and between cultures, disciplines, and subdisciplines. Broadly speaking, there are three perspectives of health (Ereshefsky, 2009). "Naturalists" regard health as normal and natural biological functioning. A "normative" view regards health as shared judgements about what we value and what an individual (or group) can do. The hybrid view declares something is healthy when there is no detectable harm or deprivation of benefit as judged by the standards of a culture and when that thing can perform its normal functions. Each of these perspectives includes the ideas of normal and values. This makes health a subjective concept created and accepted by the people in a society rather than an objective, irrefutable biological state determined by biology alone.

Health is a "humpty-dumpty" problem. Like the old English nursery rhyme of the shattered anthropomorphic egg that could not be put back together again by "all the King's horses and all the King's men," many pieces of biomedical, sociological, ecological, and individual knowledge can be assembled to define health, but there is no guarantee that different groups of people will assemble all the components together in the same way. As such, there can be several true, but conflicting, perspectives of health held by different individuals in the same circumstances. Take for example the case of a captive whale housed in an aquarium. This animal may show no pathophysiological abnormality, be normal in form and function, be able to reproduce, and can fully exploit the resources offered to her. Yet she is neither able to fulfil her evolutionarily destiny nor exhibit all behaviours typical of her wild peers. Whereas one group may conclude that she is biologically healthy, other would see impediments to her ability to express normal behaviour an affront to her welfare and therefore, not healthy. In this case, individual animal health concerns clash with the desire to learn more about the species and foster public goodwill that promotes conservation of the wild cohort (Minteer and Collins, 2013). As another example, social conflicts and disagreements over health have been one of the biggest impediments to salmon aquaculture securing its social license to operate (Stephen and Wade, 2019). Proponents see aquaculture contributing to healthy communities and ecosystems by reducing exploitation of wild stocks and contributing to food security and employment. Opponents see the potential transmission of infectious disease to free-ranging wild animals, habitat alterations, and imposition on aboriginal rights for sustainable ocean resources combining to create an unhealthy situation.

Patience, persistence, and flexibility are critical skills for the health promotion at the junction of differing views and values of health. Interdisciplinary and interspecies health promoters need to be able to negotiate a common agenda for action, rather than impose their health definitions on partners, stakeholders, and rights holders.

THERE IS NOT ONE HEALTH, THERE ARE MANY

Health across Cultures and Time

The definition of health has been changed by, adapted to, and rooted in our worldviews. Some of the earliest conceptions of health failed to see a distinction between our health and the health of the world around us. Understanding the land and the sustainable use of its resources were essential skills for providing for families and communities for many generations. Relationships with nature provided key contributors to health, including basic needs like food, water, shelter, riches, and happiness from deeply held cultural and spiritual connections with nature (Summers et al., 2012). These relationships and elements, when put together, are what health promotion specialists now call environmental and social determinants of health.

Some Indigenous peoples see health not as physical well-being of an individual but as the well-being of the whole community from a whole-life perspective. The tradition of taking a long-term view to health aligns with dedication to sharing and protecting the land to preserve its benefits for future generations. The interconnection of land, language, and culture are the foundations of health and wellness for many Indigenous peoples (see Chapter 14 for a specific example). Their land-health connections are formed from specific localized knowledge coupled with layers of personal and family experience (Johnston et al., 2007). Indigenous ways of knowing and being generally place environmental determinants on par with social determinants as positive contributions to health and wellness (Scott, 2005). Research and management that separate land use and health promotion are inconsistent with many Aboriginal and Indigenous people's perspectives of health being the result of a complex interplay between environmental, social, and individual characteristics.

The idea that health is a product of our interactions with the world around us can also be found in the writings of the ancient Greeks who recognized that good health was determined by natural rather than supernatural causes and that health could not be dissociated from physical and social environments (Tountas, 2009). The need for harmony between the individual, social, and natural environments is reflected in Hippocrates' writing. Hippocrates advocated for physicians to tailor treatments by paying attention to the characteristics of each person, his or her daily habits, the place he or she lived in, and the season of the year. Understanding the natural and built environment was essential to a proper health investigation. Several Asian traditions similarly view health as a harmonious equilibrium between internal elements, environmental conditions, and external sources of harm (Chan et al., 2002; Tai, 2012). Health, in these traditions, is not just about the condition of the body but also about the relationship between the complete person, his or her social environment, and the natural environment.

The path to separating an individual's health from his or her environment began in Western cultures in earnest when taboos against dissection opened new opportunities to understand anatomy, physiology, and pathology. With the invention of the microscope, the doors to microbiology began to open. These advances heralded a shift in emphasis in the health professions from keeping people healthy to reducing suffering and aiding in recovery from specific diseases. By the late 18th and early 19th centuries, it was now possible to cure people of certain afflictions. At the same time, public health gains started coming from interventions directed at people's relationship with their world, such as what they ate and where and how they lived. Health as the absence of disease developed its primacy in this phase of Western medicine.

The advent of statistics further shaped our view of health. The use of population data on trends and patterns of disease helped to identify circumstances and situations that predisposed to disease. Coupled with the germ theory, Western medicine was now adept at linking disease to specific causal factors. Statistics redefined what we meant as normal. No longer was normal what we usually saw in our daily existence, but rather where an individual fell within the range of an attribute with respect to the rest of the population. Volumes were published that began to medicalize health by classifying people as healthy or unhealthy based on the comparison of their clinical or sub-clinical signs of change or possession of a risk factor in comparison to a statistically expected normal.

Health took on a new social meaning in the Industrial Revolution, where a healthy workforce was defined as one that enjoyed good conditions and working ability and suffered few lost workdays due to illness. Darwinism began to tie the meaning of life to physical survival. The idea that health allowed people to tolerate and resist not just biological hazards but also environmental influences began to emerge. Coupled with earlier success in social engineering to reduce diseases such as tuberculosis, these insights began to re-tie health to our relationships with the world around us (Svalastog et al., 2017). In 1948, the World Health Organization (WHO) declared that health was not merely the absence of diseases but rather a state of complete physical, mental, and social well-being. This definition was applauded for recognizing the coexisting physical, mental, and social domains of health, but it was criticized on two fronts. Firstly, the ideal of complete well-being in each of those domains would be hard to achieve and even harder to recognize. Secondly, it unintentionally medicalized society by expanding the scope of factors for medical practice to measure. It unintentionally allowed health professionals to categorize more people as unhealthy most of the time because they were unable to attain full satisfaction of all physical, social, and environmental determinants of health (Huber et al., 2011).

The WHO's Ottawa Charter for Health Promotion (1986) and the concept of salutogenesis, introduced in the 1970s–1980s, emphasized health as the capacities and resources needed to adapt to, respond to, or control life's challenges and changes. The salutogenesis concept asks why an individual, group, or community stays well despite stressful situations and hardships (Eriksson and Lindström, 2006). The Ottawa Charter advocated for resources, capacities, and processes that enable people to have the opportunity to lead a good life. The Ottawa Charter guided the development of health promotion and shaped modern public health practice. It focuses on building healthy public policy, creating supportive environments for health, strengthening community actions, developing personal skills, and reorienting health services to enable people to increase control over and improve their health by realizing aspirations, satisfying needs, and coping with their environments (WHO, 1986). The health of individuals, communities, and environments were once again connected through the concept of reciprocal care outlined in the Ottawa Charter. The Charter declared that the "overall guiding principle for the world, nations, regions and communities alike, is the need to encourage reciprocal maintenance – to take care of each other, our communities and our natural environment." Explicit attention to the links between people and their environment became the basis for a socio-ecological approach to health.

TABLE 2.1
Comparing Human Determinants of Health with Those Proposed for Fish and Wildlife

Human Determinants of Health	Fish and Wildlife Determinants of Health
Biological endowment	Biological endowment
Physical environment	Abiotic environment
Health services	Human expectations and policies
Healthy child development	
Education	
Personal health practices	
Social environment	Inter- and intra-specific social environment
Social support and social status	
Gender	
Culture	
Income and employment	Access to the resources for daily living

Source: Adapted from Wittrock et al., 2019. See Figure 1.1 for additional details.

Population health became a foundation of public health policy and practice in the 1990s. This approach assesses health status and inequities over the lifespan at the population level. It captures not only adverse outcomes like disease but also the positive dimensions of health. The concept of determinants of health is fundamental to the population health practice. The determinants of health are those individual and collective factors and conditions that enable a person or population to be healthy. For people, these include income and social status, employment and working conditions, education and literacy, childhood experiences, physical environments, social supports and coping skills, healthy behaviours, access to health services, biological and genetic endowment, gender, and culture (WHO, 2020). Similar determinants have been found for fish and wildlife (Table 2.1).

Population health considers the entire range of determinants of health, as well as their interconnections, when planning population or community-level interventions to promote and protect health before a disease occurs. This approach recognizes health as a cumulative effect requiring a combination of health protecting and promoting actions. It aims to manage the root causes of health problems or benefits by working on the social or environmental drivers of vulnerability and resilience. The philosophy guiding population health is that action on root causes has greater potential for health gains even if the root causes are difficult to change.

With the advent of health promotion and population health, we have seemingly come full circle to Indigenous ways of knowing and the lessons of Hippocrates that saw health as a product of our interactions with the world around us. Health is not just about the state of an individual but also about the state of his or her environmental, social, and economic conditions and of the community, social, and political processes that shape those conditions. Each of the perspectives briefly described above can still be found today. The prevailing Western notion of health is but one of many ways people describe their health, their community's health, and the health of the world around them. As we expand beyond the realm of human health, the concept of health becomes even more diverse.

Animal Health

The history of animal health parallels that of human health in many ways. Different sociopolitical situations have been associated with differences in how animal health and welfare are viewed. Buddhism and Hinduism, for example, do not share the Judeo-Christian view that people have dominion over animals. This fundamental difference influences attitudes towards animal use, abuse, health, and wellness. With globalization and increasing affluence around the world, these differences seem to be changing. Chinese concepts of animal welfare, for example, are beginning to merge with those of North America and Europe (Lu et al., 2013).

Our attitudes towards animal health and welfare not only vary across cultures but also between species. It would be rare to find a person who found it acceptable to hunt a deer by snagging the animal on a hook, dragging it behind a vehicle, and suffocating it before butchering. But that is basically what we do when we go fishing. Rare would be a person who deemed his or her pet house cat to be terminally unhealthy because it was unable to produce offspring, but it would be a common practice to cull a dairy cow that was unable to conceive a calf and produce milk. Substantial investments are made in creating the circumstances that will help sustain healthy charismatic wildlife like whales and pandas, but efforts to protect and promote the health of non-charismatic species, such as the lamprey discussed in Chapter 19, are less popular and poorly funded. Social factors influence our attitudes towards animals, including (i) the extent to which we are responsible for harming them, (ii) the extent to which the harmed animals are under our stewardship, (iii) the severity of the problems that cause the harms, and (iv) the cultural and economic factors, including the popularity of the species (Kirkwood and Sainsbury, 1996). Regulations, legislations, and expectations for animal health differ based on social uses of the species, i.e. whether they are farm animals, pets, laboratory animals, wild animals, or pests. Our views of animal health are, therefore, closely tied to how we regard the animal.

Health is rarely defined in veterinary textbooks. It is largely defined as the absence of specific infectious diseases in national legislation and international agreements. Gunnarsson (2006) recognized five categories of health in the veterinary literature: (i) health as normality, (ii) health as biological function, (iii) health as homeostasis, (iv) health as physical and psychological well-being, and (v) health as productivity, including reproduction. A higher proportion of textbooks written for non-veterinarians contained definitions of health or disease than those written for veterinarians. Western veterinary medicine still largely regards health as a dichotomous state reflected by the absence of disease or ability to produce economically valuable products like meat or milk in profitable amounts. Companion animal medicine tends to emphasize disease prevention, treatment, and recovery. Herd health of animals used for agriculture or aquaculture extends the idea of health beyond the absence of disease to include external measures of physiological or economic performance. Herd health considers not just animal determinants of health but also environmental and social characteristics. Wildlife lie somewhere in between. Literature explicitly dealing with wildlife health almost exclusively deals with diseases, with a recent emphasis on infectious and parasitic diseases. However, when the literature is more broadly searched to include domains involved in wildlife management, like resilience, wildlife health is seen as a cumulative effect involving multiple factors that extend beyond the disease and pathogen focus of many wildlife health studies and legislation. Six themes have been identified as determinants of fish and wildlife health: (i) the biologic endowment of the individual and the population, (ii) the animal's social environment, (iii) the quality and abundance of resources providing the animal's needs for daily living, (iv) the abiotic environment in which the animal lives, (v) sources of direct mortality, and (vi) changing human expectations (Wittrock et al., 2019) (see Chapter 1, Figure 1.1). These parallel the determinants of health recognized for people (Table 2.1). Yet, legislatively, animal health is still largely defined as the absence of a specific subset of infectious diseases of trade, public health, or economic importance.

Animal welfare and health are allied concepts. Some people similarly conceive health and welfare as normal functioning and freedom from disease. Others conceive welfare more like the salutogenesis concept of a sense of coherence between the capacity to identify, benefit, and use resources to deal with stress and the reality of current living conditions (Stephen and Wade, 2018). This parallels the viewpoint that animal welfare is compromised when adaptations possessed by the animal make an imperfect fit to the challenges it faces in the circumstances in which it lives (Fraser et al., 1997). Animal welfare legislation and regulations generally address commonly expressed ethical concerns that animals should (i) lead natural lives through the development and use of their natural adaptations and capabilities, (ii) be free from prolonged and intense fear, pain, and other negative states, and experience normal pleasures, and (iii) have satisfactory health, growth, and normal functioning of physiological and behavioural systems (Fraser et al., 1997). The five freedoms and the 3 Rs are well-known guiding concepts of animal welfare (Table 2.2).

Organizations governing experimental use of animals have often developed animal welfare guidelines that are damage focused and intend to reduce harm by minimizing stress to individuals and discouraging procedures that have lasting negative individual or population effects, or affect the species' existence. New animal farming methods, emerging experimentation technologies, ongoing exploitation of wildlife, new understanding of animal needs, and increasing public awareness are inspiring a growth in animal welfare laws in the European Union that recognize animals as sentient beings. These regulations tend to place priority on meeting the physical and psychological needs of animals to prevent unnecessary pain, suffering, or injury (Caporale et al., 2005). The UK's 2007

TABLE 2.2
The Five Freedoms and 3 Rs of Animal Welfare

The Five Freedoms of Animal Welfare	The 3 Rs of Laboratory Animal Welfare
Freedom from hunger and thirst by ready access to fresh water and a diet to maintain full health and vigour	Replace – use alternatives which avoid or replace the use of animals in an area where animals would otherwise have been used
Freedom from discomfort by providing an appropriate environment, including shelter and a comfortable resting area	Reduce – decrease the number of animals being used to answer a research question, or maximize the information obtained per animal without compromising animal welfare
Freedom from pain, injury, or disease by prevention or rapid diagnosis and treatment	Refine – modify husbandry or experimental procedures to minimize pain and distress and to enhance animal welfare from the time of birth until death
Freedom to express normal behaviour by providing sufficient space, proper facilities, and company of the animal's own kind	
Freedom from fear and distress by ensuring conditions and treatment which avoid mental suffering	

Animal Welfare Act placed a duty of care on animal owners to provide for their animals' basic needs, such as adequate food and water, access to medical care, and an appropriate environment to live in. Such regulatory approaches to animal welfare are not found all around the world. Rapid economic changes and greater access to information from around the world is shifting societal awareness of animal welfare issues but with varying degrees of consistency and uptake (Lu et al., 2013).

Wellness and Well-Being

Like health, there is no generally accepted definition of well-being, nor is there an agreement on how to measure it. One definition of community well-being that resonates with many of the themes in this book is well-being as a state of being with others and the natural environment that arises where needs are met, where individuals and groups can act meaningfully to pursue their goals, and where they are satisfied with their way of life (McCrea et al., 2014). The concept of well-being provides a link between health and society. Health provides the raw materials for well-being. Wellness is the state of living that leads to health. Well-being is the result of health and wellness. It is the unimpaired flourishing, free of obstacles, to live in a way that conforms with expectations, opportunities, and abilities. Well-being not only encompasses basic health needs such as adequate food, safety, and lack of disease but also considers how people think and feel about their life situation, or the situation of the animals or environments they care for. Well-being implies successful biological function, positive experiences, and freedom from adverse conditions. As we learn more about the mental capacity and complexity of animals, we need to recognize their ability to feel emotions and to have needs and a degree of consciousness that may determine their well-being. Well-being may be a relevant aspiration in animal and environmental health management in that it reflects how people feel about the circumstances and states that provide the capacities and resources for animal or ecosystem health. The American conservationist Aldo Leopold recognized this in his conception of the land health in that he noted: We see a thing as being right when it tends to preserve the integrity, stability, and beauty of a biotic community (Leopold, 1989).

Ecosystem Health

Shared cultural beliefs and attitudes function as root causes of our attitude towards ecosystem and environmental changes (Stern et al., 1992). The over-exploitation of ecosystem services and resources has been traced back to the Judeo-Christian tradition that separates mankind and nature. The bias towards economic growth can be traced to Protestant teachings, while humanistic values from the Enlightenment put human wants ahead of nature and presumed that humanity could solve all problems. Western materialism divorced the consumer from awareness of the realities of production, which further impacted ecosystems. As the environmental movement gained popularity in the 1960s–1970s, the public grew increasingly aware that human population growth and economic development was depleting raw materials for our future health. A strong interest in measuring ecosystem health for planning, management, and public reporting emerged.

Aldo Leopold's idea of land health was an influential parent of ecosystem health concepts that gained prominence in the 1970s. Leopold's idea of land health combined productive use, self-renewal, and stewardship of the land with the ability to act for conservation and environmental justice (Berkes et al., 2012). As the idea of ecosystem health flourished, so did the perspectives used to define and measure it (Lu et al., 2015). Some viewed ecosystem health as the ability of the ecosystem to function within acceptable limits, becoming "diseased" when inadequate homeostatic repair mechanisms existed. Others closely linked the idea of ecosystem health with sustainability or resilience. Still others viewed an ecosystem to be healthy if it could sustainably convert solar energy and cycle nutrients. The concept of ecological health is usually tacitly understood to be undefinable in a rigorous sense. Lancaster (2000) believed the "notion that the ecological health of the environment can be assessed is a ridiculous notion in a scientific context because there can be no objective definition of 'health' or method for defining degrees of health. Ecological health is a value judgement." Ecosystems are context-specific entities because they cannot be delimited without a specific social, science, or policy context. Health, analogously, is not a biological state but rather a set of capacities and expectations defined within one's social circumstances. Therefore, the idea of ecological or ecosystem health is normative because someone must decide what ecosystem conditions or functions are good (Lackey, 2001).

There have been substantial scientific efforts towards establishing indices that measure ecosystem health. Most attempts have been heavily criticized. Many of the criticisms focus on (i) the challenge of extrapolating indices across scales, gradients, and species; (ii) the oversimplification and generalization of biological processes resulting from indices; (iii) problems in calibrating and validating indices; and (iv) challenges in linking measures such as abundance and distribution to outcomes such as productivity (Niemi and McDonald, 2004). Principle challenges to identifying suitable indicators include (i) practical constraints that restrict monitoring to a small number of indicators that fail to adequately consider the complexity of the ecosystem, (ii) vague management goals and objectives clouding the choice of indicators, (iii) failure to use a defined, consistent, and rigorous protocol for identifying indicators (Dale and Beyeler, 2001), and (iv) the complexity of dynamic ecological systems complicating prediction, thus reducing the value of indicators as forward-looking tools. Concepts linked to ecosystem health, such as resilience and well-being, have subjective, relational, and context-specific aspects in addition to more objective measurements (Brown and Westaway, 2011). There are, therefore, no universally accepted indicators of ecosystem health, nor is there a single definition of the concept.

LOOKING AT HUMAN-ANIMAL AND ENVIRONMENTAL HEALTH AS ONE

One Health

One Health is in the title of this book. But just like the other concepts of health described above, there is no single, universally recognized definition of the term (see Table 1.2). One Health, as it is commonly used today, was initially conceived by the Wildlife Conservation Society as an integrated, holistic, and preventative approach to diseases at the human-animal-environment interface. What started as a plan to reduce overexploitation of wildlife became a framework for collaboration to combat emerging infectious diseases. One Health grew to focus on improving or supporting multidisciplinary communication and collaboration with a major goal of addressing key public health outcomes of integrated disease prevention and surveillance, food safety, and food security. This characterization of One Health overlaps significantly with the scope of practice of veterinary public health. As enthusiasm grew, One Health began to further expand its scope of practice to a wider array of health issues for people and animals (Table 2.3) (Oura et al., 2017). In 2010, the World Animal Health Organization (OIE) noted the importance of co-managing the relationships between animal production systems, human influence on the environment, climate change, and emerging diseases. This paralleled the concept of healthy public policy that recognizes that health promotion goes beyond health care and therefore, health is a team activity

TABLE 2.3
Three Emphases of One Health

Emphasis	Inspiration	Focus
Connecting diseases across species	Emerging zoonotic diseases in people emerging from animal sources lead to efforts at preventing these diseases at their animal origins	Early efforts focused on building new capacity to find bacteria and viruses that could jump between species. Efforts to control infectious diseases of animal origin inspired reinvestment in veterinary public health around the world
Moving beyond infections	Getting rid of the age-old zoonotic and livestock diseases was recognized as a critical component of poverty reduction and food security	Eliminating livestock diseases to produce more affordable and available food to combat food insecurity and malnutrition. Sustaining wildlife populations to maintain ecological services and protect traditional uses of wildlife by indigenous populations
Building resilience	Global megatrends like climate change, urbanization, globalization, and landscape alterations concurrently threatening people, animals, and environments	One Health ideas were extended to protect animal and environmental health as a moral good and for long-term human benefit

rather than the purview of any one agency or discipline (see Chapter 10 for more on healthy public policy).

The One Health emphasis on multi- and interdisciplinary approaches extended veterinary public health from its emphasis on biomedical interventions targeting single hazards to a team approach that examines multiple dimensions of a shared problem. This in turn opened lines of investigation and action derived from the socio-ecological model of health. Although public health concerns predominate One Health activities, the idea that there are reciprocal relationships between people, animals, and their shared environment affecting animal health outcomes is not lost. One Health initiatives are increasingly aiming for mutual benefits for people, agriculture, and wildlife. Some One Health activities are starting to focus on human and environmental determinants of fish and wildlife health in recognition that the survival and persistence of wild animals is entirely dependent on how we decide to exploit their populations and compete for the services provided by their habitats. Attention to the human dimensions of conservation is needed to produce robust and effective conservation policies, actions, and outcomes (Bennett et al., 2017). One Health is increasingly recognizing that when health is the subject of interest, the social dimensions of human-animal interactions are important (Wolf, 2015).

The lack of a single definition of health, or of One Health, provides a flexibility to work across perspectives, species, and disciplines but can complicate program evaluation or implementation due to a lack of shared vision of success. It can also lead to unanticipated effects. For example, culling rats in Vancouver, Canada, to reduce some zoonotic disease and mental health risks from a burgeoning rat population unexpectedly or even paradoxically increased the prevalence of one zoonotic pathogen while decreasing another (Lee et al., 2018). Similarly, culling bats in Uganda to remove a host of Marburg virus may have eliminated immediate human exposure risk but subsequently increased risk as bats that repopulated the area had a higher seroprevalence of the virus, likely due to the recolonizing bat's naïve immunological status (Amman et al., 2014).

One Health is not without critics. Protecting and improving access to the fundamental social and environmental determinants of health such as water security, biodiversity, social justice, equitable access to resources, pollution, and land use planning have remained beyond the scope of most One Health programs despite their profound impacts on human and animal health and welfare (Stephen and Karesh, 2014). The increased knowledge of the role of animals in human diseases increased political and public awareness of risk interdependencies, thus fostering fear of animals, which in turn can lead to calls for their exclusion or elimination rather than improved understanding of how people and animals can safely live together (Decker et al., 2009). One Health has placed more effort on traditional public health actions to attack hazards, most often of infectious origin, rather than on creating supportive environments to enable people, animals, or environments to lead healthy lives.

Ecosystem Approaches to Health

Many health issues, whether in public health, conservation, or agriculture, are complex and best understood by paying attention to the interdependencies of human health, animal health, and the health of the ecosystems in which they live. EcoHealth (also known as ecosystem approaches to health) was pioneered at the International Development Research Centre to connect ideas of environmental and social determinants of health with ecology and systems thinking within a context of social and economic development. EcoHealth emphasizes the development of capacity and circumstances that enable individuals to make healthy choices and systems thinking to promote well-being and quality of life. EcoHealth is an explicit attempt to bring people into ecology, ecology into health, and health into community well-being.

EcoHealth recognizes that health is the result of the complex, dynamic interplay between the determinants of health and the conditions of ecosystems, often resulting from social and economic activities of people (as illustrated throughout Charron, 2012). By embracing ecology and complexity, EcoHealth admits it is deviating from the normal biomedical and epidemiological approach to health. Complex systems, like socio-ecological systems, are unpredictable and not well suited to research seeking a mechanistic truth (Holling, 1996). The systems thinking focus of EcoHealth encourages us to expand the boundaries of what is being studied as well as the set of people involved in generating that knowledge. EcoHealth tries to level the field between the external expert and the communities, shifting them from a researcher and research subject relationship to partners in discovering actions to make incremental improvements in health. The intent to solve problems, particularly those at the interface of human and natural systems, is shared by many EcoHealth projects. EcoHealth's scope of practice is characterized by a work that "(i) is undertaken in the setting where results will be applied under conditions experienced by the populations of concern; (ii) strives to inspires effective, ethical, and sustainable actions in 'real-world' settings to address complex problems important to people in those settings, rather than discover underlying mechanisms of causation; and (iii) focuses on accelerating the integration of research with policy, and practices to promote action to solve problems" (Stephen et al., 2016).

FINDING COMMON GROUND

Imagine a scene. You are looking at a marsh or a chicken farm or an urban park. Within that same space, there are many types of health. There is the health of individuals, populations, and ecosystems. There is the health of the plants, pests, valued animals, and their human caretakers. There is health as seen by the animal rights activists and health as seen by the social justice advocates. All these types of health coexist in the same place at the same time. The art of health promotion at the interface of animals, society, and environments is to find the common ground from which actions can be launched to promote and protect the health of one component of this system without harming the health of others.

Many types of health can be seen in Figure 2.1. At the bottom left, disease, in the form of trauma, can be seen – a California sealion with a neck wound due to entanglement with marine debris. The group of sealions was both promoting human community health by creating ecotourism jobs but at the same time reducing access to commercially and culturally important seafood species, thereby negatively impacting jobs as a determinant of health for others in the community. The eagle in the photo is vulnerable to lead poisoning, due to its habit of scavenging on carcasses killed by hunters using lead bullets while at the same time finding it harder to secure the needs for daily living due to competition for salmon with people and other animals and damage to their critical habitat in the foreshore. Human waste management and urbanization were changing the diets of the gulls, leading to larger population sizes and larger individual sizes of the gulls but with concurrent increased exposure to contaminants in their diets. The trees in the background were struggling with the effects of drought conditions. All these health effects were being influenced by anthropogenic changes in the local coastal ecosystem health that was affecting species distribution, habitat quality and connectedness, and climate change.

Insufficient data and a limited understanding of the cross-linking relationship between human, animal, and environmental health continue to challenge

FIGURE 2.1 There is not One Health, there are many. (Photo credit Craig Stephen.)

attempts to find a universally applicable definition of health. But, at their core, health, well-being, and resilience each deal with situations, decisions, and actions that enable coping and even thriving in a changing world. As stated in the Ottawa Charter for Health Promotion, health can only be created and sustained when we encourage reciprocal maintenance – to take care of each other, our communities, and our natural environment.

Health is not like physics where we can measure and weight different attributes to find universal laws and make definitive proclamations on a state of nature. Health is an aspiration and a cumulative effect of interdependent personal, social, and environmental factors that change and evolve over a life course. There will, therefore, never be a single obvious opening into which the person working at the animal-health-society interface can step in and readily see the one true target of success. Health is a social construct rather than a biologically determined state. In other words, the act of characterizing and describing health is a uniquely human activity. Health is also a wicked problem. Wicked problems are characterized by uncertainty, complexity, and divergent values. There is no indisputable point at which one can declare objectively something is healthy in a way that everyone will agree. There is no ultimate state of health as health goals shift as social expectations change. Health is strongly stakeholder dependent, and there can be a wide variety of opinions on how to best recognize and produce a healthy situation. Health does not sit conveniently within the responsibility of any one organization or individual. Once we start comparing health across species and settings, this wicked problem gets even messier.

A key to seeking solutions to wicked problems is through mediated dialogue aimed at finding common ground about goals and directions (Head, 2008). This requires a broad systemic perspective that looks across boundaries, incorporates many types of evidence, and engages a diversity of interests to co-produce goals, plans, and actions. The willingness and ability to have "back-and-forth" relationship between context, values, and evidence form the basis for successful research and action at the animal-health-society nexus. Interspecies health promotion and harm reduction is about working across disciplines to listen and learn. Without understanding how to frame a health problem across species and generations so that it resonates with those people who need to act to protect the many types of health in a place, cooperative, collaborative action that protects the health of one species without harming the health of others cannot be achieved.

REFERENCES

Amman, Brian R., Luke Nyakarahuka, Anita K. McElroy, Kimberly A. Dodd, Tara K. Sealy, Amy J. Schuh, Trevor R. Shoemaker et al. "Marburg Virus Resurgence in Kitaka Mine Bat Population after Extermination Attempts, Uganda." *Emerging Infectious Diseases*, 20, no. 10 (2014): 1761.

Bennett, Nathan J., Robin Roth, Sarah C. Klain, Kai Chan, Patrick Christie, Douglas A. Clark, Georgina Cullman et al. "Conservation Social Science: Understanding and Integrating Human Dimensions to Improve Conservation." *Biological Conservation*, 205 (2017): 93–108.

Berkes, Fikret, Nancy C. Doubleday, and Graeme S. Cumming. "Aldo Leopold's Land Health from a Resilience Point of View: Self-Renewal Capacity of Social–Ecological Systems." *EcoHealth*, 9, no. 3 (2012): 278–287.

Brown, Katrina, and Elizabeth Westaway. "Agency, Capacity, and Resilience to Environmental Change: Lessons from Human Development, Well-Being, and Disasters." *Annual Review of Environment and Resources*, 36 (2011): 321–342.

Caporale, V., B. Alessandrini, P. D. Villa, and S. Del Papa. "Global Perspectives on Animal Welfare: Europe." *Revue Scientifique et Technique-Office International des Epizooties*, 24, no. 2 (2005): 567.

Chan, Cecilia, Petula Sik Ying Ho, and Esther Chow. "A Body-Mind-Spirit Model in Health: An Eastern Approach." *Social Work in Health Care*, 34, no. 3–4 (2002): 261–282.

Charron, Dominique F., "Ecohealth Research in Practice," in *Ecohealth Research in Practice* (New York: Springer, 2012), 255–271.

Dale, Virginia H., and Suzanne C. Beyeler. "Challenges in the Development and Use of Ecological Indicators." *Ecological Indicators*, 1, no. 1 (2001): 3–10.

Decker, Daniel J., Kirsten M. Leong, and Darrick T.N. Evensen, "Perceptions of Wildlife-Associated Disease Risk: A Challenge or Opportunity for "One Health" in National Parks," in *Proceeding of the George Wright Society Conference*, 2009 (2009): 101–106.

Ereshefsky, Marc. "Defining 'Health' and 'Disease'." *Studies in History and Philosophy of Science Part C: Studies in History and Philosophy of Biological and Biomedical Sciences*, 40, no. 3 (2009): 221–227.

Eriksson, Monica, and Bengt Lindström. "Antonovsky's Sense of Coherence Scale and the Relation with Health: A Systematic Review." *Journal of Epidemiology & Community Health*, 60, no. 5 (2006): 376–381.

Fraser, David, Daniel M. Weary, Edward A. Pajor, and Barry N. Milligan. "A Scientific Conception of Animal Welfare that Reflects Ethical Concerns." *Animal Welfare*, 6, no. 3 (1997):187–205.

Gunnarsson, Stefan. "The Conceptualisation of Health and Disease in Veterinary Medicine." *Acta Veterinaria Scandinavica*, 48, no. 1 (2006): 20.

Head, Brian W. "Wicked Problems in Public Policy." *Public Policy*, 3, no. 2 (2008): 101.

Holling, Crawford Stanley. "Surprise for Science, Resilience for Ecosystems, and Incentives for People." *Ecological Applications*, 6, no. 3 (1996): 733–735.

Huber, Machteld, J. André Knottnerus, Lawrence Green, Henriëtte van der Horst, Alejandro R. Jadad, Daan Kromhout, Brian Leonard et al. "How Should We Define Health?" *British Medical Journal*, 343, (2011): d4163.

Johnston, Fay H., Susan P. Jacups, Amy J. Vickery, and David MJS Bowman. "Ecohealth and Aboriginal Testimony of the Nexus between Human Health and Place." *EcoHealth*, 4, no. 4 (2007): 489–499.

Kirkwood, J.K. and Sainsbury, A.W., 1996. Ethics of interventions for the welfare of free-living wild animals. *ANIMAL WELFARE-POTTERS BAR-*, 5, pp. 235–244.

Lackey, Robert T. "Values, Policy, and Ecosystem Health: Options for Resolving the Many Ecological Policy Issues We Face Depend on the Concept of Ecosystem Health, but Ecosystem Health Is Based on Controversial, Value-Based Assumptions that Masquerade as Science." *BioScience*, 51, no. 6 (2001): 437–443.

Lancaster, Jill. "The Ridiculous Notion of Assessing Ecological Health and Identifying the Useful Concepts Underneath." *Human and Ecological Risk Assessment: An International Journal*, 6, no. 2 (2000): 213–222.

Lee, Michael J., Kaylee A. Byers, Christina M. Donovan, Julie J. Bidulka, Craig Stephen, David M. Patrick, and Chelsea G. Himsworth. "Effects of Culling on Leptospira Interrogans Carriage by Rats." *Emerging Infectious Diseases*, 24, no. 2 (2018): 356.

Leopold, Aldo. *A Sand County Almanac and Sketches Here and There: Outdoor Essays and Reflections* (New York: Oxford University Press, reprint 1989).

Lu, Jiaqi, Kathryn Bayne, and Jianfei Wang. "Current Status of Animal Welfare and Animal Rights in China." *Alternatives to Laboratory Animals*, 41, no. 5 (2013): 351–357.

Lu, Yonglong, Ruoshi Wang, Yueqing Zhang, Hongqiao Su, Pei Wang, Alan Jenkins, Rober C. Ferrier et al. "Ecosystem Health Towards Sustainability." *Ecosystem Health and Sustainability*, 1, no. 1 (2015): 1–15.

McCrea, Rod, Andrea Walton, and Rosemary Leonard. "A Conceptual Framework for Investigating Community Wellbeing and Resilience." *Rural Society*, 23, no. 3 (2014): 270–282.

Minteer, Ben A., and James P. Collins. "Ecological Ethics in Captivity: Balancing Values and Responsibilities in Zoo and Aquarium Research Under Rapid Global Change." *Ilar Journal*, 54, no. 1 (2013): 41–51.

Niemi, Gerald J., and Michael E. McDonald. "Application of Ecological Indicators." *Annual Review of Ecology, and Systematics*, 35, (2004): 89–111.

Oura, Chris, Adana Mahase-Gibson, and Craig Stephen, eds. *Caribbean Resilience and Prosperity through One Health* (St. Augustine: University of the West Indies, 2017), 108.

Scott, K. 2005. "Population Health: Risk and Resistance." *Journal of Aboriginal Health*, 2, no. 1 (2009): 2–5.

Stephen, Craig, Theresa Burns, and Ana Riviere-Cinnamond. "Pragmatism (or Realism) in Research: Is There an Ecohealth Scope of Practice?" *EcoHealth*, 13, no. 2 (2016): 230–233.

Stephen, C., and W.B. Karesh. "Is One Health Delivering Results? Introduction." *Revue Scientifique et Technique-Office International des Épizooties*, 33, no. 2 (2014): 375–392.

Stephen, Craig, and Joy Wade. "Wildlife Population Welfare as Coherence Between Adapted Capacities and Environmental Realities: A Case Study of Threatened Lamprey on Vancouver Island." *Frontiers in Veterinary Science*, 5, (2018): 227.

Stephen, Craig, and Joy Wade. "Testing the Waters of an Aquaculture Index of Well-Being." *Challenges*, 10, no. 1 (2019): 30.

Stern, Paul C., Oran R. Young, and Daniel Ed Druckman. *Global Environmental Change: Understanding the Human Dimensions* (Washington, DC: National Academy Press, 1992), 320.

Summers, J. K., L. M. Smith, J. L. Case, and R. A. Linthurst. "A Review of the Elements of Human Well-Being with an Emphasis on the Contribution of Ecosystem Services." *Ambio*, 41, no. 4 (2012): 327–340.

Svalastog, Anna Lydia, Doncho Donev, Nina Jahren Kristoffersen, and Srećko Gajović. "Concepts and Definitions of Health and Health-Related Values in the Knowledge Landscapes of the Digital Society." *Croatian Medical Journal*, 58, no. 6 (2017): 431.

Tai, Michael Cheng-Tek. "An Oriental Understanding of Health." *Tzu Chi Medical Journal*, 24, no. 2 (2012): 92–95.

Tountas, Yannis. "The Historical Origins of the Basic Concepts of Health Promotion and Education: The Role of Ancient Greek Philosophy and Medicine." *Health Promotion International*, 24, no. 2 (2009): 185–192.

Wittrock, Julie, Colleen Duncan, and Craig Stephen. "A Determinants of Health Conceptual Model for Fish and Wildlife Health." *Journal of Wildlife Diseases*, 55, no. 2 (2019): 285–297.

WHO. "The Ottawa Charter for Health Promotion." World Health Organization, 1986. https://www.who.int/healthpromotion/conferences/previous/ottawa/en/

WHO. "The Determinants of Health." World Health Organization, 2020. https://www.who.int/hia/evidence/doh/en/

Wolf, Meike. "Is There Really Such a Thing as "One Health"? Thinking About a More than Human World from the Perspective of Cultural Anthropology." *Social Science and Medicine*, 129, (2015): 5–11.

3 Health Equity in One Health

Maya Gislason and Craig Stephen

CONTENTS

The World Health Organization's (WHO) 1986 Ottawa Charter for Health Promotion acknowledges the inseparable links between people and their environments. Reciprocal care occurs when we take care of each other, our communities, and our natural environment to create and sustain social and ecological options to reduce vulnerability, avoid unintended consequences, and promote resilience. A reciprocal care approach enables humans and animals to reach their full health potential. It strives to ensure that efforts to protect the health of individuals in one species, and in one generation, do not disadvantage, threaten, or lead to the extinction of other species and generations. Humanity is free to pursue long-term social and economic development if that pursuit is sustainable and developed through practices which do not fracture reciprocal care by depleting ecological services, generating great social injustices, or crossing critical thresholds for essential planetary processes. Reciprocal care requires us to strive for interspecies and intergenerational health equity.

Health equity exists when there is fair access to the resources and opportunities needed for health. Equity is a notion rooted in the principle of justice and fairness and has been enshrined in law in the Western world for over seven centuries (Holdsworth, 1914). While a widely embraced moral principle, equity is a complex idea to coherently embody in health practice, advance in policy, and robustly manifest as a society. Health equity is concerned with socially produced advantages or disadvantages that are not the result of natural biological differences yet affect the distribution of health resources and outcomes. Health equity is a term that has been almost exclusively used in reference to human health. However, at

BOX 3.1 INEQUITY VERSUS INEQUALITY

Inequality and **inequity** sound similar but mean different things.

Inequality: This refers to the uneven distribution of health or health resources as a result of genetic or other factors or the lack of resources.

Inequity: This refers to unfair, avoidable differences arising from poor governance, corruption, or exclusion.

its root, equity calls for all beings to be able to access what they need not only to survive but also to thrive.

The capacity to cope with the challenges of life is not distributed evenly between individuals, populations, species, or ecosystems. Inequalities in health are a fact of life. Ecological competition, the struggle between two organisms for the same resources within an environment, describes differences in capacity that ultimately advantages certain individuals or species over others, leading to differences in opportunities, health outcomes, and ultimately survival. Examples of inequalities can be seen in agriculture where some breeds of domestic animals fare better than others under modern farming conditions as well as in nature where, for example, some species have innate genetic traits that make them more adaptable to varying environmental conditions than others. Inequalities in health outcomes can be seen in different groups of the same species, depending on how they are managed. For example, the causes of death, longevity, and productivity are measurably different between subpopulations of the highly endangered Vancouver Island marmot. The patterns of diseases and causes of death in marmots born in the wild differ from those born in captivity, which differ from those released to the wild after captive rearing (McAdie, 2018). The simple passing of time can create inequalities. The health status of an octogenarian, for example, will be different than that of a teenager. Variations in biological endowment, environmental quality, and demographics will inevitably lead to health inequalities. Inequities, in contrast, refer to differences in health that are not only avoidable but also unfair and unjust (Braveman and Gruskin, 2003). Power, authority, and privilege drive health inequity. As we move towards interspecies and intergenerational health promotion, and work in the interplay between inequalities and inequities, the types and influences of power, privilege, and authority become more numerous and increasingly complex.

Equity is a useful concept to invoke when thinking through the determinants of health for all species and generations. Intergenerational equity is a guiding principle of sustainable development. The United Nations (UN) Brundtland Commission (1983) defined sustainable development as "development that meets the needs of the present without compromising the ability of future generations to meet their own needs" (United Nations World Commission on Environment and

Development, 1987). Interspecies equity, the consideration of non-human animals and their inherent self-interests alongside humans, together with intergenerational equity informs approaches to equitable sustainable development (Earnshaw, 1999). The United Nations Educational, Scientific and Cultural Organization (UNESCO) described interspecies equity as when people treat all creatures decently and protect them from cruelty and avoidable suffering (UNESCO, 2004): a concept not far removed from animal welfare (see Chapter 2 for more discussion on animal welfare). A new generation of work around the 2015 UN Sustainable Development Goals is seeking to redress the lack of consideration of the important roles of animal health to sustainability through integrated human, animal, and environmental frameworks. Nuanced thinking on interspecies equity necessitates careful consideration of their shared environments and of the myriad ways that health is generated, protected, connected, and distributed across social and ecological spaces, places, and times. This chapter explores how the concept of health equity might be adapted to One Health and allied fields in order to foster reciprocal care at the junction of animals, health, and society.

HEALTH EQUITY CORE CONCEPTS

Equity is fairness in treatment, access, opportunity, and advancement for all. It is promoted by identifying and eliminating barriers to allow the full realization of opportunities by all groups. Health inequities arise when social, environmental, or ecological injustices prevent an individual, group, or population from meeting their potential. Health equity is tightly allied with notions of the universal right to health enshrined in international law. Factors that make it difficult to translate this moral notion into action include the reality that ideas about equity and health are context-dependent, informed by social and cultural values and traditions, and often constrained by political practices and economic priorities (Braveman, 2010). "In operational terms, and for the purposes of measurement, equity in health can be defined as the absence of disparities in health (and in its key determinants) that are systematically associated with social advantage/disadvantage" (Braveman and Gruskin, 2003). The idea of health equity is mostly used in reference to disparities in health between different people due to differences in their access to the social determinants of health. Improving equity, therefore, most often involves increasing justice and fairness within the procedures and processes of institutions or systems, as well as in the fair distribution of resources. Health equity-informed work has an explicit focus on how inequities are often institutionalized and socially sanctioned (Krieger, 2014). Tackling equity issues requires an understanding of the root causes of disparities within our social world (Jones et al., 2009; Kapila, Hines, and Searby, 2016).

Differences in health status across human populations emerge because of ongoing marginalization and exploitation through discrimination based on categorized differences such as socio-economic status, gender, citizenship, geographic location, racialization, and sexuality. Discrimination is cumulative across the life course and is observable in the health gradient (Braveman and

Gruskin, 2003). Patterns of power and privilege are relational and many of the factors that disadvantage some groups simultaneously advantage others, for example, in the case of white privilege, poverty, and colonization. To successfully tackle human health inequalities requires working upstream to address the "causes of the causes of disparities and the causes that underlie the causes of the causes" (Greenwood and de Leeuw, 2012). Insights from research and practice around social inequities in human population health offer a way to think about determinants working across human-animal health continuums as well as to identify social, political, and economic factors that are impacting animal health. Protecting and promoting health determinants can advance equity work when patterns of discrimination, which may have played out across generations, are identified and redressed so that their impacts on current and future generations are stopped (Marmot et al., 2008). Table 3.1 summarizes some current health determinants thinking – with the recognition that this is an active and continuously evolving area of scholarship.

Our understanding of health determinants, as well as health equity, is constantly deepening. One way to progress our learning is to explicitly consider the philosophical and theoretical assumptions that inform current thinking, evidence building, policy formation, and practice. For example, scholars and practitioners are considering ways in which health determinants thinking both benefits from and is limited by its origins in Western philosophical and medical traditions. The Western worldview is anthropocentric and frames nature as separate from

TABLE 3.1
Examples of Different Conceptualizations of Determinants of Health

Conceptualization	Description
Social Determinants of Health (SDoH)	Describes a range of personal, social, economic, and environmental factors that contribute to one's state of health, such as educational attainment, employment, gender, and income (Marmot, 2005)
Indigenous Determinants of Health (IDoH)	Indicates that for Indigenous peoples in addition to being impacted by the full range of SDoH, these populations are also impacted by a history of colonization and that health and well-being require attachment and access to land and place and the right to cultural continuity (Greenwood and de Leeuw, 2012; King et al., 2009)
Environmental Determinants of Health (EDoH)	Draws attention to how the physical environment, such as the built and natural environment (e.g. water or air quality), impacts health.
Ecological Determinants of Health (EcoDoH)	Involves the inextricable interdependence that human survival has on the existence of a healthy Earth system, including the valuing of resilient ecosystems and the health of other species (Parkes et al., 2020)

culture, and it prioritizes humans over animals and economic growth over ecological resilience. The medical cosmology embodied in allopathic medical traditions tends to focus on illness and disease rather than on health and wellness; medical expertise is based on atomized rather than holistic approaches to the body, and the links between human and ecological health are often left unexplored. Other knowledge systems do not begin from these foundational assumptions as they are produced through different worldviews and offer frameworks for valuing the lives of humans, animals, and Earth, which are highly integrated. Indigenous approaches to health, for example, have for millennia viewed human health as inextricably bound to the health of family and community, the natural and the non-human world, and to land and water (FNHA, 2020). Human health is not of *a priori* importance; instead, a key guiding value is that human health is related to the health and well-being of the whole (Blackstock, 2007). Understanding the links between worldviews, social values and medical practices are important areas to reflect upon, particularly in the current context, where significant whole system problems are increasingly impacting health. For example, global pandemics, many of them zoonotic, are illustrating the inextricable links between human and animal health; climate change illuminates human dependence on Earth, while environmental degradation reminds us that health is dependent upon ecosystem services, such as the filtration of clean air and water. Issues of power, privilege, racism and other forms of discrimination add additional layers that need to be considered to understand what forces are driving and perpetuating social and environmental harms to health and well-being as well as where interventions are best targeted.

EQUITY AND ONE HEALTH

While it is important to include health determinants and health equity insights into human health, applying equity thinking to animal health is also highly relevant for a range of intersecting reasons. The extinction crisis of the Anthropocene is drawing attention to the inequitable treatment of animals and the degradation of biodiversity because of economic and human population growth. This is in part because of practices which subordinate the health and well-being of animals to humans, whether it is through direct and planned actions (such as overfishing) or indirect actions (such as habitat encroachment or destruction), or long-term processes (such as global climate change). There are diverse yet changing public perspectives on the idea that current practices of conceiving of animal health as a relational state vis-à-vis the utility value of animals *for* humans (such as for food or as companions) should be discontinued. These perspectives influence food choices, farming practices, and even the allocation of rights to some animals (Shaw, 2018). The application or adoption of these perspectives fluctuates with crisis events that can retract the extension of equitable treatment to animals.

Re-evaluating human's approach towards animals is important in the current extinction crisis, which represents the sixth mass extinction on the planet and is predicted to result in 30–50% of all species moving towards extinction by the mid-century (Center for Biological Diversity, 2020). The current ratio of humans to mammals on the planet is deeply skewed where the numbers of domestic species and humans on Earth vastly outnumber wild mammals. How we influence animal's access to determinants of health have consequences for the animals as well as for people. Chapter 23, for example, highlights how modern farming practices contribute to the global antimicrobial resistance crisis. Chapter 15 on climate change draws our attention to how animal agricultural practices contribute to climate change, which in turn is impacting farm animal, wildlife, and human health. Inequitable approaches to animal health can lead to unsustainable circumstances that are dangerous to both human and animal health now and in the future. The deep social inequities which play out across human populations also influence animal health and well-being. In North America, one consequence of settler colonialism was the systematic desecration of animals that Indigenous peoples relied on for food, shelter, spiritual practice, and tradition, such as the eradication of the plains and wood bison populations (Center for Biological Diversity, 2020). Today, threats to caribou and salmon are placing contemporary pressure on Indigenous communities in these same contexts. Colonial relationships also play out in the domain of conservation (Garland, 2008).

Social determinants such as race, gender, education, and socio-economic status are factors which drive patterns of discrimination within the social world, which in turn shape human-animal interactions. For example, slaughterhouses in some countries employ racialized, low-income, and immigrant populations who can find themselves working in unsafe environments, leading to high rates of workplace injuries (Grzywacz et al., 2007; Nibert, 2014). Over the long term, there can also be impacts to mental health derived from the routinized mass killing and processing of animals as well as from witnessing or being forced to engage in the cruel treatment of animals (Grzywacz et al., 2007). In another example, research is beginning to develop better tools for quantifying the links between animal abuse, child abuse, and domestic violence where perpetrators of animal abuse have often been found to go on to harm women and/or children (Pendergrass, 2017; Canadian Veterinary Medical Association, 2020; Febres et al., 2014). At the scale of populations, in contexts of pandemics, civil unrest, and war, animal welfare tends to diminish, and in regions impacted by increased social strain, such as economic recession, incidences of animal neglect and abuse also tend to rise. These are few examples of the myriad interlocking ways that human and animal inequities are linked and serve as an invitation for further investigation to better understand the many ways that human and animal health are intertwined with justice, fairness, and equity.

Around the world, governance bodies are calling for more concerted action on tackling health inequities. Examples include the WHO's Commission on the Social Determinants of Health report (2008), the mainstreaming of Gender-Based Analysis Plus approaches to evidence building and policy formation, the UN Sustainable Development Goals (2015), the UN Declaration of the Rights of Indigenous Peoples (2007), and the recommendations of the Truth and Reconciliation Commission of Canada. This work now needs to be strengthened by a more deeply integrated understanding of the interdependence between human, animal, and environmental health. Work on sustainability and equity offer hope and clear pathways for building more socially just, environmentally resilient, and fair societies. Given the political will to usher in change, this is a powerful time in history to be working on issues of equity, diversity, and inclusion in the human and animal health spheres.

EQUITY AS AN ESSENTIAL INGREDIENT IN COMBATING GLOBAL CHALLENGES

The current moment is being shaped by interacting global environmental, social, and species changes, such as climate change; habitat degradation and loss; overexploitation of animals; global spread of invasive species, pathogens, and parasites; and an exponentially growing human population (Ceballos et al., 2015). Despite considerable attention to the problem of health inequalities since the 1980s, striking differences in health still exist. Given that inequalities are products of the interplay between biological and social conditions as well as economic and political processes, they are complex phenomena to tackle with the greatest leverage points found where practices are unfair, unjust, and unnecessary. In other words, we refer to inequity as "unfair, avoidable differences arising from poor governance, corruption or cultural exclusion" (Tehrani et al., 2019), which often leads to unnecessary disease and death. Health differences adversely affecting socially disadvantaged groups are particularly unacceptable because ill health can be an obstacle to overcoming social disadvantage (Braveman et al., 2011).

Social inequality and imbalances of power are at the heart of environmental degradation. More unequal societies tend to have more polluted and degraded environments. Some proposed mechanisms for this relationship include the tendency for the wealthy to both consume more and hold more political power; the attempt for the less well-off to emulate the wealthy by consuming more; and the erosion of the necessary trust and social cohesion needed for environmental stewardship because of growing inequities (Cushing et al., 2015). The current extinction rate (which has been attributed to converging global stressors) is estimated to be 1,000 times higher than natural background rates (De Vos et al., 2015). These trends suggest we are failing to address the health promotion imperative to create supportive environments by protecting natural environments and conserving natural resources. Future health promoters need to ask the question: "Does the satisfaction of human health and wellbeing have to inevitably lead to the long-term degradation of ecosystems and loss of biodiversity that diminish capacity for future generations and other species to have fair access to the resources and opportunities needed for health?" The emphasis on promoting human development through highly consumptive lifestyles seems to favour creating deprivation rather than creating and preserving opportunities found in healthy biodiversity as well as sustainable social and planetary processes.

Rapidly growing demands for food, freshwater, timber, fibre, and fuel have resulted in a substantial and largely irreversible loss in the Earth's biodiversity.

The nature, amount, and variability in biodiversity often determine the sustainability of and access to ecosystem services that are key determinants of a community's capacity to adapt to future challenges, whether they arise from novel pathogens, emerging non-communicable diseases, or deprivations of the needs for daily living (Keune et al., 2013). It is more difficult for communities to recover from disasters in situations where natural resources have been degraded (Miller et al., 2006). This is reflected in the UN Sendai Framework for Disaster Risk Reduction, which recommends sustainable use and management of ecosystems to preserve ecological relationships and functions that reduce risks and support resilience. When we strive to promote health without taking nature into account, well-being comes at the cost of diminished ecological integrity and eroded environmental contributions to health. Our use of natural resources has resulted in substantial net gains in human well-being and economic development in the past two generations, but these gains have come at growing costs of the degradation of many ecosystem services and diminished access to their benefits by future generations.

At the Fifth Ministerial Conference on Environment and Health in Parma, Italy in 2010, the Member States of the WHO European Region recognized that systematic processes and shared frameworks for identifying and redressing inequitable and interconnected impacts of social and environmental harms were required and that they needed to be grounded in work that addressed the underlying mechanisms or the *causes of the causes* of these harms. Given that "both distributive justice and procedural justice often characterize sociodemographically disadvantaged groups," such a framework would need to account for both the places and populations being affected as well as the processes and procedures through which risks and harms are produced (Kruize et al., 2014). Relatedly, the conditions within which reciprocal care relationships can exist are nested within social relations of power through which discourses and practices produce environmental degradation.

Access to ecological services is affected by decisions that influence access to and the quality and quantity of habitat that is protected. Conservation efforts not only impact wildlife and ecosystem health but also human well-being and social justice. The 1993 Convention on Biological Diversity includes a requirement for protected areas to be equitably governed and managed. Conservation efforts can be stalled or blocked by people claiming them to be unfair for one group for some reason. Resentment and a sense of injustice among those inequitably affected by protected areas can lead to conflicts that impede conservation efforts. Should, for example, a sheep farmer accept restrictions on access to grazing lands for his or her sheep in order to avoid the transmission of diseases from farmed sheep to wild sheep, as in the case in western North America? How can the direct economic hardship felt by the shepherd be compared to the public good of wild sheep conservation or the community economic impacts of hunting? When local people are empowered and there is more equitable sharing of benefits, the likelihood of effective conservation action increases for issues such as these (Pascual et al., 2014). A guiding principle for linking social equity and conservation is: "no human should infringe on the well-being of others (including other species) any more than is necessary for a healthy, meaningful life" (Vucetich et al., 2018). The application of this principle is heavily influenced by prevailing (and sometimes conflicting) values and social priorities. Vucetich et al., (2018) turned to three questions that can help consider conservation from an equity perspective: (i) is there equality of opportunity; (ii) is something necessary for realizing a healthy, meaningful life, and (iii) are those involved being treated as they deserve given their ability and situation? The links between social and environmental justice are becoming increasingly clear as is the imperative to consider issues of interspecies (human and animal) and intergenerational ecosocial justice.

TOWARDS INTERSPECIES AND INTERGENERATIONAL HEALTH EQUITY

Public health leaders have been adept at advocating for investment in the social determinants of health, but there is less evidence of similar public health leadership in protecting ecological resources and services critical to individual and community resilience (Stephen, 2020). Public health's refocus on the environment has almost exclusively been on the built environment to the exclusion of the natural environment (Bracken et al., 2008). Too often, the environment is characterized as a source of public health hazards rather than of resilience. However, both the salutogenesis and socio-ecological perspectives see health in terms of the congruence of peoples needs with the structure and quality of their environment. Similar perspectives exist for domestic animal and wildlife welfare (Stephen and Wade, 2018). Despite a growing literature on the need for socio-ecological approaches to preparedness in the Anthropocene, too often management actions are *business as usual*. Health promotion programs that promote socio-ecological thinking need to consider who should be involved, whose experiences and knowledge should influence decision-making processes, and who defines and determines how determinants of health approaches will be used to promote health across species and generations.

The UN Universal Declaration of Human Rights (UN General Assembly, 1948) asserts that "recognition of the inherent dignity and of the equal and inalienable rights of all members of the human family is the foundation of freedom, justice and peace in the world." A question arising from this declaration is: "Are these rights only applicable to the current generation of the human family?" The family of one of the chapter's author currently spans 90 years and four generations. After the birth of a grandchild, the prospect of a great-grandchild becomes less of an abstract concept and more of a possibility. There may be someone in that family alive in another 90 years, and it is not unimaginable that a great-grandchild will be alive 120 years from now. A second question arising from the Universal Declaration of Human Rights is: "Are people the only members of the human family?" The question needs to be asked within the context of the Western world where dichotomous constructions have imagined humans to be distinct from animals and society to be separate from nature. Modern science understands that humans are also animals and share several biological and social characteristics that are homologous to those exhibited in other animals, even in relation to the use of languages, tools, and social lives. Perhaps what sets humans apart from other organisms are characteristics such as our ability to speak as a result of a permanently descended larynx; the range, refinement, and scope of our tool use; cognitive abilities that enable full-blown language capacity as well as reasoning and planning abilities; and significant adaptability to a range of environments (Stockholm University, 2017). If we consider the human family less as a nuclear family and more of an extended family, one can start to imagine a different way of being in the world. Ancient non-Western worldviews have centrally integrated humans and animals into a family constellation. These cosmologies offer profound insight into conceptualizations and practices grounded in the principles of reciprocal care. Fulfilling the Universal Declaration of Human Rights, it seems, needs us to consider who and what is part of the family and over how long of a time.

Integral both to Indigenous ways of knowing and being and to Western work on sustainability is the notion of intergenerational justice wherein reciprocal care considers the needs both of current and future generations. One often cited articulation of this approach is the importance of cultivating relationships and societies wherein each decision involves a consideration of past, present, and future generations. Many Indigenous nations around the world are guided by and require decision-makers to pay attention to the knowledge and decisions of past generations and consider the impacts of our actions on the future generation (Tonmyr and Blackstock, 2010). Employing an intergenerational approach to equity requires a balance to be struck between acting now so as not to compromise future generations while at the same time distributing current resources and opportunities to ensure the well-being of all in the present generation. The challenge is to find ways to work together when there is a wide diversity of needs, interests, and knowledge amongst the members of the extended human family. As discussed in Chapters 6 and 9, a key to encouraging collaborative action in situations where different needs, knowledge, and values exist is to find a shared problem. Declining ecological health is a massive problem shared by people, animals, and ecosystems that might serve as an entry point to find win-win-win solutions by reducing inequitable access to the resources needed by all of the extended human family to thrive in a changing world.

JUSTICE AND CITIZENSHIP IN PROMOTING INTERSPECIES AND INTERGENERATIONAL HEALTH EQUITY

The UN International Covenant on Economic, Social and Cultural Rights (in force in 1976) confers "the right of everyone to the enjoyment of the highest attainable standard of physical and mental health," but societies often fail to ensure this right is equitably and justly fulfilled (Gostin, 2011). Justice is concerned with equalizing relations between those who have power and those who do not. The current generation of people wield the most power in intergenerational and interspecies issues. Conservation, sustainable development, global health, One Health, and health promotion all need to confront three issues of justice: (i) justice between different people of the present generation, (ii) justice between people of different generations, and (iii) justice between humanity and other species. This leads to three questions a One Health practitioner should ask: Will achieving a health objective for one target group

1. Influence the chances for another group to also achieve their objectives?
2. Make it easier for another group to also achieve their objectives?
3. Make it more difficult for another group to also achieve their objectives?

When considering these questions, one needs to consider both whether there is equitable access to the determinants of health and whether the opportunities to benefit from these determinants are equitable. Intragenerational and interspecies justice is concerned with (i) the control of access to resources and capacity across time and species so the benefits produced by ecosystems can be enjoyed, (ii) the duties to conserve ecosystems, and (iii) the expectations to compensate for the harms caused by ecosystem degradation.

The unprecedented changes and threats arising in the Anthropocene require us to contemplate the social contract between the state and the individuals in health citizenship. Health citizenship combines the responsibilities of individuals to achieve healthy living with the responsibilities of the state to help all citizens achieve this goal (Spoel et al., 2014). No one person or group has an exclusive claim on ecosystems or the services they provide because natural ecosystems are not created by any person or group of people (Glotzbach and Baumgartner, 2012), yet every living thing depends on sustainable ecosystems – they are common pool resources. If we accept that current and foreseeable environmental change will constrain the distribution of the proceeds of economic growth as the primary means for achieving health equity, we need to conceive of health citizenship in socio-ecological terms and thus consider ecological citizenship as a core value of health promotion (Dobson, 2004).

Most health promotion efforts are directed at changing people's behaviours to promote healthy living within their social environments, empowering people to make choices and confidently take responsibility for their health (Petersen et al., 2010). The WHO's Jakarta Declaration on Leading Health Promotion into the 21st Century (1997), however, also recommended that decision-makers firmly commit to their social responsibility to protect the environment and ensure sustainable use of the resources. Health promotion and ecological citizenship are, therefore, fully compatible. Ecological citizenship entails the duties and responsibilities to ensure that ecological footprints are sustainable and provide a just distribution of ecological resources to human and non-humans alike. It entails the ability to recognize environmental issues when they arise and assume responsibility for preventing and resolving problems through individual and collective choices. The principal task to incorporating ecological citizenship into health promotion is to connect personal well-being to the ecological

BOX 3.2 CORE CONCEPTS INFORMING INTERGENERATIONAL, INTERSPECIES HEALTH EQUITY

Social justice: All people deserve and should have access to the same rights and resource regardless of race, socio-economic status, gender, or other characteristics.

Environmental justice: All people have the same degree of protection from environmental health hazards and equal access to the decision-making process to have a healthy environment in which to live, learn, and work regardless of race, socio-economic status, gender, or other characteristics.

Ecological justice: Human beings and the rest of the natural world are provided enough protection and opportunity to allow them to live according to their own forms of life and live the fullness of their natural existence.

Ecological citizenship: There are duties and responsibilities to ensure that ecological footprints make a sustainable impact and provide a just distribution of ecological space and components.

services and environmental endowments that influence health. This thinking is reflected in the 1993 UN Convention on Biological Diversity's objective to conserve biological diversity and the sustainable use of its components with the fair and equitable sharing of the benefits arising out of its use. Because of the interdependence of the human-nature community, the idea of separating social justice from the environment seems inappropriate. There are, however, legal, philosophical, and religious debates about how far the community of justice can or should be extended to include future generations as well as non-human entities.

A first step in incorporating ecological citizenship into health promotion is to evolve and expand the concept of health literacy. The holistic health literacy described by Rask et al., (2014) includes environmental awareness and interest in the state of the world. As society increasingly urbanizes, the direct connections with the land, the sea, and the rest of nature are becoming more tenuous for a growing proportion of the population. Empowering people, organizations, and institutions to act for personal, intergenerational, and interspecies health in the 21st century goes beyond documenting and disseminating alarms about the un-sustainability of the current trajectory of humanity (Ansari and Stibbe, 2009). McMichael (as quoted by Fleming et al., 2009) noted that

> Until the public health community highlights the centrality to the overall sustainability project of long-term population health, and particularly its dependence on maintaining Earth's life-support systems, society will continue to miss the real point – namely, that 'ecological sustainability' is not just about maintaining the flows from the natural world that sustain the economic engine nor maintaining iconic species and iconic ecosystems. It is about maintaining the complex systems that support health and life.

Health promoters will need the ability to access, understand, integrate, and use information about health-related ecological effects to evolve and deliver health literacy services that help communities and individuals become aware of and concerned about the environment and its associated impacts on personal, intergenerational, and interspecies health, as well as help them gain the knowledge, skills, and motivations to work towards the solution of current problems and the prevention of new ones.

PROMOTING RECIPROCAL CARE IN HEALTH EQUITY

Identifying Vulnerabilities, Barriers, and Opportunities

A systems approach is needed to map the systemic barriers to health equity. Health equity impact assessment methods were developed to help consider the cumulative effects (positive and negative) of a specific policy, program, or project on vulnerable groups of people. These tools have a distinctly anthropocentric perspective. Identifying cumulative impacts is increasingly seen as a best practice in conducting environmental assessments. This approach strives to assess the incremental impact of the action when added to other past, present, and reasonably foreseeable future actions, but it tends to have a focus on biophysical rather than social impacts. Integrated environmental impact assessment aims to combine dimensions of human determinants of health with biological and physical environments to provide a holistic understanding of the interrelationships between the human and the natural environment to help identify the unintended effects of initiatives on human health and/or on the environment. They tend to have more of a focus on sustainability than on health equity. There is a large variety of impact assessment tools and approaches but none have satisfactorily integrated the concerns of equity with human, animal, and ecosystem health. One Health practitioners will need to borrow from and adapt the various tools to best suit the context of specific issues. Figure 3.1 proposes a rubric that might guide an assessment of the interspecies, intergenerational health equity implications of a policy, intervention, or program.

Figure 3.1 considers three equity questions: (i) How will the proposed activities or decisions increase or decrease unnecessary, avoidable, and/or unjust impacts or health outcomes? (ii) How will the proposed activities or decisions increase or decrease the access, quality, or quantity of social or environmental resources or ecological services that are needed to fulfil the determinants of health for humans, animals, and the environment? (iii) How will the policies or practices needed to implement the activities or decisions systematically prevent humans or animals the opportunity to benefit from a determinant of health? Each of these questions will have both spatial dimensions (i.e. over what geographic scale does one consider these questions?) and temporal dimensions (i.e. are these questions answered in the present, over the life course of individuals, other generations, or all of these time periods?). These questions are asked for people, animals, and ecosystems that will be influenced by the decisions and activities being reviewed.

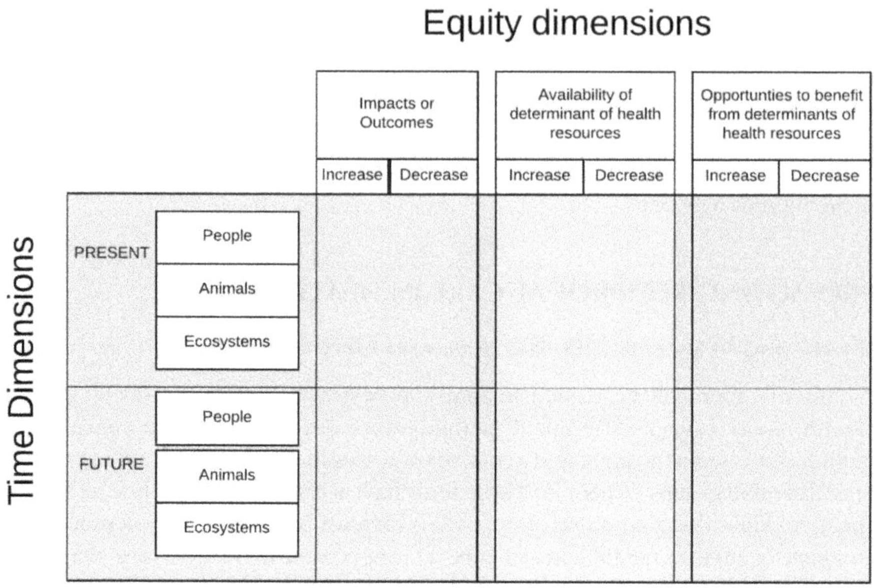

FIGURE 3.1 Conceptual framework to guide an interspecies, intergenerational health equity assessments.

Identifying which people, animals, or ecosystems are vulnerable to changes due to the proposed activities will require a collaborative, interdisciplinary, and equity informed set of approaches to appraising the situation. The success of such analytical exercises will be influenced by who is involved and what knowledge they bring to the process, underscoring the importance of considering cultural safety and related equity, diversity, and inclusion guidelines. The goal is not to produce an exhaustive conceptualization of the issues and relationships at play because this is rarely possible due to the complex, dynamic nature of ecosystem relationships in a changing world as well as due to intersectoral and disciplinary challenges to integrating knowledge. Rather, the goal is to identify who or what might be vulnerable to socially produced advantages or disadvantages derived from the proposed activities, which in itself will not be a simple or straightforward task. Identifying the right set of variables that allow one to directly measure the impact of a decision on the health of all species and forecast future impacts involves dealing with (i) difficulties detecting changes in ecological drivers of vulnerability and resilience for all species within an effected ecosystem, (ii) insufficient evidence and agreement on the identification of thresholds of impacts that are un/acceptable, (iii) the lack of methods that can integrate multiple socio-ecological scales, (iv) the effects of socio-ecological processes may take decades before changes can be seen, making it hard to recognize relationships, and (v) institutional inertia or barriers to co-managing data, intersectoral or interagency relationships, and interdisciplinary knowledge production.

There is a clear urgency to fully address health inequity as we plan strategies and interventions to cope with the threats and changes acting at the interface of animals, health, and society. One Health actions need to be attentive to barriers preventing individuals and populations, be they human or animals, from accessing and benefiting from the conditions needed to reach their full health potential. Paying attention to the "causes of the causes" of poor health and how they are distributed across species can both help prevent unanticipated consequences and ensure that we create circumstances that concurrently protect and promote the health of people, animals, and their shared environment in a just and fair manner.

REFERENCES

Ansari, Walid El, and Arran Stibbe. "Public Health and the Environment: What Skills for Sustainability Literacy – and Why?" *Sustainability*, 1, no. 3 (2009): 425–440.

Blackstock, Cindy. "The Breath of Life Versus the Embodiment of Life: Indigenous Knowledge and Western Research." *World Indigenous Nation's Higher Education Consortium Journal*, (2007): 67–79.

Bracken, Paul, Ian Bremmer, and David Gordon, eds. *Managing Strategic Surprise: Lessons from Risk Management and Risk Assessment* (Cambridge: Cambridge University Press, 2008).

Braveman, Paula. "Social Conditions, Health Equity, and Human Rights." *Health and Human Rights*, 12, no. 2 (2010): 31–48.

Braveman, Paula, and Sofia Gruskin. "Defining Equity in Health." *Journal of Epidemiology and Community Health*, 57, no. 4 (2003): 254–258.

Braveman, Paula A., Shiriki Kumanyika, Jonathan Fielding, Thomas LaVeist, Luisa N. Borrell, Ron Manderscheid, and Adewale Troutman. "Health Disparities and Health Equity: The Issue Is Justice." *American Journal of Public Health*, 101, no. S1 (2011): S149–S155.

Canadian Veterinary Medical Association. "The Link: Animal Abuse, Child Abuse, and Domestic Violence." CVMA, 2020. https://www.canadianveterinarians.net/policy-advocacy/link-between-animal-child-domestic-abuse.

Ceballos, Gerardo, Paul R. Ehrlich, Anthony D. Barnosky, Andrés García, Robert M. Pringle, and Todd M. Palmer. "Accelerated Modern Human-Induced Species Losses: Entering the Sixth Mass Extinction." *Science Advances*, 1, no. 5 (2015): e1400253.

Center for Biological Diversity. "Halting the Extinction Crisis." 2020. https://www.biologicaldiversity.org/programs/biodiversity/elements_of_biodiversity/extinction_crisis/ (accessed June 2020).

Cushing, Lara, Rachel Morello-Frosch, Madeline Wander, and Manuel Pastor. "The Haves, the Have-Nots, and the Health of Everyone: The Relationship Between Social Inequality and Environmental Quality." *Annual Review of Public Health*, 36 (2015): 193–209.

De Vos, Jurriaan M., Lucas N. Joppa, John L. Gittleman, Patrick R. Stephens, and Stuart L. Pimm. "Estimating the Normal Background Rate of Species Extinction." *Conservation Biology*, 29, no. 2 (2015): 452–462.

Dobson, Andrew. "Ecological Citizenship," in *Citizenship and the Environment* (Oxford: Oxford Scholarship Online, 2004).

Earnshaw, Gwendellyn Io. "Equity as a Paradigm for Sustainability: Evolving the Process Toward Interspecies Equity." *Animal L.* 5 (1999): 113.

Febres, Jeniimarie, Hope Brasfield, Ryan C. Shorey, Joanna Elmquist, Andrew Ninnemann, Yael C. Schonbrun, Jeff R. Temple et al. "Adulthood Animal Abuse Among Men Arrested for Domestic Violence." *Violence Against Women*, 20, no. 9 (2014): 1059–1077.

Fleming, Mary Louise, Thomas Tenkate, and Trish Gould. "Ecological Sustainability: What Role for Public Health Education?" *International Journal of Environmental Research and Public Health*, 6, no. 7 (2009): 2028–2040.

FNHA. "First Nations Perspective on Health and Wellness." First Nations Health Authority, 2020. https://www.fnha.ca/wellness/wellness-and-the-first-nations-health-authority/first-nations-perspective-on-wellness (accessed June 2020).

Garland, Elizabeth. "The Elephant in the Room: Confronting the Colonial Character of Wildlife Conservation in Africa." *African Studies Review*, 51, no. 3 (2008): 51–74.

Glotzbach, Stefanie, and Stefan Baumgartner. "The Relationship Between Intragenerational and Intergenerational Ecological Justice." *Environmental Values*, 21, no. 3 (2012): 331–355.

Gostin Lawrence A. "Health and Social Justice." *Bulletin of the World Health Organization*, 89, no. 1 (2011): 224.

Greenwood, Margo Lianne, and Sarah Naomi de Leeuw. "Social Determinants of Health and the Future Well-Being of Aboriginal Children in Canada." *Paediatrics and Child Health*, 17, no. 7 (2012): 381–384.

Grzywacz, Joseph G., Thomas A. Arcury, Antonio Marín, Lourdes Carrillo, Michael L. Coates, Bless Burke, and Sara A. Quandt. "The Organization of Work: Implications for Injury and Illness Among Immigrant Latino Poultry-Processing Workers." *Archives of Environmental and Occupational Health*, 62, no. 1 (2007): 19–26.

Holdsworth, William Searle. "Early History of Equity." *Michigan Law Review*, 13 (1914): 293.

Jones, Camara Phyllis, Clara Yvonne Jones, Geraldine S Perry, Gillian Barclay, and Camille Arnel Jones. "Addressing the Social Determinants of Children's Health: A Cliff Analogy." *Journal of Health Care for the Poor and Underserved*, 20, no. 4 (2009): 1–12.

Kapila, Monisha, Ericka Hines, and Martha Searby. "Why Diversity, Equity, and Inclusion Matter." Independent Sector, 2016. https://independentsector.org/resource/why-diversity-equity-and-inclusion-matter/ (accessed May 2020).

Keune, Hans, C. Kretsch, G. De Blust, Marius Gilbert, Lucette Flandroy, Karel Van den Berge, V. Versteirt et al. "Science-Policy Challenges for Biodiversity, Public Health and Urbanization: Examples from Belgium." *Environmental Research Letters*, 8, no. 2 (2013): 025015.

King, Malcolm, Alexandra Smith, and Michael Gracey. "Indigenous Health Part 2: The Underlying Causes of the Health Gap." *Lancet*, 374, no. 9683 (2009): 76–85.

Krieger, N. (2014). Discrimination and Health Inequities. *International Journal of Health Services*, *44*(4), 643–710. https://doi.org/10.2190/HS.44.4.b

Kruize, Hanneke, Mariël Droomers, Irene Van Kamp, and Annemarie Ruijsbroek. "What Causes Environmental Inequalities and Related Health Effects? An Analysis of Evolving Concepts." *International Journal of Environmental Research and Public Health*, 11, no. 6 (2014): 5807–5827.

Marmot, Michael. "Social Determinants of Health Inequalities." *Lancet*, 365, no. 9464 (2005): 1099–1104. doi: 10.1016/S0140-6736(05)71146-6.

Marmot, M., Friel, S., Bell, R., Houweling, T. A., and Taylor, S. (2008). Closing the Gap in a Generation: Health Equity through Action on the Social Determinants of Health. *Lancet*, *372*(9650), 1661–1669. https://doi.org/10.1016/S0140-6736(08)61690-6

McAdie, Malcolm. Indicators of Individual and Population Health in the Vancouver Island Marmot (*Marmota vancouverensis*). Masters of Science Dissertation, Thompson Rivers University, 2018. https://www.tru.ca/__shared/assets/Malcolm_McAdie_thesis44601.pdf

Miller, Fiona, Frank Thomalla, Tom Downing, and Matthew Chadwick. "Resilient Ecosystems, Healthy Communities: Human Health and Sustainable Ecosystems After the December 2004 Tsunami." *Oceanography*, 19, no. 2 (2006): 50–51.

Nibert, David. "Animals, Immigrants, and Profits: Slaughterhouses and the Political Economy," in *Critical Animal Studies: Thinking the Unthinkable*, ed. John Sorenson (Toronto: Canadian Scholars' Press Inc, 2014), 3–17.

Parkes, Margot W., Blake Poland, Sandra Allison, Donald C. Cole, Ian Culbert, Maya K. Gislason, Trevor Hancock et al. "Preparing for the Future of Public Health: Ecological Determinants of Health and the Call for an Eco-Social Approach to Public Health Education." *Canadian Journal of Public Health*, 111, no. 1 (2020): 60–64.

Pascual, Unai, Jacob Phelps, Eneko Garmendia, Katrina Brown, Esteve Corbera, Adrian Martin, Erik Gomez-Baggethun et al. "Social Equity Matters in Payments for Ecosystem Services." *Bioscience*, 64, no. 11 (2014): 1027–1036.

Pendergrass, JoAnna. "Veterinarians: Protecting Pets and People from Abuse." American Veterinarian, 2017. https://www.americanveterinarian.com/news/veterinarians-protecting-pets-and-people-from-abuse (accessed May 2020).

Petersen, Alan, Mark Davis, Suzanne Fraser, and Jo Lindsay. "Healthy Living and Citizenship: An Overview." *Critical Public Health*, 20, no. 4 (2010): 391–400.

Rask, Maija, Satu Uusiautti, and Kaarina Määttä. "The Fourth Level of Health Literacy." *International Quarterly of Community Health Education*, 34, no. 1 (2014): 51–71.

Shaw, Rosemary. "A Case for Recognizing the Rights of Animals as Workers." *Journal of Animal Ethics*, 8, no. 2 (2018): 182–198.

Spoel, Philippa, Roma Harris, and Flis Henwood. "Rhetorics of Health Citizenship: Exploring Vernacular Critiques of Government's Role in Supporting Healthy Living." *Journal of Medical Humanities*, 35, no. 2 (2014): 131–147.

Stephen, Craig. "Rethinking Pandemic Preparedness in the Anthropocene." *Healthcare Management Forum*, 33, no. 4 (2020): 153–157.

Stephen, Craig, and Joy Wade. "Wildlife Population Welfare as Coherence Between Adapted Capacities and Environmental Realities: A Case Study of Threatened Lamprey on Vancouver Island." *Frontiers in Veterinary Science*, 5 (2018): 227.

Stockholm University. "Memory for Stimulus Sequences Distinguishes Humans from Other Animals." Science Daily, June 2017. https://www.sciencedaily.com/releases/2017/06/170620200012.htm (accessed May 2020).

Tehrani, Shadi O., Shuling J. Wu, and Jennifer D. Roberts. "The Color of Health: Residential Segregation, Light Rail Transit Developments, and Gentrification in the United States." *International Journal of Environmental Research and Public Health*, 16, no. 19 (2019): 3683. doi: 10.3390/ijerph16193683.

Tonmyr, Lil, and Cindy Blackstock. "Commentary: Indigenous Health Special Issue." *International Journal of Mental Health and Addiction*, 8, no. 2 (2010): 135–144.

UN General Assembly. "Universal Declaration of Human Rights." *UN General Assembly*, 302, no. 2 (1948).

UNESCO. "Technical and Vocational Education and Training and the 'Triple Bottom Line' of Sustainable Development." Supplement to UNESCO-UNEVOC Bulletin, No. 9, September 2004. https://unevoc.unesco.org/fileadmin/user_upload/pubs/bulletin/Supplement-04-e.pdf (accessed May 2020).

United Nations Strategic Imperative. "Report of the World Commission on Environment and Development: Our Common Future." United Nations, 1987. Print. Pg 16. https://sustainabledevelopment.un.org/content/documents/5987our-common-future.pdf (accessed June 2020).

Vucetich, John A., Dawn Burnham, Ewan A. Macdonald, Jeremy T. Bruskotter, Silvio Marchini, Alexandra Zimmermann, and David W. Macdonald. "Just Conservation: What Is It and Should We Pursue It?" *Biological Conservation*, 221 (2018): 23–33.

4 Health Promotion as a Foundation for Reciprocal Care and Collective Action

Craig Stephen and Christa Gallagher

CONTENTS

The World Health Organization's (WHO) World Health Assembly declared in 1977 that a major goal of governments around the world should be to ensure that all people can attain a level of health that would lead them to live a socially and economically productive life (WHO, 1981). This was the start of a Health for All movement. Health for All set out to evenly distribute resources for health and make essential health care accessible to everyone. Health promotion, the process of enabling people to increase control over and improve their health, is a cornerstone of Health for All. One Health has reminded society that Health for Us (people) cannot be achieved without attending to the Health of Them (animals and ecosystems). The UN 2030 Sustainable Development Goals affirmed that healthy environments and healthy animals are essential for human health and human rights. Other sectors are also responding to the calls for healthy equity across people, animals, and environments. Agriculture is acting on consumer concerns about the inherent value of farmed animal health while attending to the economic health of farms and protecting human health through on-farm food safety and security. Processes enlarging people's choices to lead long and healthy lives and enjoy decent standards of living (known as the human development agenda) are realizing that functional and sustainable ecosystems are needed for human health, wellness, and cultural well-being. The Health for All movement can no longer only include people in the category of "all" if its founding goals are to be achieved. Health for All needs new principles and new ways of knowing and measuring health as an interdependent resource shared by people, animals, and ecosystems. This chapter introduces some core concepts of health promotion and explores how they may be applied in a One Health world.

HEALTH PROMOTION GUIDING PRINCIPLES AND VALUES FOR AN INTERCONNECTED WORLD

The pace of social and ecological changes, the interconnections and interdependencies of species and ecosystems, and the problems this complexity creates require prompt attention. In a world of concurrent interacting harms, tackling one disease in one species at a time is neither practical nor efficient. We cannot always rely on the next new problem to inspire action. Health promotion needs to focus on promoting health across species in advance of harm and ensure that action to promote the health of one species does not adversely impact the health of others (see Chapter 3 for more discussion on interspecies health equity).

The concept of health is conceived and applied differently depending on the species, context, and culture (see Chapter 2). Promoting the health of chickens in a subtropical extensive farm, for example, will involve different goals, priorities, and partners than promoting the health of marginalized poor people in an urban inner-city environment. Health promoters must be sensitive to the variety of health goals that arise from the differing opinions of what constitutes health and its determinants. There are, however, commonalities between how one could approach health promotion in such diverse situations. Health promotion generally sees health as more than being physically and mentally well. It includes the ability to use skills, knowledge, and resources that give individuals or populations the ability to manage and even to change their surroundings and circumstances to feel well and live a fulfilling life. Health promotion builds capacity to cope with the circumstances of everyday living.

Health promotion was founded on principles of justice, equity, empowerment, and self-determination (Davies, 2013). Health promotion has the dual task of improving health and increasing opportunities to have more control over it. It tries to make it easier for people to make good choices that lead to healthier conditions by equipping them to make healthy choices while also creating conditions that make those choices easier. Promoting health means improving, advancing, supporting, encouraging, and placing health higher on personal and public agendas. Health promotion that uses a true socio-ecological perspective must mediate relationships between people, animals, and their environments by linking the actions individuals take in their own interest with outcomes that help the other domains. In doing so, it creates the social and environmental conditions conducive to health for all.

Health promotion primarily takes place outside of the health care sector. Let us clarify that sentence. There is a difference between most of clinical care (which can be conceived as disease service) and health promotion. Health promotion can be interested in the question of why some people or animals have better access to and benefit from clinical care services. This might involve questions of why some marginalized people fail to receive the care they require. It may involve asking why non-charismatic wildlife species fail to get the protective services they need or why people in poverty have less access to veterinary care for their animals. Many of the impediments to access to care, such as cultural biases, economic opportunities, or geographic location, fall outside of the diseases service sector. Broadly speaking, health promotion takes place outside of the health care clinical service sector because the personal, social, and environmental determinants of health are generally also found outside of that sector.

Health promotion tries to understand and influence social and environmental capacity to be healthy rather than target biomedical or technical interventions to take people, animals, or environments out of a diseased state. Health promotion perspectives and tools fall across a spectrum ranging from understanding human behaviour that influences the need for, access to, and use of clinical services to work on the moral and political environments that influence health equity (Davies, 2013). Regardless of where health promoters sit in this spectrum, they serve the needs of the population as a whole in the context of their daily lives. They seek opportunities to promote action to protect and sustain the root sources of health.

Once we advocate for a model of health promotion that sees Health for All as an interspecies and intergenerational activity, it becomes increasingly complex to understand, implement, and evaluate multifaceted health promotion efforts. There is no guarantee that all the authors of this book share the same perspective of health promotion, nor do they recognize a shared set of values and principles that drive their actions at the animal-health-society interface. Despite these differences, there are shared core values, perspectives, and methods that resonate across fields, disciplines, and approaches that work at the human-animal-environment nexus (Table 4.1).

A growing number of fields look at the complex of physical, social, economic, and environmental relationships that determine health, but they emphasize different aspects of those relationships. EcoHealth, for example, goes beyond health promotion's "traditional" emphasis on the social determinants of health and uses a transdisciplinary approach to examine the ecosocial aspects of health. Global health research uses equity-centred, problem-focused, systems-based approaches to find upstream determinants that could make people more resilient to social and ecological factors impacting their health (Stephen and Daibes, 2010). For example, significant attention has been placed on women's empowerment as a path to food security. This can be illustrated by a study in Bangladesh that found increases in women's empowerment were positively associated with calorie availability and dietary diversity at the household level (Sraboni et al., 2014). While little is written about health promotion in veterinary medicine, the field has adapted and applied some core concepts and methods familiar to health promotion in its efforts to promote sustainable food systems, deliver disease control programmes, and protect wildlife health. Conservation science applies some of the values, perspectives, and tools found in health promotion in order to achieve conservations goals. Community-based conservation projects, for example, need to pay attention to how people, and especially poorer people, are enabled to take more control over their own lives and secure a better livelihood in the face of conservation

TABLE 4.1
Ten Perspectives Shared by Many Allied Health Disciplines Trying to Promote and Protect Health of Different Species in Different Situations

People and animals exist in multiple scales, from the individual to group, community, population, and ecosystem level, and from local to regional to global scales. Health is affected by and affects relationships within and across these scales
The knowledge, culture, resources, skills, and processes people bring with them to a health promotion situation are valued. Those being asked to make a choice or implement an action need to actively participate in the health promotion process
People construct their knowledge and worldviews differently based on their culture, experience, and expertise. There is not one truth that will work for everyone, so health promotion needs to be pragmatic and adaptive
New knowledge has its own merits but without the capacity to mobilize knowledge to the people who can act to make the necessary changes, little will come of it
Health outcomes like longevity or lack of disease are not the end goals. The goal is to sustainably allow individuals or populations to fulfil their purpose, whether social, ecological, or otherwise, and therefore have a good life
Relationships between those who have power and those who do not should not result in differences in health that are unnecessary, avoidable, unfair, or unjust
Health promotion is context specific. Activities and interventions need to account for local circumstances and conditions
Health improvements depend on the development of an environment that is conducive to and supportive of healthy choices and actions
Social forces and social relationships are critical determinants of what we value as being healthy and why we undertake health-promoting activities
Health promotion directs actions towards the socio-ecological determinants of health. It requires co-operation across multiple sectors and combines diverse but complementary approaches

Sources: Adapted from Gregg and O'Hara, 2007; WHO, 1984.

actions (Berkes, 2004). Although the term health promotion may not be widely used outside of the human population health and public health fields, its core values and concepts are found wherever the goal is to help people take actions to protect health by proactively addressing the social and environmental obstacles to making and implementing healthy choices.

EMPOWERMENT – A KEY TO HEALTH PROMOTION

The word empowerment comes up a lot in health promotion literature. People are empowered when they have access to and can use their knowledge, self-esteem, self-confidence, self-efficacy, and autonomy to act (Tengland, 2007). Empowerment helps people gain control over and influence events and outcomes important to them. In the case of this book, empowerment means gaining mastery over the knowledge needed to influence circumstances that protect and promote the health of animals and ecosystems along with the health of people and their communities. Feelings of empowerment fuel action.

The path to empowerment begins by identifying personally meaningful goals related to a person's or his or her communities' power to make and implement decisions (Cattaneo and Chapman, 2010). Goals leading to empowerment cannot be imposed upon a person or a group. Imposing a goal, when it differs from another person's, will not facilitate that person's empowerment process. Empowerment requires a person to believe in his or her abilities to act when he or she has the power to do so (this is known as their self-efficacy). A person's perceptions of the success or failure of an action will strongly influence his or her perceptions of self-efficacy, and thus the ability or willingness to keep acting. Once empowered, people can have some control or influence over the next steps to accomplish their goals. Helping people become empowered requires developing an understanding of the relevant social context, power dynamics at play, the possible means to achieve control over attaining goals, and the resources needed to set a course of action.

Empowering is not without its challenges. Power dynamics play a role. For example, when we give an individual more power over his or her own choices, we may decrease the power of someone else (or some organization) to determine those choices. This tension can manifest itself differently across various sectors as well as in different cultures. Actions that empower one community in one situation could disempower another community in another situation. These types of tensions have been manifested as previous colonial powers negotiate new ways to manage wildlife or natural resources with Indigenous governments in co-management systems. Co-management provides direct involvement for Indigenous governments and organizations in wildlife management, sharing management responsibility between governments, users, and renewable resource boards rather than retaining power in one government agency. Trying to change power dynamics can have unintended and negative consequences if it triggers actions from a repressive authority. Individuals or groups that are marginalized as power dynamics shift may not equitably benefit from actions intended to empower the entire community.

Further challenges can arise by a choice to make top-down or bottom-up decisions on priorities and actions. What if a health promoter, for example, decides the priorities for action should focus on a disease-prevention programme but the community members prioritize food security actions? Who decides who needs to be empowered and for what? While there is significant literature exploring these tensions, questions, and challenges, there is no one best approach to empower people or communities. A careful understanding of needs, power dynamics, and the local circumstances is essential before making any decisions on who to empower, for what, and how.

STRATEGIES AND COMPETENCES FOR CROSS-SECTORAL HEALTH PROMOTION

The WHO's Ottawa Charter for Health Promotion (1986) set five strategies that are essential for successful health promotion (Table 4.2).

TABLE 4.2
Five Strategies for Health Promotion Established in the Ottawa Charter for Health Promotion (WHO, 1986)

Build healthy public policy
Create supportive environments
Strengthen community action
Develop personal skills
Reorient health services

The first strategy, the healthy public policy approach, refers to the need to be concerned with the implications of all public policies on the social, personal, biological, and ecological influences on health (known as the determinants of health; see Chapter 1) and not just health policy. The aim is to shine a health light on all policies that can influence social and natural environments affecting health and the ability to make healthy choices. (See Chapter 10 for a more detailed discussion of healthy public policy.) This supports the creation of situations and circumstances that allow people to be healthy. The Ottawa Charter recognized the need not only to create social circumstances for health but also the criticality of protecting the natural and built environments and natural resources when constructing supportive health environments. Community empowerment and actions are at the heart of health promotion, making strengthened community action a health promotion pillar. Translating community support into personal action requires people to develop the personal skills to exercise more control over their own health and over the environments where they live, work, and play. In doing so, they can create and access options to be healthy. The fifth strategy encouraged us to see health systems not as doctors, nurses, and hospitals but as a collaboration of individuals, community groups, professionals, institutions, and governments which connect the health sector with broader social, political, economic, physical, and natural environmental components, the goal is to move beyond providing clinical and curative services to partnerships in the pursuit of health. Intersectoral action, interorganizational partnerships, community engagement and participation, creating healthy settings, political commitment, funding and infrastructure, and awareness of the socio-environmental context are all essential for successful health promotion.

Health promoters strategically work with others to influence behaviours, decisions, and determinants of health across sectors (WHO, 1998). There have been and continue to be efforts to define essential health promotion skills, attitudes, and knowledge. Table 4.3 synthesizes some of these efforts and generalizes them for work across species.

The ideals and principles of health promotion tend to be shared across jurisdictions, but their emphasis and organization can differ with different target audiences (Hyndman, 2009). Patrick et al., (2012), for example, emphasized the need for health promotion to re-engage with the interconnectedness of humans and the

TABLE 4.3
Cross-Sectoral Health Promotion Competencies for Reciprocal Care

Attitude or Perspective	Domains of Thinking or Doing
Systems thinking	Understands the socio-ecological model of health and the determinants of health framework and can apply them to problem assessment and programme planning
	Integrates, assesses, and communicates complex and diverse evidence and types of knowledge in an understandable manner
	Can identify the people and settings involved in an issue as well as their interrelationships and current situations
Facilitates change	Can identify, engage, and work with communities and individuals seeking or requiring help to act or make health-promoting decisions
	Can develop strategic action plans that are feasible, acceptable, and effective in the realities encountered in the involved circumstances, communities, and individuals
	Supports and enables incremental change to a healthier situation using available knowledge and resources while promoting longer-term gains coming from new knowledge and changed situations
Evidence-based and action-oriented	Can systematically and effectively bridge the knowing-to-doing gap in people and organizations
	Uses evidence-based approaches to planning, recognizing the diverse types of knowledge and perceptions of evidence relevant to health promotion
Adapts to local context and circumstances	Can build partnerships, alliances, and coalitions to develop accessible, feasible, and effective actions sensitive to and relevant to the local setting and local people
	Can identify locally relevant structural and social barriers and opportunities to equitable and just distributions of health determinants, resources, and services
	Can identify critical social, environmental, or behavioural factors that promote or compromise meeting health goals within a local socio-ecological system

Sources: Adapted from Allegrante et al., 2009; Hyndman, 2009.

natural environment and adopt ideas from the sustainability field while remaining attentive to core competencies, if it aims to rise to the challenge of climate change. Hancock (2015) similarly recommended more emphasis on the "eco" side of the socio-ecological approach if population health and health promotion want to deal with the new planetary challenges emerging in the Anthropocene. Although a common language by which health promotion and its meaning can be communicated beyond the human health sector is still lacking, there is substantial overlap in lists of competencies for One Health and EcoHealth (e.g. Charron, 2012; Stephen and Stemshorn, 2016) with those of health promotion. Competencies for partnering, collaborating, leading, systems thinking, strategic planning, communicating, evaluating, and enabling action across disciplines arise whenever people seek to mobilize knowledge and resources for sustainable change.

Health promoters need leadership skills to imagine and implement strategic actions for developing policy, mobilizing, and managing resources for health promotion, and building capacity. Advocacy with and on behalf of individuals and communities or other species is part of the health promoters' purview. Being able to assess the needs and assets in communities and systems will help to develop measurable goals and objectives that are acceptable to the community as well as target research and interventions to develop the resources, capacity, and knowledge to achieve those goals. Health promoters must be able to determine the effectiveness of actions and policies in order to enable evidence-based, culturally sensitive, and ethical actions that can effectively produce the greatest improvements of health within the available resources. In addition to these skills, an effective health promoter needs to be able to work well with others. Working collaboratively across disciplines, sectors, and partners is essential to enhance the impact and sustainability of health promotion programmes and policies (Fertman et al., 2016).

THE SOCIO-ECOLOGICAL APPROACH

The socio-ecological model of health was introduced in Chapter 1. It was first used in the 1970s to understand human development (Kilanowski, 2017). At its essence, the socio-ecological model recognizes that there is a nested hierarchy of interacting factors that determine the resources and opportunities that influence an individual's well-being. Key messages for One Health from the socio-ecological approach are provided in Table 4.4. The model can be thought of as

TABLE 4.4
The Key Messages of the Socio-Ecological Model of Health

Key Messages
No single factor can explain why one individual (or community, or species, or systems) is healthy and others are not
There are many social, built, and natural environments with which an individual's biological endowment interacts to result in healthier outcomes
There are multiple levels of influence to shape the opportunities for healthy choices. Directing action at roots causes found in the outer circles of the socio-ecological model (known as upstream determinants) can have broad and significant influences on downstream outcomes
The interrelations among various circles in the socio-ecological model are dynamic and context-specific
The most effective approach to health promotion uses a combination of interventions at all levels of the model

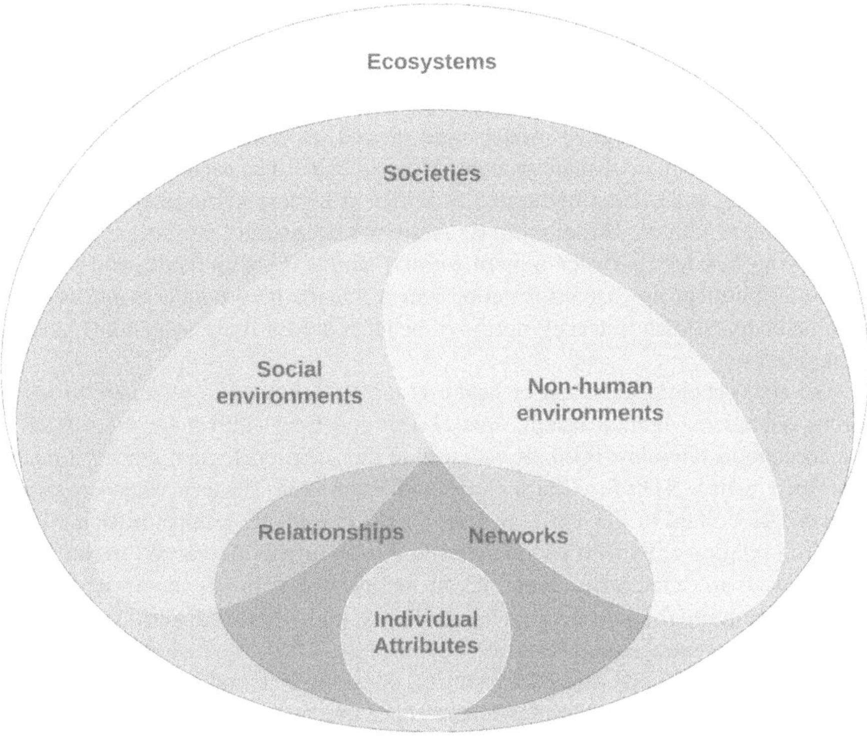

FIGURE 4.1 An example of nested hierarchy to represent the socio-ecological model of health. The model emphasizes how relationships and forces beyond an individual level influence health of individuals, populations, and communities. The model represents how behaviours, decisions, and actions to influence health can be found beyond the level of the individual.

a series of nested relationships, with an individual at the core (Figure 4.1). The inner core contains an individual's attributes and interactions and relationships with his or her immediate surroundings. It contains the individuals' biological endowment, skills, and knowledge. The social and non-human environments include interactions with other individuals and the needs for daily living (like food and water) and hazards. The third level exerts social forces that positively or negatively influence relationships that are subordinate in this nested hierarchy. Examples here include social organizations and policies. Next come the capacities and services that ecosystems give us, that provide the "raw materials" found in our relationships with our immediate world that enable social decisions. Beyond this are global and planetary forces, both historic and current, that influence all aspects of the system, such as climate change. This model can be readily adapted to different circumstance and species as all types of health (individual, species, systems, communities) are influenced by higher and lower levels of organization.

There are many examples of socio-ecological thinking in the One Health and EcoHealth. Parkes and Horwitz (2009), for example, used watersheds as an integrating concept to more effectively link human well-being, health promotion, and disease prevention with ecosystem management at the local, regional, and global scales. The socio-ecological model underpinned calls for systems-based public health approaches to a climate change (Morris, 2010). The theory and practice of understanding and managing human activities as a socio-ecological system were used extensively in the Millennium Ecosystem Assessment (Zinsstag et al., 2011). Identifying linkages between human, animal, and ecosystem health and strategically communicating that information can influence how people consider their own behaviours such that they optimize benefits across these systems (Lapinski et al., 2015).

The socio-ecological model of health is most often used in reference to social forces influencing human health. One Health and EcoHealth more often refer to socio-ecological systems thinking rather than the socio-ecological model of health described in the UN Ottawa Charter for Health Promotion. The term socio-ecological system was coined in the 1970s to refer to an ecological system's intricate links with and relationships with social systems. It was further developed in the 1990s to find ways to "match the dynamics of institutions with the dynamics of ecosystems for mutual socio-ecological resilience and improved performance" for local resource management (Colding and Barthel, 2019). Agriculture, sustainable development, infectious disease dynamics, conservation, and veterinary services delivery are just a few examples where there are problems being conceived of and managed as socio-ecological systems. Rüegg et al. (2017) noted that "many health decisions are linked to dilemmas between scales, namely, individual versus social or global ecological interests. The solution lies in a continuous process of negotiation that includes all stakeholders and results in benefits from the interaction between different sectors. One Health acknowledges that people's choices are made within the context of economic, social and cultural values." Chapter 7 provides more detail on systems thinking in a health context.

Systems thinking can be hard and it can be messy, but there are three core systems concepts that can guide your thinking. First, a system is made up of parts, relationships, functions, and products. Understanding the relationships is as much or more important than knowing the parts. Second, because these relationships can be complex and dynamic, and because there are feedbacks, synergies, and antagonisms within relationships, systems do not act in a linear predictable fashion. Surprises are to be expected. Third, the goal of applying a systems perspective is not to elucidate the intricacies of all parts, products, functions, and all relationships. The goal is to know them well enough to find entry points that allow you to start to work in the system to find the leverage points where a small intervention can have a meaningful impact. Systems thinking has been applied to many different fields from health, to computers, to biology, and more. Interested readers will find few challenges in locating additional books and publications on systems thinking, but we encourage you to start exploring systems thinking in EcoHealth to bridge systems science with health management.

ANIMALS, HEALTH, AND SOCIETY

Rudolf Virchow, the 19th-century physician, pathologist, and biologist, is often quoted as saying, "between animal and human medicine there is no dividing line – nor should there be." There is abundant literature on the influences of animals on human health and on the dependencies of animal health and ecosystems' integrity on human decisions. As seen in the diverse examples throughout this book, "animals are part of the fabric of life, all around the world" (Rock et al., 2009). Animals are sources of many determinants of human health. Lack of attention to the interdependence of human and animal health neglects opportunities for reciprocal care in health promotion.

There is no shortage of examples in recent years exemplifying the need to work across the species divide to find shared solutions to shared problems through collective, cross-sectoral action. Antimicrobial resistance is an archetype of the co-benefits of a collaborative and cross-sectoral approach to a health threat. A One Health assessment of antimicrobial resistance would not only examine the implications for clinical care of people but also wider societal costs, impacts on animal health and welfare, higher costs of animal-origin food production, and reduced consumer confidence in food safety and international trade (Queenan et al., 2016). A greater understanding is required by the medical and veterinary sectors alike of how health-seeking behaviour – for ourselves, our families, and our animals – influences the pandemics of multi-drug-resistant pathogens (Queenan et al., 2016) (see Chapter 23 for a more detailed consideration of this issue). Romenalli et al. (2014) argued that a better understanding of the relationships between biodiversity, health, and disease presents "major opportunities for policy development and can enhance our understanding of how health-focused measures influence biodiversity and how conservation measures affect health." The dynamic and complex interdependence of people, animals, and the environment, and how they are collectively impacted by disasters, underscores the need for a cross-species health promotion approaches to disaster management (Gallagher et al., in press). A comprehensive, systems thinking approach that incorporates human, animal, and environmental considerations into each of the continuous and overlapping disaster management phases of mitigation, preparedness, response, and recovery is essential for inclusive and multi-sectoral actions to mitigate a disaster's impacts. As a final example, the similarities between obesity in people and companion animals speak to the need to address shared environmental and lifestyle elements of this multifactorial disease (Bomberg et al., 2017).

There are even more examples of how health of one species can promote the health of another. Livestock contribute to health security either directly through consumption or indirectly through sales. Without healthy livestock, income and nutritional benefits would evaporate for livestock farmers (which include many of the world's poor people), hence the focus on livestock disease control for international organizations such as the World Organization for Animal Health (also known as the OIE). Chapter 18 illustrates the important role healthy animals play in poverty reduction, food security, and farm sustainability in Sri Lanka.

Livestock provide cash for farmers and their families, fertilizer in the form of manure, draft power, and they can boost sustainable crop production in mixed crop-livestock systems (Smith et al., 2013). Livestock production can transform human well-being by enhancing food and income security to pay for education, basic needs, and other health care. Wildlife and fish provide consumptive and non-consumptive economic resources for many Indigenous, rural, and remote communities around the world (Stephen and Duncan, 2017). They provide a sense of identity and purpose for a variety of natural resource workers and Aboriginal peoples. Recreational fishing, hunting, and other nature uses can promote exercise and group membership leading to healthy lifestyles and diets. Personal and spiritual connections with animals can influence healthy childhood development, mental health, and personal coping skills. O'Bryan et al. (2018) argue that predators and scavengers have far-reaching benefits to human health through disease mitigation, agricultural production, and waste-disposal services, and at the same time they show that human decisions are the key driver to the continued existence of these species.

In return for all these benefits, human actions and decisions are the primary modifier of the determinants of animal and ecosystem health. How we use animals, house them, shape their built or natural environments, and regulate their access to vital needs for living determine if some species thrive and others disappear. Unfortunately, constructive, well-balanced, and informed practices and policies that try to co-manage animal, human, and environmental health are rare. Many fields and programmes talk about it but few equally invest in concurrent care of people, animals, and environments. The question is no longer: Should we act together?" but "With whom can we act, and how can we act together?"

Reciprocal Care Health Promotion Teams

When there is a clear and urgent goal that transcends the needs, interests, and abilities of individuals, teams are necessary. The changes being witnessed in the Anthropocene that transcend single disciplines are shifting scientific disciplines away from their own unique solutions to problems. Given the complexity of health issues at the animal-environment-society interface, various types of partnerships will be essential to ensure that investigations and interventions are relevant to the priority population for whom they are intended, and are meaningful and appropriate to the targeted communities, populations, and species.

There is a variety of terms used to describe cooperative work of different groups or disciplines. Descriptors using the prefix "multi" are assigned to situations where different professional groups or disciplines work together. Multidisciplinary approaches allow teams to address a problem by more than one perspective, but members stay rooted in their own disciplines. Building the circumstances and situations conducive to reciprocal care of human, animal, and environmental health requires a wide suite of skills, capacities, connections, and knowledge, and therefore, a reciprocal care health promotion team must, at the least, have a multidisciplinary organization and orientation. When the prefix "inter" is assigned to a team, it does more than work together. It also integrates knowledge and methods from different disciplines, striving to transfer knowledge from one discipline to another in order to synthesize knowledge and methods into a single coherent approach. Once the term "trans" is assigned to the team, such as transdisciplinary, the team is looking to transcend any disciplinary approach. Most often it is used to describe teams that go beyond academic interdisciplinarity to engage directly with the production and use of knowledge outside the academic environment to support action-oriented work. The composition of these various types of teams and the processes they use to work together will, and should, vary with the context of the problems they address. EcoHealth and, increasingly, One Health are using transdisciplinary approaches to address the health of people, animals, and ecosystems.

Health promotion uses interdisciplinary and transdisciplinary perspectives and methods. It brings together practitioners from varied disciplinary backgrounds with the communities they serve to collectively address key determinants of health in the most effective manner (Orme et al., 2007). Team members bring with them their own histories, values, and knowledge. Having a shared conceptual framework for the problem at hand is, therefore, essential for transdisciplinary teams. Without a shared conceptual framework, team members might find themselves working on the same problem but on "different paths," which could lead to disjointed outcomes and delays in research-to-action timelines (Min et al., 2013). There are several other challenges transdisciplinary teams confront such as; recognizing that a way of thinking that is appropriate in one situation can be inappropriate in another; unbalanced problem ownership; conflicts in the selection and application of methods; and even finding agreement that a transdisciplinary approach is appropriate (Norris et al., 2016). It can be challenging to overcome disciplinary perspectives in transdisciplinary teams. Dialogue at the outset of team building on how various forms of knowledge and perspective will be shared, respected, and accommodated can help avoid future team conflict.

HEALTH PROMOTION IN A ONE HEALTH WORLD

A foundational concept of health promotion is, "The overall guiding principle for the world, nations, regions and communities alike, is the need to encourage reciprocal maintenance – to take care of each other, our communities and our natural environment" (WHO, 1986). It has been rare that coordinated and collaborative efforts truly seek policy and programme outcomes that concurrently take care of each other, our communities, and the natural environment. Too often there are winners and there are losers. Conservation programmes that protect wildlife but drive marginalized communities into poverty are not health promotion in this sense. Nor are programmes protecting people from emerging infections by eliminating wildlife reservoirs of those infections. In the world of specialists, it is rare to find a person who can think across and within the relationships between ourselves, our communities, and our natural environments. Chapter 5 provides some thinking on how we might get better at doing so, but elements of this trans-species thinking can be found throughout many of the case studies presented in this book. The goal of this book is to provide some foundational concepts and case studies to encourage future practitioners of true reciprocal care who are willing to become "specialized generalists": that is, those people skilled at looking across disciplines, values, species, and generations; able to inspire incremental improvements in health that provide mutual benefits to animals, societies, individuals, and ecosystems.

REFERENCES

Allegrante, John P., Margaret M. Barry, Collins O. Airhihenbuwa, M. Elaine Auld, Janet L. Collins, Marie-Claude Lamarre, Gudjon Magnusson et al. "Domains of Core Competency, Standards, and Quality Assurance for Building Global Capacity in Health Promotion: The Galway Consensus Conference Statement." *Health Education and Behaviour*, 36, no. 3 (2009): 476–482.

Berkes, Fikret. "Rethinking Community-Based Conservation." *Conservation Biology*, 18, no. 3 (2004): 621–630.

Bomberg, E., L. Birch, N. Endenburg, A. J. German, J. Neilson, H. Seligman, G. Takashima et al. "The Financial Costs, Behaviour and Psychology of Obesity: A One Health Analysis." *Journal of Comparative Pathology*, 156, no. 4 (2017): 310–325.

Cattaneo, Lauren Bennett, and Aliya R. Chapman. "The Process of Empowerment: A Model for Use in Research and Practice." *American Psychologist*, 65, no. 7 (2010): 646.

Charron, Dominique F., "Ecohealth Research in Practice," in *Ecohealth Research in Practice* (New York: Springer, 2012), 255–271.

Colding, Johan, and Stephan Barthel. "Exploring the Social-Ecological Systems Discourse 20 Years Later." *Ecology and Society*, 24, no. 1 (2019): 423–432.

Davies, John Kenneth. "Health Promotion: A Unique Discipline?" Health Promotion Forum of New Zealand, 2013. http://hauora.co.nz/assets/files/Occasional%20Papers/Health%20Promotion%20is%20a%20Discipline%2027%20Nov(1).pdf (accessed May 2020).

Fertman, Carl I., Diane D. Allensworth, and M. Elaine Auld. "What Are Health Promotion Programs?" in *Health Promotion Programs: From Theory to Practice*, eds. Carl I. Fertman and Diane D. Allensworth (John Wiley & Sons, Inc., 2016), 3–28.

Gallagher, Christa, Barbara Jones, and Jimmy Tickel. "Towards Resilience: The One Health Approach in Disasters" in *One Health: The Theory and Practice of Integrated Health Approaches*, 2nd Edition, eds. Jakob Zinsstag, Esther Schelling, Lisa Crump, Maxine Whittaker, Marcel Tanner, and Craig Stephen (CABI Boston, in press).

Gregg, Jane, and Lily O'Hara. "Values and Principles Evident in Current Health Promotion Practice." *Health Promotion Journal of Australia*, 18, no. 1 (2007): 7–11.

Hancock, Trevor. "Population Health Promotion 2.0: An Eco-Social Approach to Public Health in the Anthropocene." *Canadian Journal of Public Health*, 106, no. 4 (2015): e252–e255.

Hyndman, Brian. "Towards the Development of Skills-Based Health Promotion Competencies: The Canadian Experience." *Global Health Promotion*, 16, no. 2 (2009): 51–55.

Kilanowski, Jill F. "Breadth of the Socio-Ecological Model." *Journal of Agromedicine*, 22, no. 4 (2017): 295–297.

Lapinski, Maria Knight, Julie A. Funk, and Lauren T. Moccia. "Recommendations for the Role of Social Science Research in One Health." *Social Science and Medicine*, 129 (2015): 51–60.

Min, Bernard, Lisa K. Allen-Scott, and B. Buntain. "Transdisciplinary Research for Complex One Health Issues: A Scoping Review of Key Concepts." *Preventive Veterinary Medicine*, 112, no. 3–4 (2013): 222–229.

Morris, George P. "Ecological Public Health and Climate Change Policy." *Perspectives in Public Health*, 130, no. 1 (2010): 34–40.

Norris, Patricia E., Michael O'Rourke, Alex S. Mayer, and Kathleen E. Halvorsen. "Managing the Wicked Problem of Transdisciplinary Team Formation in Socio-Ecological Systems." *Landscape and Urban Planning*, 154 (2016): 115–122.

O'Bryan, Christopher J., Alexander R. Braczkowski, Hawthorne L. Beyer, Neil H. Carter, James EM Watson, and Eve McDonald-Madden. "The Contribution of Predators and Scavengers to Human Well-Being." *Nature Ecology and Evolution*, 2, no. 2 (2018): 229–236.

Orme, Judy, Nick De Viggiani, Jennie Naidoo, and Teri Knight. "Missed Opportunities? Locating Health Promotion Within Multidisciplinary Public Health." *Public Health*, 121, no. 6 (2007): 414–419.

Parkes, Margot W., and Pierre Horwitz. "Water, Ecology and Health: Ecosystems as Settings for Promoting Health and Sustainability." *Health Promotion International*, 24, no. 1 (2009): 94–102.

Patrick, Rebecca, Teresa Capetola, Mardie Townsend, and Sonia Nuttman. "Health Promotion and Climate Change: Exploring the Core Competencies Required for Action." *Health Promotion International*, 27, no. 4 (2012): 475–485.

Queenan, Kevin, Barbara Häsler, and Jonathan Rushton. "A One Health Approach to Antimicrobial Resistance Surveillance: Is There a Business Case for it?" *International Journal of Antimicrobial Agents*, 48, no. 4 (2016): 422–427.

Rock, Melanie, Bonnie J. Buntain, Jennifer M. Hatfield, and Benedikt Hallgrímsson. "Animal–Human Connections, "One Health," and the Syndemic Approach to Prevention." *Social Science and Medicine*, 68, no. 6 (2009): 991–995.

Romanelli, C., H. D. Cooper, and B. F. De Souza Dias. "The Integration of Biodiversity into One Health." *Revue Scientifique et Technique-Office International des Epizooties*, 33, no. 2 (2014): 487–496.

Rüegg, Simon R., Barry J. McMahon, Barbara Häsler, Roberto Esposito, Liza Rosenbaum Nielsen, Chinwe Ifejika Speranza, Timothy Ehlinger et al. "A Blueprint to Evaluate One Health." *Frontiers in Public Health*, 5 (2017): 20.

Smith, Jimmy, Keith Sones, Delia Grace, Susan MacMillan, Shirley Tarawali, and Mario Herrero. "Beyond Milk, Meat, and Eggs: Role of Livestock in Food and Nutrition Security." *Animal Frontiers*, 3, no. 1 (2013): 6–13.

Sraboni, Esha, Hazel J. Malapit, Agnes R. Quisumbing, and Akhter U. Ahmed. "Women's Empowerment in Agriculture: What Role for Food Security in Bangladesh?" *World Development*, 61 (2014): 11–52.

Stephen, Craig, and Ibrahim Daibes. "Defining Features of the Practice of Global Health Research: An Examination of 14 Global Health Research Teams." *Global Health Action*, 3, no. 1 (2010): 5188.

Stephen, Craig, and Colleen Duncan. "Can Wildlife Surveillance Contribute to Public Health Preparedness for Climate Change? A Canadian Perspective." *Climatic Change*, 141, no. 2 (2017): 259–271.

Stephen, Craig, and Barry Stemshorn. "Leadership, Governance and Partnerships are Essential One Health Competencies." *One Health*, 2, (2016): 161–163.

Tengland, Per-Anders. "Empowerment: A Goal or a Means for Health Promotion?" *Medicine, Health Care and Philosophy*, 10, no. 2 (2007): 197.

WHO. "Global Strategy for Health for All by the Year 2000." World Health Organization, 1981: 90.

WHO. "Health Promotion: A Discussion Document on the Concept and Principles: Summary Report of the Working Group on Concept and Principles of Health Promotion, Copenhagen, 9–13 July 1984." No. ICP/HSR 602 (m01), 1984. Copenhagen: WHO Regional Office for Europe.

WHO. "The Ottawa Charter for Health Promotion." World Health Organization, 1986. https://www.who.int/healthpromotion/conferences/previous/ottawa/en/ (accessed March 2020).

WHO. "Health Promotion Glossary." World Health Organization, 1998. www.who.int/healthpromotion/about/HPR%20Glossary%201998.pdf (accessed May 2020).

Zinsstag, Jakob, Esther Schelling, David Waltner-Toews, and Marcel Tanner. "From "One Medicine" to "One Health" and Systemic Approaches to Health and Well-Being." *Preventive Veterinary Medicine*, 101, no. 3–4 (2011): 148–156.

Part 2

Core Concepts

5 Working Together for WHOLE Systems: Approaching Well-Being and Health While Oriented to Living Systems and Equity

Margot W. Parkes

CONTENTS

INTRODUCTION

Drawing on precedents and convergence across disciplines, sectors, and cultures, this chapter identifies, examines, and profiles key characteristics of an integrative shift in how health is protected, promoted, and planned for. A preoccupation with specialization is starting to be balanced with a re-emphasis on integrative approaches that include and are informed by Indigenous knowledges that have fostered this awareness for many thousands of years (Durie, 2004; McGregor, 2004; Parkes, 2011; Redvers, 2018). This chapter examines the value of addressing the WHOLE (**W**ell-being and **H**ealth **O**riented to **L**iving Systems and **E**quity) as a foundation for a healthy, equitable, and ecologically sustainable future. Whether the focus is on individuals, or groups, of humans, animals, or other species, any approach to protecting, promoting, and planning for health and well-being will fall short if it is not oriented to equity as well as ecosystems (living systems) within our planetary home. A WHOLE perspective recognizes that without the living systems we depend on, health, equity, and life itself are simply not possible. A living systems view also demands attention to equity, across cultures, generations, species, and society and across scales from the local to the global (see Capra and Luisi, 2014; Wahl, 2016; Horwitz and Parkes, 2019).

In a time when our health systems are being confronted by what can seem to be competing and converging imperatives ranging from climate change to the COVID-19 pandemic, WHOLE approaches are more necessary than ever. An emphasis on WHOLE systems resonates with other efforts paying attention to transformative, creative, collective, and integrative approaches to health in social and ecological context, especially those that foster regenerative rather than degenerative processes (Waltner-Toews, 2004; Brown, 2007; Poland and Dooris, 2010; Fullerton, 2015; Bartlett et al., 2015; Benne and Mang, 2015; Wahl, 2016; Gislason et al., 2018; Ratima et al., 2019; Redvers, 2018; White, 2019). Orientation to WHOLE systems demands that we work together in new ways and focus our attention to both knowledge(s) and action(s). A WHOLE approach draws on many efforts to appraise health in relation to both ecological and social contexts, spanning, for example: ecosocial approaches; socio-ecological health; social-ecological systems and health; ecohealth; planetary health; and One Health[1] (Zinsstag et al., 2011; Hallström et al., 2015; Hancock et al., 2017; Buse et al., 2018; Harrison et al., 2019; Parkes et al., 2020). Attention to the WHOLE brings to the fore interrelationships between knowledge, participation, and action from the outset of our efforts. This requires awareness of what we hold in mind as we enter into a WHOLE-systems approach, and the questions we ask as we progress.

A primary point of reference and starting point, therefore, is the need for attention to both content and process, where *what* we know is directly in relation to *how* we know it and will inform *how* responses are designed (Parkes et al., 2012; Cole et al., 2018). An *integrative turn*, in relation to concepts, content and knowledge, has increased potential for impact if it is combined and "matched" with an overtly *collaborative turn*, in relation to processes, practices, and actions. A combined integrative and collaborative turn leads to an emphasis on asking different kinds of questions (see, for example, Brown, 2010; Wahl, 2016). Along these lines, the WHOLE acronym can also stand for "Who and How are we Open to Listen, and Engage/exchange?" This question extends attention beyond the act of integrating knowledge(s) about a specific content or topic area, towards the processes of interaction, exchange, and potential collaboration among knowledge holders (Parkes et al., 2012). This combined content and process emphasis underscores an active, dynamic, and oftentimes complex relationship between issues being observed (the objects and subjects under consideration) and the interactions of the observer(s) of the issue, including those who may be working together to do the observing, and how this will influence what is addressed. Awareness of interrelationships between observations and observer(s), between knowledge and action, and the nuances of integration, participation, and collaboration are important considerations when seeking to address complex health, ecological, and equity concerns (see Brown, 2010; Parkes et al., 2012; Parkes, 2015; Brown et al., 2019).

An implication of the WHOLE view is the respect for, and relevance of, crossing boundaries between worldviews and cultures, between knowledge and action, between research and policy, and between theory and practice. A focus on application and implementation (and learning about both) should be familiar to many working in human health care and veterinary practice. It is also increasingly important when health and well-being are considered beyond individuals or single species, and across the range of personal, public, population, and planetary health.

The WHOLE approach asks us to shift from describing and analyzing "problems" and "solutions" (e.g. how can I describe this problem to select the "right" solution?) towards strength-based orientations that privilege contextually nuanced, regenerative, restorative, and emergent principles and perspectives (e.g. what do we need to know and understand about this system in order to enhance conditions for thriving?). A WHOLE view also highlights axes of equity that need more attention, specifically in relation to future generations and those who have less or no voice within our existing knowledge generation and decision-making processes, including other species beyond humans (see de Groot, 1998; Masuda et al., 2008; Poland and Dooris, 2010; Masuda et al., 2014; and Chapter 3 for more on healthy equity in One Health). A WHOLE perspective prompts the question: Who or what is this work serving? and drives a need to respond in ways that identify, create, and promote co-benefits for the health of humans, animals, and other species.

This chapter presents ideas and processes that can enhance and operationalize a WHOLE approach through strength-based and appreciative processes. The chapter begins by considering the role of the "specialist generalist" in undertaking or engaging with this kind of work, with attention to learning cycles (described in Brown, 2005, 2010) relevant to creating healthy WHOLE systems change. The rest of the chapter is structured around the four questions within Brown's collective learning cycle, where each learning stage poses the questions: "What should be?" "What is?" "What could be?" and "What can be?" (Brown, 2010). The chapter concludes by revisiting the role of the "specialized generalist" in understanding and responding to contemporary health and well-being challenges while oriented to living systems and equity for the benefits of both people and other species.

ADDRESSING WHOLE SYSTEMS AS A "SPECIALIZED GENERALIST"

The idea of a "specialized generalist" (Kates, 1967) is a phrase that is neither new nor clearly defined. The idea tends to arise in relation to strengthening capacity for applied knowledge and innovation, including the dynamic capabilities of "T-shaped professionals" and a combination of special expertise and collaborative capacities needed to respond to complex 21st-century challenges, spanning health, environmental, design, and systems contexts (Barile and Saviano, 2013; Boumeester, 2014; Donofrio et al., 2018; Ashhurst, 2019; Lisefski, 2019). The idea of a "specialized generalist" has particular relevance when addressing well-being and health oriented to both living systems and equity. This is because a WHOLE approach requires a *generalist orientation* (e.g. attention to and synthesis across *general principles, patterns, and practices*) supported by a *specialized suite of knowledge, attitudes, and skills* (*including specific tools and processes*), required to put general principles into practice in different contexts, settings, and circumstances. A sense of reciprocity is important for this framing: A generalist foundation that enables bridging across expertise may be supported by specialization, and specific expertise may be complemented by capabilities in communicating and working across different types of knowledge. In either case, an essential feature of this perspective is a forward-looking, applied, and collaborative emphasis on design, application, implementation, and change that is considered to be equally important, if not more, to specific expertise, research, analysis, and prediction. In the language of Boyer (1990), while the specialized generalist may value and benefit from the "scholarship of discovery," emphasis and value are also focused on the knowledge and practice advances arising from the "scholarship of teaching and learning," with the "scholarship of integration," "scholarship of application," and what Woollard (2006) describes as the "scholarship of engagement."

The role of both understanding *and* working across and among different knowledge systems has been proposed in a variety of other health-related contexts. Adapting from the WHO (2000) partnership orientation of "Towards Unity for Health," Woollard (2006) emphasized the value of working across the "partnership pentagram" of policymakers, health professionals, academic institutions, communities, and health administrators. In the context of emerging infectious diseases, Parkes et al. (2005, Figure 1) underscore the importance of applying transdisciplinary thinking to connecting not only across different disciplines but also among different types of knowledge (including communities and cultures, practitioners and field workers, disciplines, sectors, and units of governance). These efforts resonate with ongoing and expanding work that focuses on transdisciplinary research as a form of knowledge generation that transcends disciplinary boundaries and values other knowledge systems, often in relation to complex and wicked problems (see, for example, Brown, 2010; Pohl, 2011; Lang et al., 2012).

Brown's work elaborates on the need to identify and value different kinds of knowledge, in ways that combine a big-picture, integrative (also holistic) generalist perspective with the specialized skills required to actively engage different knowledge cultures. Brown recognizes "Individual" (personal lived experience), "Community" (shared place-based lived experience), "Specialized" (academic disciplines and professions), "Organizational" (strategic agendas, regulations, policies), and "Holistic" (synthesis, metaphors, images) knowledge cultures, noting that individuals are often engaged with several of these (Brown, 2010; Brown et al., 2019). Brown's "collective learning cycle" (sometimes described as a "social learning cycle") provides questions that were refined across multiple contexts, and proposed as a way to benefit from the collective knowledge that can arise working across these five knowledge cultures, in order to better address complex issues at the interface of environment, health, and community (Brown, 2007, 2010).

The four questions of Brown's (2010) collective learning cycle commence with an aspirational and asset-based posture, considering "What should be?" underscoring the point that "how we begin matters." The next part of the cycle is guided by objective orientation to "What is?": grounding in the reality, constraints, enablers of what is happening, and the parameters that can be used to assess and describe this. The question "What could be?" focuses on the potential arising when considering how to move from the current state to the desired state (e.g. from "What is?" to "What should be?") and how it could be possible to move ideas into practice to do so. A final pragmatic step focuses on "What can be?" and the practicalities arising from specific action planning associated with questions of who, how, when, where, and associated responsibilities. By completing this cycle, there is a final step to consider moving from the practicalities of "What can be?" to the development and revision of principles in relation to "What should be?" This aligns with the important question of "Now what?" (Rolfe et al., 2001) to inform whether and how the sense of direction needs to be adapted prior to embarking on the next iteration or cycle of learning.

Brown's collective learning cycle can guide the ongoing learning processes required by a "specialized generalist" informed by a WHOLE-systems perspective on health of animals, humans, and others. It offers guidance on how to proceed when addressing specific issues. The learning cycle also provides a big-picture overview of priorities, patterns, and questions that are also seen in other forms of assessment and practice (Table 5.1).

Brown's learning cycle resonates with other cycles of learning and action in practice-based contexts spanning health, education, and ecosystem management. Table 5.1 presents the stages of some of these cycles, underscoring patterns of similarity across different contexts relevant to the "general" practice of a "specialized generalist." The examples in Table 5.1 profile a series of similar questions and stages that span Brown's collective learning cycle (2010), and they can be aligned to the features of the "clinical method" or the "clinical diagnostic approach" for humans and veterinary medicines (Shah, 2005a, 2005b; Waltner-Toews and Kay, 2005), professional learning, action research cycles (Zuber-Skerritt, 2015; Fletcher, 2015), and adaptive (ecosystem) management (Allen et al., 2011; Williams, 2011). The examples in Table 5.1 also resonate with related learning and iterative cycles emphasized in Disaster Risk Reduction (Phibbs et al., 2016) and integration across the Sustainable Development Goals (Stafford-Smith et al., 2017; Allen et al., 2019). Although the rows in Table 5.1 do not align exactly, the table is intended to depict a series of patterns, commonalities, and relationships that are relevant to a specialist generalist seeking to adopt a highly contextual and learning-oriented approach to health embedded within WHOLE systems.

Other authors have observed the patterns in Table 5.1 and have proposed overall approaches that can be applied in multiple contexts such as the Adaptive Methodology for Ecosystem Sustainability and Health (AMESH) (Waltner-Toews and Kay, 2005). Although the other examples in Table 5.1 do not share the same kind of "presenting complaint" as the "clinical diagnostic approach," there are substantive overlaps in the quality of the assessment and the nature

TABLE 5.1
Learning Patterns Relevant to WHOLE Systems Approaches: Lessons from Brown's Collective Learning Cycle (and Other Cycles of Learning)

Collective Learning Cycle, Including Four Guiding Questions[a]	The "Clinical Method" or Clinical Diagnostic Approaches[b]	Professional Learning and Action Research Cycles[c]	Adaptive Management for Natural Resource Management[d]	Associated Reflective Questions[e]
What should be? ***What is?*** ***What could be?*** ***What can be?***	***History*** ***Examination*** ***Investigations*** ***Management***	***Plan*** ***Act*** ***Observe*** ***Reflect***	***Set up,*** ***Decision-Making*** ***Monitoring*** ***Assessment***	***What?*** ***So what?*** ***Now what?***
What should be? (Principles, ideals) Collective learning draws on multiple knowledge cultures[1] to develop or revise this shared intent	Ideally, an aspiration, state, or capacity of "***health***" is negotiated between patient and clinician in the clinical encounter	***Plan:*** This phase involves learning among individuals, groups, and teams informed by different contexts	"***Set-up***" involves framing in terms of stakeholders and objectives. Technical learning involves iterative learning through decision-making, monitoring, and assessment	Consider intentions. Who is involved? Why? Where? Factor into reflective questions
What is? Uses parameters, facts, empirical constraints, and enablers to determine What is, now?	Appraise ***history*** of specific issue, starting with the ***presenting complaint***, and through related contextual features	***Act***: Actions are informed by planning, context, and assessment of current situation and appraisal of Where are we now?	***Structured decision-making*** requires capacity to define the problem, identify objectives, formulate evaluation criteria, estimate outcomes, and evaluate trade-offs	***What*** happened? (description and self awareness: facts and reactions)
What could be? (Potential, ideas) Explore possibilities for taking ideas into practice to get to: What should be?" from "What is?"	***Examination and investigations*** via physical examination and other diagnostic testing give more detail	***Observe***: Monitor, gather data – How are we doing? – What evidence do we have?	Decision-making needs to be complemented by higher-order learning enabled through implementation and ***monitoring and assessment***	***So what?*** (analysis, evaluation: interpretation and relationship to existing experience)

What can be? (Practicalities, actions) Action planning, taking practicalities into account to establish what can be done to complete this iteration of learning	Create ***management*** plans, and design follow-up to confirm or adjust diagnosis and plan ahead	***Reflect*** on the process and learning to date – How did we do? – Where to next?	Learning about management choices needs to be completed through ***evaluation and adjustment*** to inform next iteration of the process	***Now what?*** (synthesis, actions, application)

Notes:

[a] Brown (2010). In addition to the four guiding questions (***What should be***? ***What is***? ***What could be? What can be***?) collective learning draws on individual, community, specialized, organizational, and holistic knowledge.

[b] Shah (2005a, b) describes the clinical encounter through ***history, examination, investigations,*** and ***management***. Waltner-Toews (2005) focuses on health of people, animals, and ecosystems through ***presenting complaint,*** clinical examination, diagnosis, setting, and achieving goals.

[c] Zuber-Skerritt (2015) extends on the classic Action Research cycle of ***"Plan, Act, Observe, Reflect,"*** with connections to action learning and the important links to critical reflection, professional learning, action leadership, sustainable development, and social justice.

[d] For Allen et al. (2010), adaptive management is a "formal iterative process of resource management" combining structured decision-making and high-order learning. Williams et al (2009) links a ***set-up*** phase with iterative learning through ***decision-making, monitoring,*** and ***assessment***.

[e] Rolfe et al.'s (2001) reflective questions are ***What? So What? Now What?***

of the questions that are asked. Waltner-Toews (2004) is especially clear about the relevance of a system-informed approach to a "presenting complaint," which can serve as a starting point for learning and decision-making about issues and challenges that may be well outside of the traditional health sector, especially in the context of health and ecosystem sustainability. Of utmost importance to the "specialized generalist," Waltner-Toews focuses on the quality of the contextual appraisal, and the need to consider combined social and ecological factors when seeking the "history of presenting complaint." The quality and nuance of the initial assessment can influence all subsequent learning and decisions.

This quality of initial assessment is reflected in the other cycles presented in Table 5.1, whereby any appraisal, assessment, or framing of proposed actions and next steps needs to be done *in relation to* intended goals, directions, and objectives. The strong tendency to "start with the issue/problem" without taking time to clarify contextually relevant goals, intentions, and priorities can be deeply problematic because it limits our ability to attend to complex system issues (Brown, 2010). This is a key issue for the "specialized generalist" and others seeking a WHOLE-systems perspective on health. Accordingly, the following walk through of the four questions in Brown's collective learning cycle commences with attention to "What should be?" in the context of working together for WHOLE systems, also acknowledging the overlaps and relevance of related approaches included in Table 5.1.

WHAT SHOULD BE?

A WHOLE system health issue needs to be understood in the context of some kind of desired future – with a view to intentions, ideals, goals, aspirations. If there is no sense of what kind of "health" we are working towards or aspiring to, it becomes challenging to assess where we are at now. Any efforts to address or assess health need to be informed by an individual's or a group's perspective on health, which may help understand and assess when things are progressing well, even thriving or, alternatively when there is a problem (see Chapter 2 for more discussion on defining health).

Starting with "What should be?" (instead of "What is?") provides a sense of direction or goal to work towards it. Asking "What should be?" is not intended to provide a moral directive but, rather, to encourage recognition of the intended conditions and principles that underlie a state of thriving, including those that can guide efforts to focus on WHOLE systems. The emphasis is not health in isolation but rather health in relationship with life (living systems) and in relation to others (equity). When viewed alongside living systems, equity applies among and between people, generations, and species, fostering attention to reciprocity and regenerative approaches (see Chapter 3 for more on interspecies and intergenerational equity).

Others working in socio-ecological contexts for health have outlined principles to guide how we can understand and respond to interconnected health issues (see, for example, Table 5.2 adapted from Poland and Dooris, 2010; Charron, 2012a,b). The orientation towards principles reflects the context-specific nature of work at this interface, which is not well suited to definitive, checklist concepts or approaches. These principles reflect a distillation of patterns, recurring values, and priorities that have consistently emerged in the several decades of work seeking to refine and extend the ambitious socio-ecological orientation of the Ottawa Charter for Health Promotion (WHO, 1986). Table 5.2 offers a synthesis of converging principles to inform leadership in building a healthier more equitable and sustainable future.

The principles in the left column in Table 5.2 are ongoing efforts to strengthen the decades of work with "settings" approaches to health, through a clearer focus on building health, equity, and sustainability (see also Capon and Dixon, 2007; Parkes and Horwitz, 2009; Hancock, 2011). The principles in the right column in Table 5.2 draw on developments in ecosystem approaches to health (also known as ecohealth) as a field of research, education, and practice that adopts systems approaches to promote the health of people, animals, and ecosystems in the context of social and ecological interactions (Webb et al., 2010; Parkes and Horwitz, 2016; Buse et al., 2018). Both complementary sets of principles are relevant when considering a WHOLE-systems approach to "What should be?" The combined present and future orientation of the principles in Table 5.2 offers new ways of conceiving the role of the health sector, including expanding and unifying a perspective of "health for all" that considers equity and living systems for all species, including humans.

TABLE 5.2
Principles for Re-Engaging with the Socio-Ecological Context for Health Converging towards Health Equity for Future Populations

Principles for a Healthy and Sustainable Settings Approach (from Poland and Dooris, 2010)	Principles to Inform the Application of Ecosystem Approaches to Health (from Charron, 2012a,b)[a]
Adopt an ecological "WHOLE system" perspective	Systems thinking
Start where people are (listen to and respect lived experience, diverse forms of knowledge)	Participation
Practice rooted in place	Sustainability
Deepen the social analysis (connect lived experience to that of others and to the practices, structures that create and sustain inequity located in broader context of relations of power)	Transdisciplinary research
Asset-based/appreciative inquiry approach	Gender and social equity
Build resilience (for change) instead of efficiency (for narrow set of operating conditions)	Knowledge to action

[a] The order of Charron's principles is adapted to highlight commonalities with Poland and Dooris (2010).

The principles and WHOLE-systems considerations presented in Tables 5.1 and 5.2 will need to be adapted to specific contexts. A WHOLE-systems perspective may, for example, need to be adopted when considering "What Should be?" in relation to the health of non-human species, recognizing the challenge of identifying a goal, aspiration, or desired future for animals that are not able to directly share that using human language(s). This challenge of assessing health for those with other-than-human "voices" is explored in Chapters 2, 17, and 19. In many cases, expressing the desires or aspirations for animal health or welfare draws on a person or people whose expertise or experience to frame the heath needs and aspirations of "those that have no voice" (de Groot, 1998). This may include – for example – a pet owner, a livestock farmer, a conservation biologist, or a veterinarian. Although the challenge of "who speaks for health" is particular when considering animals, it is, arguably, equally challenging for those who are considering any form of health that is beyond individual or "personal health." In 2020, the scope of health beyond individuals (humans or animals) extends to a whole new lexicon and framing, working across scales and contexts, from public health and population health through to planetary health (Buse et al., 2019; Redvers, 2018; Harrison et al., 2019).

The synergies across principles in Table 5.2 point to the benefits of re-learning old lessons about approaches that are integrative and regenerative (as compared to disintegrative and degenerative), with implications that span how we think about governance and priorities across scales and contexts. This requires a combined recognition of both "ideas" and content (often associated with understanding "What is?"), as well as practice and processes (often associated with "What Could be?").

WHAT IS?

Those working with Brown's learning cycle may experience considerable pressure to start with "What is?" There are, indeed, situations where this may be warranted, such as situations of urgency or acute crises, for example the 2020 COVID-19 pandemic. In these cases, health-oriented professionals (including a "specialized generalist") may need to focus all efforts immediately on the demands of "What is?" (Where we are now? What is happening?) without having a sense of what we are aiming for, and where we want to get to. This may mean that a focus on the WHOLE may not be immediately relevant, or even possible.

As an analogy, when you have a broken-leg, it is time for a specialist, who can fix this particular problem with a cast or surgery. In this case, "What should be?" is the treatment of an immediate problem, but once the urgency has passed (e.g. especially in preparation for a patient with a broken leg to be discharged home), broader, contextual factors become much more relevant, including the need for an overall goal, grounded in both social and ecological context, that should be considered in the next phase of learning and assessment following initial response. Extra time may need to be spent on questions of principles, intentions, and direction, after an initial crisis is over. In this analogy, an initial crisis of "What is?" needs to be followed by learning about and paying attention to the upstream drivers that influence access to orthopaedic and rehabilitation services, or create settings conducive to injury or, conversely that increase opportunities for healing and recovery.

Brown argues that whether in situations of urgency and crisis or as part of intentional learning cycles, problem-oriented appraisals and analyzes of "What is?" will be improved and enhanced through overt, collective attention to "What Should be?" Having a sense of where we are heading provides a sense of where we wish to go, and therefore a better position to consider and interpret the patterns and parameters of "What is?" in terms of what we are currently experiencing or have experienced. This emphasis echoes related examples in Table 5.1 to plan before we act (Zuber-Skerritt, 2015) and to set-up and frame our decision-making processes (Williams, 2011) in ways that are consistent with WHOLE systems. In the same spirit, Box 5.1 provides a case to think through how the principles and concepts presented thus far can be applied in response to a presenting complaint that has both personal and collective implications for a WHOLE-systems approach to health.

Considering the learning cycles presented in Table 5.1, what would you suggest as a WHOLE-systems approach to answering this student's questions in Box 5.1? The hypothetical presenting complaint in Box 5.1 likely has a familiar ring to those interested in the themes of this chapter. The WHOLE-systems orientation introduced earlier challenges us to resist the temptation to define and describe the problem in order to select a solution and instead adopt a strength-based

BOX 5.1 A CASE STUDY OF "WHAT IS?": A PROMPT FOR LEARNING CYCLES AND WHOLE-SYSTEM PRINCIPLES

PRESENTING COMPLAINT ("WHAT IS?")

A 27-year-old graduate student with a clinical health background is completing their Masters research addressing a complex issue that raises ecological, equity, and health concerns for humans and animals in New Zealand and that has necessitated a mixed methods study. The student's experience to date has made them feel that to have the desired impact, their future work will require collaborative effort across difference disciplines and knowledge. The student has been looking into new approaches to understanding health in the context of social and environmental change and has become unclear about how to proceed.

This student approaches you for guidance and direction about the best way to build on their background, and presents you with a series of questions:

- I am thinking of doing a PhD, but should I gain practical "real-world" experience prior to undertaking doctoral studies?
- I am really interested in understanding health in relation to both ecosystems and equity, not one or the other. There are so many new fields claiming to address aspects of this. I have seen eco-health, One Health, planetary health, environmental justice, ecological public health. Which of these is best?
- Some people say studying for a PhD is all about becoming a specialist about one thing, but when I was doing clinical work, I was always more interested in and better at being a generalist – understanding the big picture. I want to do a PhD that gives me this big-picture view, but how can I become a specialist in the big picture?
- If I do pursue doctoral studies, I am worried about becoming further disconnected from my clinical and applied origins, because I really want to be able to make a difference and contribute to a better world. What do you recommend?

orientation that prompts questions such as "what do we need to know and understand about this system in order to enhance conditions for thriving?" The intent is not to demonstrate the "right" answer to these questions. Rather, responding to the questions in Box 5.1 draws on the ideas presented above in ways that address the student's questions and also point to the wider challenge of strengthening capacity to support WHOLE-systems perspectives on ecosystems, equity, and health. Several notable challenges and dilemmas raised by the example in Box 5.1 are explored here.

Each of the questions in Box 5.1 presents the challenge to *overcome dichotomous thinking.* Dichotomies in the questions above are evident in the implied tone of the questions that it is necessary to make a choice between clinical versus academic and applied versus theoretical knowledge, or that there is an implied either/or needed when considering the fields of ecohealth or One Health, or planetary health. Related and false dichotomies embedded in each of the questions include the idea of a choice between specialization and integration, or between nature and society, between individual and collective work, between the ecological and the social, between environment and equity, between the global and the local, and to some extent between highly populated (urban) and less populated (rural, remote) regions. A key consideration here is that "What is?" (in terms of career choices, and strengthening capacity through training) is often presented as binary either/or choice between priorities or parameters of concern, which misses opportunities to see the important relationships and connections between these factors (Horwitz and Parkes, 2019; Waltner-Toews et al., 2019). Combatting these false dichotomies needs to be accompanied by alternative ways of understanding and respecting complementary knowledge and priorities that, at its simplest, require both/and thinking, engagement with paradox, and the potential of holding contradictory ideas in our minds at the same time. This idea is explored by multiple authors in Brown et al. (2019) and revisited in the "What can be" section to follow, including Box 5.2.

The questions in Box 5.1 also underscore a related dilemma: Although many integrative approaches to health are being proposed, overt *emphasis on both social and ecological context for health remains especially rare.* The student described in Box 5.1 has interests in health (of humans and animals) in relation to both ecological and social context (living systems and equity). Despite the aspiration of socio-ecological approaches to health promotion (WHO, 1984), concerted efforts to take stock of combined social and ecological context for health remained sparse in the decades following the Ottawa Charter (Cole et al., 1999; Parkes et al., 2003; Hancock, 2011; Hallström et al., 2015).

The first two decades of the 21st century have seen considerable increase of options for the kind of student described in Box 5.1 to consider, including the rapid expansion of integrative efforts that focus on health relation to environment and ecosystems, with particular emphasis on what can be seen as a trio of emerging (and sometimes competing) fields of ecohealth, One Health, and planetary health (see Buse et al., 2018; Oestreicher et al., 2018; Harrison et al., 2019). Yet, despite ongoing developments, and associated calls for a healthy, just, and ecologically sustainable future (Capon and Dixon, 2007), concerted attention to both the social and ecological context for health remains an area which requires more attention in almost all integrative approaches. The sets of principles presented in Table 5.2 can be seen as an example of this, where calls for "Greening Healthy Settings" by Poland et al. (2019) is a direct response to the need for a more overt attention to the ecological and environmental perspectives, which are ironically often absent within what are often claimed as socio-ecological approaches (see Hancock, 2011; Hallström et al., 2015; Horwitz and Parkes, 2019). Similarly, although the ecological can be seen as integral to ecosystem approaches to health, it is possible to apply Charron's six principles in ways that attend to social context, processes, and engagement across all six principles, and yet overlook the fundamental relationship between health and ecosystems in a living systems sense (see Parkes and Horwitz, 2016; Horwitz and Parkes, 2019).

Although responses to these dilemmas tend to flip-flop from one to another (with emphasis, for example, on social and equity issues, that override an emphasis on living systems; or vice versa), there is a growing return to language that overtly brings these considerations together in the form of ecosocial approaches and various forms of socio-ecological approaches, including socio-ecological systems and health (Zinsstag et al., 2011; Hancock, 2016; Horwitz and Parkes, 2019; Parkes et al., 2020). A conversation with the student in Box 5.1 might acknowledge that the English language is especially clumsy in this regard (resulting in multiple syllables and hyphenated words), and that Indigenous languages often have a suite of words, metaphors, and terminologies that better reflect these interrelationships (see Panelli and Tipa, 2007; Bartlett et al., 2015; Moewaka Barnes et al., 2017; Redvers, 2018), and which warrant increasing respect and attention.

Finally, the options available in response to the student's questions in Box 5.1 are heavily contingent on the proposed desired future or "What should be?" Identifying what is/are options for this student will be influenced by the student aspirations and intent to make a difference and contribute to a better world. Notions of difference and "better" could involve, for example, a shift in focus from understanding degenerative driving forces of change (creating cascades of effects and impacts that flow between ecological degradation, social inequities, and short-, medium-, and long-term health impacts) towards WHOLE-systems practices. To foster this, Brown's collective learning cycle is overt about the need to shift from ideals, parameters, and content about integrative and regenerative perspectives ("What should be?" and "What is?") towards a clear focus on potential processes and practicalities needed to adopt these integrative and regenerative approaches ("What could be?" and "What can be?").

WHAT COULD BE?

In the context of WHOLE systems, the purpose of "What is?" is not to provide a definitive description of the situation but to create a basis to synthesize and interpret the insights to create a better understanding of how to respond and move forward. Brown's collective learning cycle calls upon those learning to acknowledge the differences and disconnects (even dissonance) between the current situation (What is?) and the desired future (What should be?). Learning about the current situation creates new opportunities to address this, initially by designing "What could be?" This shift also resonates with Rolfe et al.'s (2001) reflective questions of "So What?" and "Now What?" (Table 5.1).

Orienting to "What could be?" in a WHOLE-systems approach can be the most creative phase of the collective learning cycle. Designing for what is possible can fuel potential for transformation and change towards (and even beyond) the desired future and goals you have identified. When working together for WHOLE systems, this creates an expansive opportunity to emphasize processes that focus on regeneration and life, rather than death, disease, and disability (and related degenerative processes).

One way of fostering the transformative potential of "What could be?" from a WHOLE-systems perspective is through a focus on co-benefits: designing solutions that proactively enhance health, equity, and the living systems we depend on. Co-benefits present a simple but transformative idea: focusing on opportunities to prioritize policy, actions, and interventions that achieve multiple benefits (for ecosystems, for equity, for the earth's diverse inhabitants) and pay attention to the kinds of processes required to do so. In the context of synergies across health, biodiversity, and climate objectives, "co-benefits can only be achieved, however, through joined-up, collaborative, cross-sectoral and transdisciplinary working" (Marselle et al., 2019).

Focusing efforts on attaining co-benefits is also in keeping with the shift from a problem-oriented to a strength-based approach focused on assets. This shift has far-reaching implications in how we think about both the integration challenge of WHOLE-systems approaches and the pathways to change required to get there. Take, for example, a focus on the cumulative environmental, community, and health impacts of climate change, resource development, or other drivers of social and ecological change. Attention to cumulative impacts responds to calls for attention to the "integration imperative" (Gillingham et al., 2016), by addressing combined health, ecosystem, and equity impacts across space, time, and multiple drivers of change. Yet such approaches can be at risk of limiting their emphasis on understanding and analyzing the problem from multiple perspectives, and not focusing sufficiently on actions to response to these issues (Parkes et al., 2019).

A WHOLE-systems perspective encourages a focus on cumulative processes to be reframed with a focus on potential co-benefits. It turns the emphasis towards the possibilities created when responses are designed to achieve multiple converging positive objectives and to meet different co-benefit criteria. Co-benefits (or, cumulative benefits) can, therefore, be seen as a flip side to cumulative impacts, shifting from "disease" and "degenerative" approaches to an emphasis on promoting and protecting health and regenerative approaches to equity and living systems. Arguments for collaborative, multilevel "harm reduction" approaches to animal health that "looks throughout the socio-ecological system at drivers of harm to find strengths, possibilities, and opportunities for solutions in the face of a prevailing challenges and uncertainties" has some similar features (Stephen et al., 2018) (see Chapters 6 and 23 for more on harm reduction).

Examples of "What could be?" in terms of co-benefits include interventions, actions, and programs that are designed to have positive, regenerative impacts across spatial and temporal scales. This requires an ability to see the connections and co-benefits at different levels, spanning macro/global-level processes through to the intermediate mesoscales of landscapes and regions down to micro-level interactions for specific communities or individuals. Co-benefit work has, largely, focused on health and climate goals in urban contexts, targeting built environments, public transportation, housing, and public health (Walpole et al., 2009; Karlsson et al., 2020). However, calls for recognition of the interactions between health and, for example, biodiversity, the natural environment and related Sustainable Development Goals are increasing (Sandifer et al., 2015; Fleming et al., 2018; Nilsson et al., 2018). Viewing co-benefits from a WHOLE-systems perspective can shift attention to interacting health, ecosystems, and equity-related benefits and challenges within mesoscale living systems such as catchments and watersheds (through which we see upstream-downstream dynamics and flows of water, energy, people, animals, etc.,) and also in other mesoscale socio-ecological systems such as bioregions, specific landscapes (forests, deserts), and oceanscapes.

Table 5.3 presents examples of "What Could be?" in terms of co-benefits for WHOLE systems, reflecting interventions and actions that operationalize

TABLE 5.3
"What Could Be?" Examples of Co-Benefit Thinking Using a WHOLE-Systems Perspective

	Co-Benefits That Are Good for:		
	Planetary Health (e.g. climate for all species)	**Living Systems** (e.g. catchments, watersheds)	**Intergenerational Equity** (communities across generations)
Examples	Macro/global	Mesoscale/regional	Micro/local
The health benefits of Indigenous protected areas and stewardship initiatives	Protected areas prevent deforestation, limiting carbon-release and providing carbon offsets	Protection of landscapes and oceanscapes support intact habitat for native forest and endemic biota	Nearby Indigenous and other communities gain multiple benefits from social, ecological and economic resilience, as well as contact with nature
Landscape-level management to address the ecosocial impacts of wildfires on human and animal health	Soil, vegetation, and water management provide options for climate mitigation (carbon sequestering) and adaptation (drought)	Landscape-level soil, vegetation, and water management attuned to climate aims to increase biodiversity and habitat values	Prevention of extreme wildfires prevents morbidity and mortality for affected human communities and animals (wildlife and domestic)
Youth-based initiatives that connect climate, catchments, and community through regenerative land-based practices	Fosters both climate mitigation (habitat restoration as carbon offset) and adaptation (drought, flood, etc.)	Riparian planting and corridors of habitat restoration in waterways and wetlands throughout catchment areas	Youth and "educational" engagement in downstream communities, create well-being benefits, with links to other species
Shifting from extractive to regenerative economies	Reduction of carbon, net-zero orientation, emphasize mitigation but also include adaptation opportunities	Prevent climate impacts in land – water living systems, including drought and flood	Principles, processes, and practices that build economic and political power across rural, remote, and urban communities

thinking across diverse literatures that connect benefits across climate, ecosystems, equity, and health (see, for example, Walpole et al., 2009; Romanelli et al., 2015, Horwitz and Finlayson, 2011; Horwitz and Kretsch, 2015; Jenkins et al., 2018; Vandyk et al., 2018; Marselle et al., 2019).

The WHOLE-systems co-benefit examples presented in Table 5.3 may also resonate with the long-standing traditions of Indigenous knowledges that have emphasized these interrelationships for millennia and have also fostered ongoing practices that prioritize a more holistic orientation (Panelli and Tipa, 2007; Moewaka Barnes et al., 2017; Jenkins et al., 2018; Ratima et al., 2019; Redvers, 2018). Connecting different ways of knowing and traditions is an integral part of conceiving "What could be?" (see Bartlett et al., 2015; Henwood et al., 2016).

WHAT CAN BE?

The question "What can be?" moves us into the practical, practice-based elements of a WHOLE-systems approach. Working together for WHOLE systems demands collaborative actions, processes, and practices attentive to the question "Who and How are we Open to Listen, and Engage/exchange?"

Moving from "What could be?" to "What can be?" shifts from designing potential options to the practical implications of doing. Given the importance of context, it would be disingenuous to delineate specific practical and operational aspects of "What can be?" for all WHOLE approaches. Even so, some overarching guidance can be offered to those aspiring towards a specialized generalist approach to WHOLE systems (Box 5.2). Getting to and operationalizing "What can be?" will benefit from ongoing attention to the questions and guidance provided in Box 5.2.

Designing for "What could be?" in relation to WHOLE systems is likely to challenge the status quo. This means that practical questions (What needs to be done? Who is going to do what? When and how is it going to happen?) may need to be linked to less familiar change-oriented questions such as "Who needs to do what, differently?" "Who talks to whom?" "Who needs to work with/listen to/learn from whom?" and "Do these right people know how to do this, and would they be valued for doing so?" The related questions "whose day-jobs are going to be different because of this?" "who is benefitting from this" and "whether these changes do or don't happen?" acknowledges the potentially far-reaching structural, institutional, systemic, or societal implications and power dynamics of moving through associated change processes. Given the boundary-crossing nature of WHOLE approaches, answering these kinds of questions rekindles the need for attention to how to strengthen individual and collective capacity for this kind of work, linking back to calls for cultivating independent thinking (Brown et al., 2019) and related transformations in education and training as part of wider systems change that aspires towards a healthy, just, and regenerative future (Parkes et al., 2012; Parkes et al., 2016; Redvers, 2018; Walpole et al., 2019).

BOX 5.2 GETTING TO WHAT CAN BE?: GUIDANCE FOR A "SPECIALIZED GENERALIST" APPROACH TO WHOLE SYSTEMS

- **How we begin matters**: The quality of the questions asked when we start influences all subsequent efforts in WHOLE-systems approaches.
- **Working and learning will happen in cycles and be iterative**: Future learning may create opportunities to deepen understanding of matters missed earlier. Celebrating different, connected phases of work recognizes the new insights possible across short, medium-, and long-term practice.
- **Context is essential**: Commitment to context is critical to all stages of a WHOLE approach to health. It requires attention to past, present, and future influences on living systems, equity, and their implications for health.
- **Benefit from "approaches"**: There will never be only one approach to gaining a contextually informed WHOLE-systems understanding or response to health issues for animals, humans, or other species. Embrace the plurality and opportunities provided by working with many approaches that share common principles and patterns of work.
- **Know when to refer and/or collaborate**: Be prepared for teamwork. A "specialized generalist" needs to leverage from their own skills to embrace working with other specialties, disciplines, or partners to complete their work. This requires the "specialized" skills of recognizing, respecting, and learning to work with, and bridge across, other knowledge and approaches.
- **Commit to overcoming false dichotomies**: Develop and cultivate a "both/and" orientation that counters the emphasis on either/or – a both/and approach that is oriented to both living systems *and* equity; people *and* nature; urban *and* rural; crisis *and* recovery; individual *and* collective; knowledge *and* action; health of human *and* other species.
- **Focus on knowledge and action**: Be attentive to the ways that a primary focus on knowledge and understanding can lead to an emphasis on describing and analyzing problems (especially the processes of disintegration and degeneration). New and valuable insights will arise when learning is also focused on how to foster, enhance, and amplify actions and interventions designed to be integrative and regenerative.
- **Continue to ask good questions**: The quality and scope of our questions will influence our work and practice towards WHOLE systems more than finding solutions. Ongoing reflection on "Who and How are we Open to Listen, and Engage/exchange?" will prompt further questions to inform the ongoing practice and cycles of work.

CLOSING SUMMARY

This chapter explored the challenge of "Working Together for WHOLE Systems." This work extends beyond individuals or singular perspectives to collective capacities for transformative change. The chapter applied Brown's collective learning cycle to harness experience, capacities, and processes for change across individual, community, specialized, organizational, and holistic knowledge cultures (Brown, 2010). While recognizing the importance of collective processes for emergent knowledge and actions, there is also an important role for specific individuals with a set of orientations that enable and encourage opportunities to work together within the complex terrain of WHOLE systems. This chapter underscored the combination of boundary-crossing perspectives required for framing, fostering, and advancing WHOLE-systems approaches, and the potential role of a "specialized generalist" to progress this work. Opportunities are created when we combine an integrative turn (where Well-being and Health is oriented to Living Systems and Equity) with a collaborative turn (that asks "Who and How are we Open to Listen, and Engage/exchange?")

In the face of crises such as the COVID-19 pandemic or climate change, there is value in framing and revisiting first principles along with the kinds of tools and processes that can ensure that our collective efforts move beyond relief and recovery to reimagined, regenerative futures, including those oriented to WHOLE systems. This chapter provided principles, questions, guidelines, and points of reference relevant to a new era where addressing health and well-being challenges can be done in a way that prioritizes both living systems and equity – for the benefit of humans, other species, and our shared planetary home.

NOTES

1. This chapter uses decapitalized versions of ecohealth, One Health and planetary health for several reasons. First, to differentiate these emerging fields from the trademarked and branded uses of One Health (e.g. One World-One Health™), *EcoHealth* (the journal, and EcoHealth Alliance) and the Planetary Health Alliance. Second, to encourage recognition of the 'maturing' of each of these fields beyond their early origins (see Buse et al., 2018), noting that capitalization is often associated with aspirational claims to novelty and exceptionalism with newly coined terms, that become less relevant as fields become more well-established. In this way ecohealth, One Health, and planetary health are treated like other academic disciplines and fields of research, education and practice, that are generally not capitalized.

REFERENCES

Allen, Craig R., Joseph J. Fontaine, Kevin L., Pope, and Ahjond S. Garmestani. "Adaptive Management for a Turbulent Future." *Journal of Environmental Management*, 92, no. 5 (2011): 1339–1345.

Allen, Cameron, Graciela Metternicht, and Thomas Wiedmann. "Prioritising SDG Targets: Assessing Baselines, Gaps and Interlinkages." *Sustainability Science*, 14, no. 2 (2019): 421–438.

Ashhurst, Craig A. "Transcoherence: Labels and Wicked Problems," in *Independent Thinking in an Uncertain World: A Mind of One's Own*, eds. Valerie A. Brown, John A. Harris, and David Waltner-Toews (New York: Routledge, 2019): 224–241.

Barile, Sergio, and Marialuisa Saviano. "Dynamic Capabilities and T-Shaped Knowledge: A Viable Systems Approach," in *Contributions to Theoretical and Practical Advances in Management. A Viable Systems Approach (VSA)*, ed. Sergio Barile (Rome: Arachne, 2013) 39–59.

Bartlett, Cheryl, Murdena Marshall, Albert Marshall, and Marilyn Iwama. "Integrative Science and Two-Eyed Seeing: Enriching the Discussion Framework for Healthy Communities," in *Ecosystems, Society and Health: Pathways Through Diversity, Convergence and Integration*, eds. Lars K. Hallström, Nicholas P. Guehlstorf, and Margot W. Parkes (Canada: McGill-Queens University Press, 2015): 280–326.

Benne, Beatrice, and Pamela Mang. "Working Regeneratively Across Scales-Insights from Nature Applied to the Built Environment." *Journal of Cleaner Production*, 109, no.16 (2015): 42–52.

Boumeester, Marc. "The Specialized Generalist: Art and Design and the Osmotic Oxymoron." *Proceedings of the Cumulus Conference: Cultural Diversity, Social Engagement and Shifting Education.* University of Aveiro, Aveiro, Portugal, May 8–10, 2014, 83–88.

Brown, Valerie A. "Collective Decision-Making Bridging Public Health, Sustainability Governance and Environmental Management," in *Sustaining Life on Earth: Environmental and Human Health Through Global Governance*, eds. Colin L. Soskolne, Laura Westra, Lewis J. Kotzé, Brendan Mackey, and William E. Rees (Washington, DC: Lexington Books, 2007): 139–154.

Brown, Valerie A. "Collective Inquiry and Its Wicked Problems," in *Tackling Wicked Problems: Through the Trandsiciplinary Imagination*, eds. Valerie A. Brown, John A. Harris, and Jacqueline Russel (London: Earthscan, 2010): 61–81.

Brown, Valerie A, John Grootjans, Jan Ritchie, Mardie Townsend, and Glenda Verrinder. *Sustainability and Health: Supporting Global Ecological Integrity in Public Health.* (Abingdon: Allen and Unwin, 2005).

Brown, Valerie A., John A. Harris, and David Waltner-Toews. *Independent Thinking in an Uncertain World: A Mind of One's Own* (Routledge, 2019): 336.

Boyer, Ernest L. *Scholarship Reconsidered: Priorities of the Professoriate*, 1st Edition (Stanford, CA:Carnegie Foundation for the Advancement for Teaching, 1990): 147.

Buse, Charles G., Maxwell Smith, and Diego S. Silva. "Attending to Scalar Ethical Issues in Emerging Approaches to Environmental Health Research and Practice." *Monash Bioethics Review*, 37, no.1 (2019): 4–21.

Buse, Chris G., Jordan Sky Oestreicher, Neville R. Ellis, Rebecca Patrick, Ben Brisbois, Aaron P. Jenkins, Kaileah McKellar et al. "Public Health Guide to Field Developments Linking Ecosystems, Environments and Health in the Anthropocene." *Journal of Epidemiology and Community Health*, 72, no. 5 (2018): 420–425.

Capon, Anthony G., and Jane M. Dixon. "Creating Healthy, Just and Eco-Sensitive Cities." *New South Wales Health*, 18, no. 3–4 (2007): 37–40.

Capra, Fritjof, and Pier Luigi Luisi. *The Systems View of Life: A Unifying Vision* (Cambridge: Cambridge University Press, 2014), 498.

Charron, Dominique F. *Ecohealth Research in Practice: Innovative Applications of an Ecosystem Approach to Health* (New York: Springer/International Development Research Centre, 2012a), 288.

Charron, Dominique F. "Ecosystem Approaches to Health for a Global Sustainability Agenda." *EcoHealth*, 9, no. 3 (2012b): 256–266.

Cole, Donald C, John Eyles, Brian L. Gibson, and Nancy Ross. “Links between Humans and Ecosystems: The Implications of Framing for Health Promotion Strategies.” *Health Promotion International*, 14 (1999): 65–72.

Cole, Donald C., Margot W. Parkes, Johanne Saint-Charles, Maya Gislason, Kaileah McKellar, and Jena Webb. “Evolution of Capacity Strengthening Approaches for Health, Ecosystems and Society: Insights Informed by Involvement in the Canadian Community of Practice in Ecosystem Approaches to Health.” *Transformative Dialogues: Teaching & Learning Journal*, 11, no. 2 (2018): 1–21.

de Groot, Wouter T. “Problem-in-Context: A Framework for the Analysis, Explanation and Solution of Environmental Issues,” in *Environmental Management in Practice. Vol. 1: Instruments for Environmental Management*, eds. Paul Compton, Dimitri Devuyst, Luc Hens, and Bhaskar Nath (London: Taylor & Francis, 1998): 22–43.

Donofrio, Nicholas, Jim Spohrer, and Hossein S. Zadeh. “Research-Driven Medical Education and Practice: A Case for T-Shaped Professionals,” in *T-Shaped Professionals: Adaptive Innovators*, eds. Yassi Moghaddam, Haluk Demirkan, and Jim Spohrer (New York: Business Expert Press, 2018).

Durie, Mason. “Understanding Health and Illness: Research at the Interface Between Science and Indigenous Knowledge.” *International Journal of Epidemiology*, 33, no. 5 (2004): 1138–1143.

Fleming, Lora E., Giovanni S. Leonardi, Mathew P. White, Jolyon Medlock, Ian Alcock, Helen L. Macintyre, Kath Maguire et al. “Beyond Climate Change and Health: Integrating Broader Environmental Change and Natural Environments for Public Health Protection and Promotion in the UK.” *Atmosphere*, 9, no. 7 (2018): 245.

Fletcher, Margaret. “Professional Learning,” in *Professional Learning in Higher Education and Communities: Towards a New Vision for Action Research*, eds. Ortrun Zuber-Skerritt, Margaret Fletcher, and Judith Kearney (Basingstoke: Palgrave Macmillan, 2015), 41–75.

Fullerton, John. *Regenerative Capitalism: How Universal Principles and Patterns Will Shape Our New Economy* (Capital Institute, 2015). https://capitalinstitute.org/regenerative-capitalism/.

Gillingham, Michael P., Greg Halseth, Chris J. Johnson, and Margot W. Parkes. “The Integration Imperative: Cumulative Environmental,” *Community and Health Impacts of Multiple Natural Resource Developments* (Switzerland: Springer International Publishing AG, 2016), 256.

Gislason, Maya K., Vanessa Sloan Morgan, Kendra Mitchell-Foster, and Margot W. Parkes. “Voices from the Landscape: Storytelling as Emergent Counter-Narratives and Collective Action from Northern BC Watersheds.” *Health & Place*, 54 (2018): 191–199.

Hallström, Lars K., Margot W. Parkes, and Nicholas P. Guehlstorf. “Convergence and Diversity: Integrating Encounters with Health, Ecological and Social Concerns,” in *Ecosystems, Society and Health: Pathways Through Diversity, Convergence and Integration*, eds. Lars K. Hallström, Nicholas P. Guehlstorf, and Margot W. Parkes (Canada: McGill-Queens University Press, 2015): 3–28.

Hancock, Trevor. “It’s the Environment, Stupid! Declining Ecosystem Health Is the Threat to Health in the 21st Century.” *Health Promotion International*, 26, no. 2 (2011): ii168–ii172.

Hancock, Trevor. “Population Health Promotion 2.0: An Eco-Social Approach to Public Health in the Anthropocene.” *Canadian Journal of Public Health*, 106, no. 4 (2015): e252–e255.

Hancock, Trevor, Anthony Capon, Mark Dooris, and Rebecca Patrick. “One Planet Regions: Planetary Health at the Local Level.” *Lancet Planetary Health*, 1, no. 3 (2017): e92–e93.

Harrison, Sarah, Lucy Kivuti-Bitok, Alexandra Macmillan, and Patricia Priest. "EcoHealth and One Health: A Theory-Focused Review in Response to Calls for Convergence." *Environment International*, 132, (2019): 105058.

Henwood, Wendy, Helen Moewaka Barnes, Troy Brockbank, Waikarere Gregory, Kaio Hooper, and Tim McCreanor. "Ko Tāngonge Te Wai: Indigenous and Technical Data Come Together in Restoration Efforts." *EcoHealth*, 13, no. 4 (2016): 623–632.

Horwitz, Pierre, and C. Max Finlayson. "Wetlands as Settings for Human Health: Incorporating Ecosystem Services and Health Impact Assessment into Water Resource Management." *BioScience*, 61, no. 9 (2011): 678–688.

Horwitz, Pierre, and Conor Kretsch. "Contribution of Biodiversity and Green Spaces to Mental and Physical Fitness, and Cultural Dimensions of Health," in *Connecting Global Priorities: Biodiversity and Human Health, a State of Knowledge Review,* eds. Cristina Romanelli, David Cooper, Diarmid Campbell-Lendrum, Marina Maiero, William B. Karesh, Danny Hunter, and Christopher D. Golden (Geneva: World Health Organization, Convention on Biological Diversity, 2015), 200–220.

Horwitz, Pierre, and Margot W. Parkes. "Intertwined Strands for Ecology in Planetary Health." *Challenges*, 10, no. 1 (2019): 20.

Jenkins, Aaron, Anthony Capon, Joel Negin, Ben Marais, Tania Sorrell, Margot Parkes, and Pierre Horwitz. "Watersheds in Planetary Health Research and Action." *The Lancet Planetary Health*, 2, no. 12 (2018): e510–e511.

Jenkins, Aaron, Pierre Horwitz, and Kerry Arabena. "My Island Home: Place-Based Integration of Conservation and Public Health in Oceania." *Environmental Conservation*, 45, no. 2 (2018): 125–136.

Karlsson, Mikael, Eva Alfredsson, and Nils Westling. "Climate Policy Co-Benefits: A Review." *Climate Policy*, 20, no. 3 (2020): 292–316.

Kates, Kobebt W. "Geography: The Case for the Specialized Generalist in a Science of Environment." *Geography*, 53 (1967): 72.

Lang, Daniel J., Arnim Wiek, Matthias Bergmann, Michael Stauffacher, Pim Martens, Peter Moll, Mark Swilling et al. "Transdisciplinary Research in Sustainability Science: Practice, Principles, and Challenges." *Sustainability Science*, 7, no. 1 (2012): 25–43.

Lisefski, Benek. "Why Designers Need to Be 'Specialized Generalists'. Being 'T-shaped' will make you more effective and more valuable." Modus, 2019. https://modus.medium.com/why-designers-need-to-be-specialized-generalists-520b020ec039.

Marselle, Melissa R., Jutta Stadler, Horst Korn, Katherine N. Irvine, and Aletta Bonn. "Biodiversity and Health in the Face of Climate Change: Challenges, Opportunities and Evidence Gaps," in *Biodiversity and Health in the Face of Climate Change* (Cham: Springer, 2019), 1–13.

Masuda, J. R., T. Zupancic, E. Crighton, N. Muhajarine, and E. Phipps. "Equity-Focused Knowledge Translation: A Framework for 'Reasonable Action' on Health Inequities." *International Journal of Public Health*, 59, no. 3 (2014): 457–464.

Masuda, Jeffrey R., Tara Zupancic, Blake Poland, and Donald C. Cole. "Environmental Health and Vulnerable Populations in Canada: Mapping an Integrated Equity-Focused Research Agenda." *The Canadian Geographer/Le Géographe canadien*, 52, no. 4 (2008): 427–450.

McGregor, Deborah. "Coming Full Circle: Indigenous Knowledge, Environment, and Our Future." *American Indian Quarterly*, 28, no. 3/4 (2004): 385–410.

Moewaka Barnes, Helen, Te Raina Gunn, Angela Moewaka Barnes, Emerald Muriwai, Margaret Wetherell, and Timothy McCreanor. "Feeling and Spirit: Developing an Indigenous Wairua Approach to Research." *Qualitative Research*, 17, no. 3 (2017): 313–325.

Nilsson, Måns, Elinor Chisholm, David Griggs, Philippa Howden-Chapman, David McCollum, Peter Messerli, Barbara Neumann et al. "Mapping Interactions Between the Sustainable Development Goals: Lessons Learned and Ways Forward." *Sustainability Science*, 13, no. 6 (2018): 1489–1503.

Oestreicher, Jordan S., Chris Buse, Ben Brisbois, Rebecca Patrick, Aaron Jenkins, Jonathan Kingsley, Renata Távora et al. "Where Ecosystems, People and Health Meet: Academic Traditions and Emerging Fields for Research and Practice." *Sustainability in Debate*, 9, no. 1 (2018): 45–65.

Panelli, Ruth, and Gail Tipa. "Placing Well-Being: A Maori Case Study of Cultural and Environmental Specificity." *EcoHealth*, 4, no. 4 (2007): 445–460.

Parkes, Margot W. *Ecohealth and Aboriginal Health: A Review of Common Ground* (National Collaborating Centre for Aboriginal Health, 2011).

Parkes, Margot W. "'Just Add Water': Dissolving Barriers to Collaboration and Learning for Health, Ecosystems and Equity." in *Ecosystems, Society and Health: Pathways Through Diversity, Convergence and Integration*, eds. Lars K. Hallström, Nicholas P. Guehlstorf, and Margot W. Parkes (Canada: McGill-Queens University Press, 2015): 184–226.

Parkes, Margot W., S. Allison, H.G. Harder, D. Hoogeveen, D. Kutzner, M. Aalhus, E. Adams et al. "Addressing the Environmental, Community, and Health Impacts of Resource Development: Challenges Across Scales, Sectors, and Sites." *Challenges*, 10, no. 1 (2019): 22.

Parkes, Margot W., Leslie Bienen, Jaime Breilh, Lee-Nah Hsu, Marian McDonald, Jonathan A. Patz, Joshua P. Rosenthal, et al. "All Hands on Deck: Transdisciplinary Approaches to Emerging Infectious Disease." *EcoHealth*, 2, no.4 (2005): 258–272.

Parkes, Margot W., Dominique F. Charron, and Andres Sanchez. "Better Together: Field-Building Networks at the Frontiers of Ecohealth Research," in *Ecohealth Research in Practice: Innovative Applications of an Ecosystem Approach to Health*, ed. Dominique Charron (New York: Springer/International Development Research Centre, 2012), 231–253.

Parkes, Margot W., and Pierre Horwitz. "Water, Ecology and Health: Ecosystems as Settings for Promoting Health and Sustainability." *Health Promotion International*, 24, no. 1 (2009): 94–102.

Parkes, Margot W., and Pierre Horwitz. "Ecology and Ecosystems as Foundational for Health," in *Environmental Health: From Global to Local*, 3rd Edition, ed. Howard Frumkin (New Jersey: John Wiley & Sons, Inc., 2016), 27–58.

Parkes, Margot, Ruth Panelli, and Philip Weinstein. "Converging Paradigms for Environmental Health Theory and Practice." *Environmental Health Perspectives*, 111 (2003): 669–675.

Parkes, Margot W., Blake Poland, Sandra Allison, Donald C. Cole, Ian Culbert, Maya K. Gislason, Trevor Hancock et al. "Preparing for the Future of Public Health: Ecological Determinants of Health and the Call for an Eco-Social Approach to Public Health Education." *Canadian Journal of Public Health*, 111, no. 1 (2020): 60–64.

Phibbs, Suzanne, Christine Kenney, Christina Severinsen, Jon Mitchell, and Roger Hughes. "Synergising Public Health Concepts with the Sendai Framework for Disaster Risk Reduction: A Conceptual Glossary." *International Journal of Environmental Research and Public Health*, 13, no. 12 (2016): 1241.

Pohl, Christian. "What Is Progress in Transdisciplinary Research?" *Futures*, 43, no. 6 (2011): 618–626.

Poland, Blake, and Mark Dooris. "A Green and Healthy Future: The Settings Approach to Building Health, Equity and Sustainability." *Critical Public Health*, 20, no. 3 (2010): 281–298.

Poland Blake, Margot W. Parkes, Trevor Hancock, George McKibbon, and Andrea Chircop. "A Changing Role for Public Health in the Anthropocene: The Contribution of Scenario Thinking for Reimagining the Future," in *Health in the Anthropocene: Living Well on a Finite Planet*, eds. Stephen Quilley and Katharine Zywert (Toronto: University of Toronto Press, 2019).

Ratima, Mihi, Debbie Martin, Heather Castleden, and Treena Delormier. "Indigenous Voices and Knowledge Systems – Promoting Planetary Health, Health Equity, and Sustainable Development Now and for Future Generations." *Global Health Promotion*, 26, no. 3 (2019): 3–5.

Redvers, Nicole. "The Value of Global Indigenous Knowledge in Planetary Health." *Challenges*, 9, no. 2 (2018): 30.

Rolfe, Gary, Dawn Freshwater, and Melanie Jasper. *Critical Reflection for Nursing and the Helping Professions a User's Guide* (Palgrave, 2001), 194.

Romanelli, Cristina, David Cooper, Diarmid Campbell-Lendrum, Marina Maiero, William B. Karesh, Danny Hunter, and C. D. Golden. "Connecting Global Priorities: Biodiversity and Human Health: A State of Knowledge Review." World Health Organization/Secretariat of the UN Convention on Biological Diversity (2015).

Sandifer, Paul A., Ariana E. Sutton-Grier, and Bethney P. Ward. "Exploring Connections Among Nature, Biodiversity, Ecosystem Services, and Human Health and Well-Being: Opportunities to Enhance Health and Biodiversity Conservation." *Ecosystem Services*, 12 (2015): 1–15.

Shah, Nayankumar. "Taking a History: Introduction and the Presenting Complaint." *Student BMJ*, 13 (2005a): 0509314.

Shah, Nayankumar. "Taking a History: Conclusion and Closure." *Student BMJ*, 13 (2005b): 0510358.

Stafford-Smith, Mark, David Griggs, Owen Gaffney, Farooq Ullah, Belinda Reyers, Norichika Kanie, Bjorn Stigson et al. "Integration: The Key to Implementing the Sustainable Development Goals." *Sustainability Science*, 12, no. 6 (2017): 911–919.

Stephen, Craig, Julie Wittrock, and Joy Wade. "Using a Harm Reduction Approach in an Environmental Case Study of Fish and Wildlife Health." *EcoHealth*, 15, no. 1 (2018): 4–7.

Vandyck, Toon, Kimon Keramidas, Alban Kitous, Joseph V. Spadaro, Rita Van Dingenen, Mike Holland, and Bert Saveyn. "Air Quality Co-Benefits for Human Health and Agriculture Counterbalance Costs to Meet Paris Agreement Pledges." *Nature Communications*, 9, no. 1 (2018): 1–11.

Wahl, David Christian. *Designing Regenerative Cultures* (England: Triarchy Press, 2016).

Walpole, Sarah Catherine, Stefi Barna, Janet Richardson, and Hanna-Andrea Rother. "Sustainable Healthcare Education: Integrating Planetary Health into Clinical Education." *The Lancet Planetary Health*, 3, no. 1 (2019): e6–e7.

Walpole, Sarah Catherine, Kumanan Rasanathan, and Diarmid Campbell-Lendrum. "Natural and Unnatural Synergies: Climate Change Policy and Health Equity." *Bulletin of the World Health Organization*, 87 (2009): 799–801.

Waltner-Toews, David. *Ecosystem Sustainability and Health: A Practical Approach* (Cambridge: Cambridge University Press, 2004).

Waltner-Toews, David, Valerie A. Brown, and John A. Harris. "Knowing Our Own Minds," in *Independent Thinking in an Uncertain World: A Mind of One's Own*, eds. Valerie A. Brown, John A. Harris, and David Waltner-Toews (New York: Routledge, 2019), 315–328.

Waltner-Toews, David, and James Kay. "The Evolution of an Ecosystem Approach: The Diamond Schematic and an Adaptive Methodology for Ecosystem Sustainability and Health." *Ecology and Society*, 10, no. 1 (2005).

Webb, Jena C., Donna Mergler, Margot W. Parkes, Johanne Saint-Charles, Jerry Spiegel, David Waltner-Toews, Annalee Yassi et al. "Tools for Thoughtful Action: The Role of Ecosystem Approaches to Health in Enhancing Public Health." *Canadian Journal of Public Health*, 101, no. 6 (2010): 439–441.

White, Damian. "Ecological Democracy, Just Transitions and a Political Ecology of Design." *Environmental Values*, 28, no.1 (2019): 31–53. https://doi.org/10.3197/096327119X15445433913569.

Williams, Byron K., Robert C. Szaro, and Carl D. Shapiro. *Adaptive Management: The US Department of the Interior Technical Guide*. US Department of the Interior, 2009.

Woollard, Robert F. "Caring for a Common Future: Medical Schools' Social Accountability." *Medical Education*, 40, no. 4 (2006): 301–313.

WHO. "The Ottawa Charter for Health Promotion." World Health Organization, 1986. https://www.who.int/healthpromotion/conferences/previous/ottawa/en/

WHO. "Towards Unity for Health: Challenges and Opportunities for Partnership in Health Development." World Health Organization, 2000. http://www.who.int/entity/hrh/documents/en/TUFH_challenges.pdf

Zinsstag, Jakob, Ester Schelling, David Waltner-Toews, David and Marcel Tannera. 'From "one medicine" to "One Health" and systemic approaches to health and well-being.' *Preventive Veterinary Medicine*, 101 (2011): 148–156.

Zuber-Skerritt, Ortrun. "Participatory Action Learning and Action Research (PALAR) for Community Engagement: A Theoretical Framework." *Educational Research for Social Change*, 4, no. 1 (2015): 5–25.

6 Harm Reduction for Reciprocal Care

Craig Stephen

CONTENTS

To harm something is to damage it or make it less effective or successful than it was. There are many types of harms to consider (e.g. physical, financial, psychological, ecological, climatological, and others) when managing animals, health, and society. The socio-ecological model of health reminds us that each of these harms can impact the determinants of health as well as each other. Health management needs to address known and existing harms while at the same time minimize the chance of future and unanticipated harms. One Health, EcoHealth, and allied fields have documented how managing one type of harm in one species can lead to harmful implications for other species, populations, or generations. This chapter discusses harm reduction from the perspective of reciprocal care, where we look after ourselves, our community, and Nature over time.

HARM REDUCTION IN A ONE HEALTH WORLD

Health promotion reduces risks, promotes healthy settings, and tends to the underlying determinants of health. Try as we might, we are unlikely to achieve a world of no risk, completely healthy settings, and a full spectrum of supportive determinants of health. Harms still occur. Prohibitions of harmful situations or eradication of hazardous agents can drive the chance of harm to zero, but in their absence, a series of actions are needed to reduce harms.

The central defining feature of the harm reduction approach is its focus on reducing harm. This intentionally obvious description is meant to situate harm reduction within the spectrum of health promotion and health protection scopes of practice. Harm reduction does not argue against eliminating hazards or for accepting them but rather it aims to decrease the negative consequences of the hazardous circumstances in the face of uncertainty and conflicting opinions that prevent eradication of the hazards. Harm reduction is consistent with the precautionary approach that states that lack of full certainty shall not be used as a reason for postponing measures to prevent harm.

Harm reduction is both a goal and a process. It is a goal shared across many health and environmental sciences. Most people interested in health ultimately want to reduce harm to individuals, species, or environments, whether through clinical care, preventive actions, or health promotion. Harm reduction, as a process, focuses on developing local relationships and collaborations for collective actions leading to incremental improvements in health. Reciprocal care harm reduction asks us to consider the interconnections of harms and how actions to reduce one harm influence harms in other species, circumstances, or generations. Harm reduction works to minimize harms through non-judgemental strategies by enhancing skills and knowledge to live safer, more sustainable, healthier lives.

The paradox of promoting human well-being in an exponentially growing human population illustrates the need for reciprocal care harm reduction. The social justice ethic of public health encourages policies that leave no one behind by ensuring all people have equitable access to the conditions and resources needed to be healthy. Political, cultural, and technical advances over the past two centuries have greatly reduced mortality and changed the material conditions needed to bring people out of poverty and improve their well-being. Linking well-being with material conditions resulted in the coupling of well-being to consumption. Increased consumption by a larger proportion of an exponentially growing human population tremendously expanded societies' ecological footprint, resulting in unprecedented rates of ecological degradation. The World Wildlife Fund's 2006 Living Planet Index concluded that Earth's regenerative capacity can no longer keep up with human demand. The drive to reduce harms to people by increasing global health equity has caused serious and, in many cases, irreversible, harms that threaten both current biodiversity and future generations of people, animals, and plants (McMichael, 2002).

Examples of unintentional impacts of harm reduction in one sector causing harm in another can be found in many situations, including the global growth of aquaculture. People in developed countries have been encouraged to increase their consumption of fatty fish as part of dietary strategies to reduce the personal harms from cardiovascular disease. These recommendations fail to consider the reality that global fisheries cannot meet the seafood demands from affluent and developing economies. The collapse of fish stocks can have profound socioeconomic effects on coastal communities, which in turn can shift supplies of local protein that, in some countries, harm terrestrial wildlife (Brashares et al.,

2004). Some countries have looked to aquaculture to reduce harm to wild fish stocks and coastal communities by providing local jobs and meeting the growing human demand for fatty fish without exploiting wild fish. The "blue revolution" of aquaculture has, however, been dogged by concerns about energy-intensive animal production, nutrient pollution, habitat destruction, water diversions that disrupt aquatic ecosystems, introduction of alien species, and the use of fish meals in feeds (Costa-Pierce, 2002). Advice to reduce personal cardiovascular disease risks cannot, in this example, be uncoupled from this web of inter-acting social and ecological harms.

Many One Health issues involve multiple concurrent harms. Chronic wasting disease (CWD) of cervids illustrates this point. This invariably fatal disease of deer, elk, and moose causes wasting, neurological dysfunction and eventual death of infected animals. The Western Association of Fish and Wildlife Agencies in North America deemed this prion disease as a significant threat to the future health and vitality of free-ranging deer and elk (WAFWA, 2017). Because of its similarities to the zoonotic prion disease causing mad cow disease (also known as bovine spongiform encephalopathy), public health agencies advised precaution in consuming cervids known to have the disease or hunted from CWD-positive areas (despite no evidence of human CWD to date). These recommendations discouraged hunting and in doing so harmed rural hunting businesses as well as lowering confidence in consuming wild game for subsistence needs. The resulting food safety concerns had implications for rural food security and on Indigenous rights to access safe and sustainable wildlife. As CWD appears in both wild and farmed cervids, international trade in cervid products was impeded by this disease. What at first was framed as a disease harming individual wild animals was later seen to cause conservations harms, inflict harms to agricultural trade, increase social conflict, cause psychological harm by increasing fear of natural places, alter ecosystem functions as deer and elk populations declined in number, cause cultural harms by discouraging traditional hunting, and create political conflicts linked to debates on how to control this epidemic. A harm reduction process would not focus on eradicating the CWD prion from populations or environments (although recognizing the importance of trying to do so) but rather would rely on existing knowledge, resources, and values to promote collaborative approaches to make incremental gains to reduce ecological, population, or social harms related to the disease.

Harm reduction recognizes that many critical drivers of harm cannot be eliminated. Take for example Japanese encephalitis: a mosquito-borne viral disease that circulates between birds, humans, and pigs. A 2012 study in Nepal sought to control this disease in humans by working with pig farmers to reduce the prevalence of the disease in pigs (Dhakal et al., 2014; Metelka et al., 2015; Robertson et al., 2013). Because of cultural biases against pig ownership, pig farmers were from poor, marginalized communities. They were often unable to purchase or afford Japanese encephalitis vaccines for their animals. Human vaccine uptake was low in these communities, and mosquito control efforts were inconsistent due to lack of financial resources and distrust of government. Many farmers did not own the land they farmed, causing them to lack infrastructure or financial capacity to house pigs indoors or away from people. Climate change was shifting the distribution of the mosquitoes higher into the mountains, exposing a new cohort of pigs and people to the virus. Policies to increase food security were creating conditions conducive to perpetuating the infection, conditions further compounded by climate change. Culture, poverty, and climate change coupled with an endemic virus were, therefore, the ultimate drivers of Japanese encephalitis harms in this system, but the likelihood that any of these drivers could be changed fast enough and extensively enough to eliminate present-day health and economic harms to families were extremely low. Actions were needed in the interim to reduce harms. In this example, the team focused on improving ecological and personal health literacy to increase uptake of free vaccines and inform agriculture and suburban planning to reduce conditions conducive to the disease but good for the farmers.

Reciprocal care harm reduction operates in a landscape of interacting influences and constantly changing conditions. It must be attentive to the interspecies and intergeneration implications of interventions to prevent or reduce the adverse consequences to all members of the One Health community rather than targeting only the hazard and its human victims. Lessons from managing other complex problems suggest that cooperative approaches that provide insight into cross-sectoral and socio-ecological systems context are critical to working in interspecies health (Head and Alford, 2015). Taking a systems approach to One Health can provide a wider perspective to discover alternative means of solving problems and avoid unintended consequences.

The upstream drivers of health found in social and ecological determinants can greatly influence the persistence, diversity, and magnitude of harms, yet they are most challenging to manage in One Health systems. The inseparable array of factors that determine and modify these determinants, coupled with tremendous challenges in eliminating hazards like pollutants, pathogens, or climate change, suggest that effective harm reduction requires multi-level interventions, ranging from influencing policy to targeted biomedical interventions.

HARM REDUCTION AS A GOAL – HOW TO GET THERE?

There are four general harm reduction goals: (1) prevent persistent, irreversible or severe harms after a harm has emerged; (2) prevent harmful consequence by early intervention; (3) reduce the likelihood that a harmful situation arises; and (4) maintain conditions that are not harmful or hazardous. Health practitioners and investigators work to reduce harm across the continuum from a premature or bad death, through disease and disability, to optimal well-being. Total harm faced by a population can, in general, be reduced by reducing the total amount of harm (which is achieved by reducing exposure or sensitivity), or by reducing the total impact of harm (achieved by increasing capacity to cope or reducing cumulative effects) (Figure 6.1) (Stephen et al., 2018).

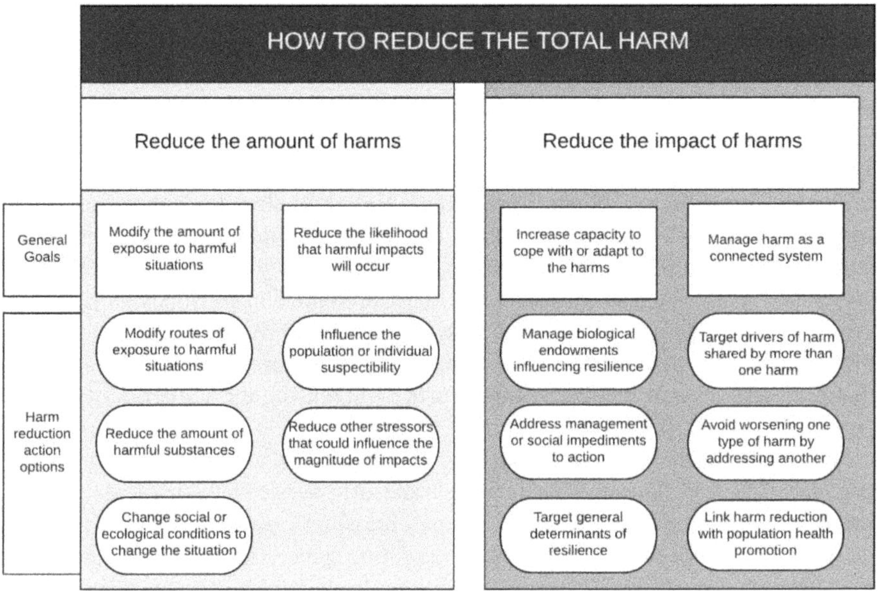

FIGURE 6.1 A taxonomy of harm reduction actions. This classification links general actions to harm reduction goals. Harm reduction programs require multi-level actions targeting more than one goal.

Goal 1: Intervene in a Harmful Situation to Prevent Persistent, Irreversible, or Severe Harms

In this scenario, harms have occurred. Recovery, mitigation, and remediation are key words in this circumstance. The goal is to lessen the impact of an ongoing harm, minimize the duration of the effects, and ensure it does not lead to irreversible harm. This is the realm of treatment and rehabilitation in clinical medicine. The health promotion practitioner may not be involved in delivering clinical services, but he or she could ensure equitable access to effective and appropriate health services to mitigate inequities in access to care. Removing socio-cultural, organizational, economic, geographic, and gender-related barriers to health care underpins equitable access to care (Evans et al., 2013). Socio-economic impediments can also deter people from accessing animal health services. For example, poverty limited farmer's access to preventive veterinary services in the Nepalese case described earlier. Subsidized or free veterinary services in developing countries or remote areas are examples of harm reduction for livestock development.

A challenging question, whether for people, animals, or ecosystems, is what to restore or recover, and how close to historical states should we aim? The question may seem easier for people or individual animals where we can strive to recover and individual back to its former state or to the norms of form and function expected by society. It becomes more challenging when thinking of ecosystems.

Take for example the 2010 oil spill in the Gulf of Mexico. Millions of barrels of oil that leaked into the ocean were met with millions of barrels of chemicals intended to disperse the oil. Political leaders vowed to restore the Gulf back to how it was before the spill. Unfortunately, the state of the Gulf was not pristine before the spill. The region had endured centuries of natural and technologically hazardous events. The impacts of the spill were immense, but so too were the challenges of pollution, habitat loss, overfishing, and climate change before the spill. Similar challenges can be found in recovering endangered species. What happens, for example, when two endangered species are engaged in a predator-prey relationship, such as sea otters and abalone in the United States, or lions and giraffes in Africa, or reindeer and wolves in Europe? Managers need to assess the acceptability and feasibility of controlling predator pressures to the extent that prey species can recover without further increasing the vulnerability of the predator.

Trade-offs in management strategies will affect restoration or recovery goals (e.g. see Box 6.1). These often involve trade-offs between values (e.g. ecological vs economic), preferences (e.g. saving charismatic species rather than cryptic unattractive species), or current versus future generations (e.g. carbon taxes). Because socio-ecological systems provide multiple health services which influence each other, decision-makers need to consider the system-wide synergies or trade-offs between ecosystem services, well-being components, and values being affected by decisions made. Ecologic, economic, and systems dynamic models can be used with participatory methods to systematically examine trade-offs in recovery and rehabilitation efforts.

BOX 6.1 CASE STUDY: FINDING WIN-WIN SOLUTIONS WHEN RECOVERY AND REHABILITATION GOALS CONFLICT — A CASE OF INVASIVE SPECIES ACTION VERSUS ENDANGERED SPECIES PROTECTION

Freshwater mussels are some of the most endangered groups of animals in North America. The Rocky Mountain ridged mussel (*Gonidea angulata*) is a bivalve species whose Canadian range is limited to the Okanagan Valley, British Columbia. Its habitat is being encroached upon by the invasive Eurasian watermilfoil (*Myriophyllum spicatum*). Milfoil control is a local priority due its impact on the recreational and real estate value of lakeshore properties. Protecting the mussels was believed to require curtailment of milfoil control in portions of the lake, a solution unpalatable to local businesses and politicians. Mapping the overlaps in distributions of these two species helps identify adaptive management strategies by proposing evidence-based buffer zones and/or tailoring milfoil control activities to site-specific mussel-milfoil overlaps (Figure 6.2). Milfoil control can help protect habitat quality for the mussels and, therefore, could be reframed as a shared conservation and invasive species goal, rather than a competitive goal.

FIGURE 6.2 Measuring the spatial overlap of Rocky Mountain ridge mussels and Eurasian watermilfoil in Okanagan Lake, British Columbia, Canada, to gather evidence to create win-win solutions for endangered species and invasive species management. (Image credit Joy Wade.)

Goal 2: Intervene Early in a Harmful Situation before the Outcomes Cause Notable Damage

Early warning is the key phrase for this scenario. The goal is to act quickly to limit the extent of harms that could occur if the harmful situation was left unattended. Actions try to prevent the early signs of harms turning into disease or loss of function. Screening tests play a major role for this form of harm reduction in clinical medicine. Tests or examinations seek sub-clinical signs of disease that inspire interventions that prevent clinical effects, like blood pressure testing to screen for pre-clinical cardiovascular disease. To be effective, this form of harm reduction needs to generate signals that inspire people to act. Confidence is needed in the predictive value of early warning signals. Despite some challenges in linking early warning signs in clinical medicine with the certainty of a harmful outcome, tremendous gains in combatting individual and community harms from many diseases have been made through health screening. Routine recording of blood pressure, cholesterol, urine sugars, and cervical cytology, for example, have become standard part of annual physical exams as part of the battle against diabetes, cardiovascular disease, and cancer, respectively.

The pace of environmental and social changes is creating a greater need for screening for upcoming threats and identifying the most vulnerable population in advance of harms. Confidence in taking preventive steps requires access to the right information fast enough from a trusted source. The use of animal or environmental signals as early warning for human diseases underlies the founding premise of One Health. Much has been written about the use of animal and ecological indicators as early warning of pandemics and emerging diseases. Less has been written about how well these indicators motivate people to act in a way that reduces harms. The frequency with which animal and ecological early warning signs are suggestive of human disease risk exist without an associated human disease outbreak far outnumber the few occasions when a risk to humans does emerge. Despite billions of dollars spent globally to predict the next emerging infectious disease, we continue to be surprised by emerging threats, such as the coronavirus pandemics (COVID-19) of 2020 and the emergence of Zika virus in the Americas before that. It is no surprise that repeated proclamations of disease outbreak early warning without concomitant impacts creates future warning fatigue.

There are methods to predict emerging diseases, but so far, they act as "the art of the possible" rather than predictive tools to direct specific risk management actions in specific locations and times. Significant knowledge gaps for even the most studied threats along with a view of causation disconnected from social and ecological context reduces the likelihood that we will be able to predict surprising harms with accuracy or regularity in the foreseeable future. The rate of health surprises is expected to accelerate in the Anthropocene and to grow beyond infectious diseases to include changes in contaminant exposure, catastrophic events like wildfire and floods, and shifts in access to natural resources that underpin economic and cultural needs. While we may not predict the next surprise, we still need early warning systems to help anticipate vulnerable situations in order to prioritize limited resources.

Surprises arise when there is a gap between what you expect and what occurs. Our expectations can be restricted when we limit what we listen to (so we are unreceptive or insensitive to warning signals outside of our usual scope of practice); who we listen to (such as when priorities and overcrowded agendas discourage collaborations that extent beyond immediate interests), and what we believe (power dynamics, too much information, and wishful thinking can lead to failure to recognize and act on very early warning signals) (Stephen, 2019). Five warning scenarios and associated foci for early warning are presented in Table 6.1. Accessing, integrating, and applying information from a socio-ecological system perspective is needed for early warning of changes at the human-animal-environment junction. But, more importantly, early warning signals need to be adapted to the community being warned and often need to be sought from more than one agency. Surprise anticipation can be enhanced by improving awareness of changes in

TABLE 6.1
Five Early Warning Scenarios and Their Focus for Early Warning

Scenario	Situation	Focus
Risk management	Health managers are fully aware of the nature of the hazard and the vulnerability of the community or ecosystem	Document how a known hazard is changing in distribution or abundance and inform local risk perception to modify protective and preventive actions accordingly
Coping assessment	There is uncertainty about how a community or ecosystem can cope with known hazards	Track changes to determinants of health that influence sensitivity to harms and coping capacity
Hazard detection and monitoring	Variation in community, individual, or system vulnerability is primarily driven by differences in exposure to known hazards	Provide reliable signals of the distribution and abundance of known hazards and information on the likelihood of exposure by examining changes in social or ecological interactions
Signals of effect	A suspected hazard is present in the environment but there is insufficient evidence of effect to warrant risk reduction actions, or there are newly introduced or previously unknown agents, the effects of which are unknown	Link the presence of the agent/hazard with a biological effect to inform recognition of a risk and help prioritize risks requiring further assessment or management
Mixed strategies	The characteristics of community or system vulnerability are largely unknown and effects of hazards unanticipated	A mixed strategy combing attributes of the four preceding scenarios is warranted

Source: Adapted from Stephen and Duncan, 2017.

distributions of and exposure to hazards and/or variations in population vulnerability or resilience to shocks (Figure 6.3) (Stephen et al., 2015).

Goal 3: Reduce the Likelihood That a Risky Situation Will Initiate Harmful Outcomes

This form of harm reduction focuses on reducing the likelihood that a harm will occur. Prevention is the keyword. Actions are promoted in advance of a measurable harm. Understanding the risks inherent in a situation allows you to prioritize and target actions to reduce risk as well as to examine any offsets between the risks and benefits of that situation.

Risk factors are any attribute, agent, or situation that affects the likelihood of harm. Risk factors are not diagnostic. They give a sense of the likelihood of

TYPE OF SURPRISE	GENERAL RESPONSE
Knowable in retrospect, but eluded detection	DIVERSIFY PARTNERSHIPS AND INFORMATION
Failed to recognize a warning signal or were unable to act	• Recognize harmful situations • Communicate warning • Influence those who need to act
Unanticipated consequence of social or ecological change	PROMOTE RESILIENCE
Previously inconceivable events	• Reduce susceptibility and exposure to harmful situations • Create conditions conducive to health

FIGURE 6.3 The relationship between the type of surprising circumstance one could encounter and general responses to prepare for the next surprise. (Adapted from Stephen et al., 2015.)

something occurring in populations like the ones from which risk estimates have been derived. Risk factors can be categorized in many ways. There are modifiable and non-modifiable risk factors. There are environmental, social, behavioural, and biological risk factors. There are individual, population, and community risk factors. Care must be taken in making inferences about the impacts of risk factors measured at one level (like an individual) on another level (like a community). Additional care needs to be taken in extrapolating the impacts of a risk factor in one species (like a cow) to another (like a deer). Risk factor analysis is contextual.

Another distinction is the difference between perceived risk and a measured risk. Measured risks are the outcome of epidemiological studies that systematically compare a study group with a control group that presents the same characteristics of the study group, except for the risk factor being studied. Perceived risk refers to individuals' "instinctive and intuitive reactions to danger" (Slovic and Peters, 2006). Both types of risk are real, and both are prone to error. Measured risks are influenced by the limitations of study design, challenges in finding adequate control groups, and problems in generalizing how multifactorial systems work in one context compared to another. The issue of finding enough true controls becomes more challenging as we move from individuals to communities to ecosystems because as systems become more complex, finding identical compositions and relationships becomes more difficult. Perceived risks are influenced by culture, personal experience, psychological state, and social norms. They can, therefore, vary significantly within and between communities. Mobilizing society and individuals to act preventatively needs one to be attentive to perceived risks.

Instead of concentrating solely on risk factors in isolation, greater attention is now paid to interacting risks. For example, obesity in pets has been linked to genetic predisposition, reproductive management, and dietary/exercise management (Bland et al., 2009). Obesity in humans has been linked to trade liberalization, economic growth and rapid urbanization that change environments, diets, and lifestyles in ways that promote positive energy balance (Malik et al., 2013). Interventions focused on changing the diet and exercise behaviour of individuals without attending to the condition leading to obesogenic conditions have had little impact on the obesity epidemic (Dehghan et al., 2005). Harm reduction focused on preventing obesity requires a variety of interventions targeting attitudes, lifestyles, and environments of people or pet owners to create the circumstances that keep normal-weight patients from becoming obese. Comprehensive prevention planning is important because of the multiple, complex individual, community and societal and ecological factors that not only drive the origins of obesity-related harms but also influence the ability and willingness to act in advance of harms.

Pandemic planning is an example where planners need not only consider risk factors that influence the likelihood of a pandemic but also those that influence motivations to act in advance of disease outbreak (Stephen, 2019). Being aware of and understanding a risk does not inevitably lead to risk avoidance. Personal experience, risk perception, trust in authorities, and exposure to false alarms, all affect willingness to act. A study of Chinese chicken farmers found that farmer personal attributes (gender, age, education, risk perception), farm production variables (animal density, feed conversion ratios, chicken weight gain), and societal variables (access to services, availability of subsidies), all affected farmer biosecurity decisions to reduce avian influenza risks (Huang et al., 2016). The innumerable reports of people violating recommendations or requirements for social distancing during the 2020 coronavirus (COVID-19) pandemic further demonstrated that acceptance of logical risk reduction recommendations is not universal, even when people are provided the same facts in the same media. Knowledge of the presence of a pathogen may be helpful but is often insufficient to motivate risk reduction actions. Risk management at this stage requires attention to facts, knowledge, and beliefs. Chapter 9 provides details on motivating health promoting and harm reduction actions.

Goal 4: Create an Environment Where the Underlying Conditions Leading to Harmful and Risky Situations Are Unlikely to Occur

Resilience, thriving, and flourishing are three words associated with this form of harm reduction. The goal at this stage is to equip people with the knowledge, resources, and circumstances that allow them to make choices that build health promoting capacities in themselves, their communities, other species, or ecosystems (e.g. see Box 6.2). The focus is on avoiding, removing, or mitigating pressures that lead to unhealthy conditions or environments that can lead to harmful situations. This can be achieved either by compulsion (such as laws or penalties) or voluntarily (by incentives or by finding win-win situations producing healthy co-benefits).

Social and economic structures along with personal choices and behaviours can determine decisions and actions that promote reciprocal care of human, animals, and environmental health. Addressing these factors can be a challenging yet effective investment in health. For example, clean indoor air laws, tobacco taxes, and reducing sodium from the food supply can have profound impacts on settings and decisions that affect cardiovascular health for a large segment of the human population (Weintraub et al., 2011). Modern dairy herd health has developed into a transdisciplinary approach, combining sociology, psychology, economics, behavioural science, and communication with classical veterinary disciplines to motivate farmers to make systems-wide decisions to protect animal health and productivity (Kristensen and Jakobsen, 2011). Ecosystem conservation without cultural considerations is not only insufficient but also risks unanticipated negative impacts to communities (Poe et al., 2014). Climate change adaptation is another proactive, health promoting form of harm reduction. It tries to create a sustainable future for all people and species through an integrated approach at all scales. It recognizes the limits of natural and human systems to adapt to climate change and advocates for a new future where people retain critical functions,

BOX 6.2 CASE STUDY: MATCHING EVOLVED NEEDS WITH LIVING CONDITIONS TO PROMOTE ANIMAL WELFARE AND CREATE COMMUNITY CONSERVATION AWARENESS

A new phenomenon of community-based, capture-hold-release aquariums is an example of the harm reduction in practice that is attentive to social, conservation, and animal welfare needs. These types of aquariums (Figure 6.4) have been growing in number in small coastal communities as ways to increase community connection to local marine environments by displaying local animals for shorter periods while avoiding negative perceptions about holding individual animals too long in captivity. They strive to have a low ecological footprint as well as promote awareness and activities to protect local marine environments through environmental health literacy and community-based action. However, their remoteness, facility design, and scale can sometimes prevent access to veterinary services and infrastructure that allow medical intervention typical of larger aquariums. They are, therefore, heavily dependent on creating the conditions that provide the captive animals access to the resources needed to stay healthy. While biosecurity and early triage of ill animals are instrumental health management tools, success is found by matching the species' adapted capacities and the realities of its holding environment. This is done by the habitat design, species composition and density, animal collection and transportation protocols, regulating and monitoring public access, and feeding protocols using accessible local foods that all match the animals' biology (Stephen et al., 2019b).

FIGURE 6.4 A giant Pacific octopus in a coastal community capture-hold-release aquarium.

resources, and opportunities that will allow health to continue. Strong and closely connected communities are better able to adapt to, withstand, or get out of harm's way in a disaster, which in turn contributes to the future well-being of individuals and communities (Wulff et al., 2015). Policies that decrease social inequities and improve social cohesion can minimize and offset the drivers of change (Bunch et al., 2011). Working upstream in a One Health world means creating the social and ecological conditions that concurrently enable healthful situations for animals, society, and environments.

Health and environmental literacy are important in this realm of harm reduction due to the reliance on affecting people's attitudes and behaviours. Health literacy supports individuals' capacity to obtain, process, and understand information and services needed to make appropriate health decisions. Environmental literacy promotes awareness of and concern for the environment and provides the knowledge, skills, and motivations to work towards solutions of current problems and the prevention of new ones (McBride et al., 2013). Ecological health literacy promotes everyday practices that support the sustainability of both non-human and human communities. It integrates concepts from environmental literacy and health literacy to develop a range of skills and competencies that people need in order to seek out, comprehend, evaluate, and use information to make informed choices that improve quality of life and protect the environment (Finn and O'Fallon, 2017). Concurrently promoting actions to protect human, animal, and environmental health requires competencies not only in communicating complex science in an integrated and accessible manner but also in mobilizing information to communities and individuals that may have diverse risk perceptions and conflicting values. Chapter 9 describes in more detail some theory and methods to support people making healthy decisions for themselves or the animals or ecosystems they care for.

HARM REDUCTION AS A PROCESS

Scientific advances, changes in technology, or regulatory changes needed to eliminate hazards or harms can take considerable time to be achieved. Many hazards cannot be quickly eliminated because of the often slow pace of scientific, social, and political change. Too often harms prevail and actions are delayed due to challenges in securing the required new knowledge, regulations, or technology. The pervasive uncertainties and simplifying assumptions that can plague One Health create a gap between what science provides and what society demands in dealing with harms. Different standards and expectations for how much and what types of harms should be attacked can discourage actions, especially when the science remains inconclusive. Harm reduction, as a process, helps inspire collective action to make incremental improvements towards a healthy situation in the presence of uncertainty and conflict. It is a perspective and a set of strategies that applies to all the determinants of health and not merely problematic risks. It involves pragmatic, multidisciplinary, approaches to remove barriers to the implementation of knowledge to protect health and promote sustainability in situations where the hazard or harmful situation cannot be eliminated in the near future.

Harm reduction is most associated with public health actions against persistent problems such as addiction and homelessness. Adapting harm reduction to One Health provides new tactics to overcome entrenched perspectives and inaction to ensure progress on shared goals. Harm reduction works by reducing the more immediately harmful consequences of an activity through pragmatic, realistic programs feasible under current conditions. It promotes relationships, structures, and processes to make gains towards safer situations by incrementally reducing the negative health, social and ecological consequences to individuals, communities, and ecosystems, without relying on elimination of the hazard.

The process of harm reduction is characterized by five features:

1. *It is a collaborative process.*
   Harm reduction seeks multidisciplinary pathways to overcome barriers to implementing recommendations, fostering collaboration on shared goals, and reducing conflict. Social and organizational factors that influence actions and opportunities to prevent, mitigate, or cope with harms are targeted. A diversity of players is engaged in finding solutions throughout the chain of causation. The process does not blame or judge the participants. Collaboration involves individual who, thorough formal or informal negotiation, find ways to act together towards a shared vision that results in mutually beneficial interactions. Collaboration helps people see different aspects of the problem and, by exploring these differences, find solutions that go beyond their own perspective of the problem. Harm reduction processes build new forms of strategic collaboration and governance that allow actions while debate remains on the scope and mechanisms of harm.
2. *It creates an enabling environment for collective actions.*
   There are four preconditions to collaboration and cooperation (Thomson and Perry, 2006; Singh and Kant, 2008). Firstly, people need to have some ownership of the problem. Rather than seeing the issue as someone else's problem, successful collaborators need to see their role and responsibilities for helping to reduce harms that may extend beyond their interests. Some degree of negotiation will be needed to create a shared vision that will help collaborators see how working towards collective interests will meet the interests of themselves or their organization. Secondly, the collaborators need a shared goal of where they want to go and a hierarchy of achievable steps that, taken one at a time, can lead to a safer and healthier situation. Harm reduction emphasizes actions that can benefit multiple parties and lead to progress on shared goals. Thirdly, participants need to understand and endorse the process for making decisions and for moving what they know into action. Fourthly, there needs to be trust. Trust can be built by being honest in negotiations, communicating purposefully and regularly, behaving in accordance with agreements, and not taking advantage of others or event when the opportunity is available. Trust, commitment, and a deeper understanding of the value of collective action are gained by focusing on series of incremental small wins towards the long-term goal.
3. *It is oriented to finding pragmatic solutions.*
   While not ruling out the longer prohibitions or elimination of hazardous situations, gains that are feasible within the current circumstances and state of knowledge are sought rather than relying on the creation of a preferred future before acting. The focus is on finding strengths, possibilities, and opportunities to reduce negative consequences rather than emphasizing discovery of the proximate cause of harms or attributing blame to others. It is about working with what we have and who we have today to make incremental improvements.

4. *It is inclusive and local.*
   Whereas many forms of risk management emphasize top-down actions, harm reduction emphasizes bottom-up, locally developed planning. This reflects its focus on working with the knowledge, resources, and relationships that are currently affecting the problems of concern in the context in which they are occurring. Harm reduction recognizes that no one approach works for everyone in all situations. It emphasizes action plans that adapt generic recommendations to local circumstances to produce gains that can be built on over time to lessen present harms while preparing for tomorrows risks. Individuals, agencies, companies, and communities affected by or affecting harms and risks need to be involved in the co-creation of harm reduction strategies tailored to a specific situation. Incorporating the context in which environmental, organizational, and personal factors interact increases the likelihood of finding shared priorities.
5. *It is integrative.*
   Social, ecological, and individual harms are interrelated and examining them together helps build consensus on goals and finds actions that may have benefits across domains. Harm reduction promotes a range of interventions but shifts the focus for change from technical and biological matters alone to include social innovations and opportunities. Seeking consensus on biological harms without accounting for social harms can increase conflict and delay actions. The usual approach of examining one type of harm in isolation from another reduces chances of finding common pathways or opportunities to reduce or eliminate risks and harms.

The pathway from knowledge to harm reduction action requires trust. People need to trust both the information provided and the information providers. To be successful in communicating and building trust, one must recognize and respect the diverse and complex value systems operating at the animal-health- society interface. Inspiring people to act in a way that addresses a suite of harms needs to do more than present the facts. The mere acquisition of new information often does little to affect risk perceptions and willingness to act (Gerrard et al., 1999). Information sharing needs to be tailored to the individual characteristics of targeted audiences, including an assessment of their readiness for change. Blending objective information with an understanding of people's emotions, values, and personal experiences is essential when promoting ecological and interspecies harm reduction actions.

The harm reduction process is starting to be used outside of the public health realm, for example, in managing monkey-human conflicts in St. Kitts and Nevis (see Chapter 22) and in recommendations to manage salmon aquaculture in British Columbia, Canada. An expert advisory panel recommended harm reduction as a new tactic to overcome entrenched perspectives and inaction associated with salmon aquaculture (Anon, 2018). To ensure progress on sustaining healthy and abundant wild salmon, irrespective of the diverse perspectives on salmon farming, the panel emphasized finding strengths, possibilities, and opportunities to reduce harms and the development of achievable small wins to build public trust through collaboration.

The Anthropocene is bringing unprecedented changes along with more far-reaching interconnections. While significant effort is and should be directed at detecting and responding to harms arising from these changes and connections, there is room for people to create the conditions that incrementally make the world healthier by focusing on circumstances and actions that reduce the total amount and total impact of harms across the human-animal-environment interface through harm reduction processes.

REFERENCES

Anon. "Minister of Agriculture's Advisory Council on Finfish Aquaculture Final Report and Recommendations." Province of British Columbia, 2018. https://www2.gov.bc.ca/assets/gov/farming-natural-resources-and-industry/agriculture-and-seafood/fisheries-and-aquaculture/minister-or-agriculture-s-advisory-council-on-finfish-aquaculture/maacfa-2017-docs/minister_of_agricultures_advisory_council_on_finfish_aquaculture_final_report_and_appendices.pdf (accessed May 2020).

Bland, I. M., A. Guthrie-Jones, R. D. Taylor, and J. Hill. "Dog Obesity: Owner Attitudes and Behaviour." *Preventive Veterinary Medicine*, 92, no. 4 (2009): 333–340.

Brashares, Justin S., Peter Arcese, Moses K. Sam, Peter B. Coppolillo, Anthony R.E. Sinclair, and Andrew Balmford. "Bushmeat Hunting, Wildlife Declines, and Fish Supply in West Africa." *Science*, 306, no. 5699 (2004): 1180–1183.

Bunch, Martin J., Karen E. Morrison, Margot W. Parkes, and Henry D. Venema. "Promoting Health and Well-Being by Managing for Social-Ecological Resilience: The Potential of Integrating EcoHealth and Water Resources Management Approaches." *Ecology and Society*, 16, no. 1 (2011).

Costa-Pierce, Barry A. "The 'Blue Revolution': Aquaculture Must Go Green." *World Aquaculture*, 33, no. 4 (2002): 4–5.

Dehghan, Mahshid, Noori Akhtar-Danesh, and Anwar T. Merchant. "Childhood Obesity, Prevalence and Prevention." *Nutrition Journal*, 4, no. 1 (2005): 24.

Dhakal, Santosh, Durga Datt Joshi, Anita Ale, Minu Sharma, Meena Dahal, Yogendra Shah, Dhan Kumar Pant et al. "Regional Variation in Pig Farmer Awareness and Actions Regarding Japanese Encephalitis in Nepal: Implications for Public Health Education." *PLoS One*, 9, no. 1 (2014): e85399.

Evans, David B., Justine Hsu, and Ties Boerma. "Universal Health Coverage and Universal Access." *Bulletin of the World Health Organization*, 91 (2013): 546–546A.

Finn, Symma, and Liam O'Fallon. "The Emergence of Environmental Health Literacy: From Its Roots to Its Future Potential." *Environmental Health Perspectives*, 125, no. 4 (2017): 495–501.

Gerrard, Meg, Frederick X. Gibbons, and Monica Reis-Bergan. "The Effect of Risk Communication on Risk Perceptions: The Significance of Individual Differences." *Journal of the National Cancer Institute Monographs*, no. 25 (1999): 94–100.

Head, Brian W., and John Alford. "Wicked Problems: Implications for Public Policy and Management." *Administration and Society*, 47, no. 6 (2015): 711–739.

Huang, Zeying, Di Zeng, and Jimin Wang. "Factors Affecting Chinese Broiler Farmers' Main Preventive Practices in Response to Highly Pathogenic Avian Influenza." *Preventive Veterinary Medicine*, 134 (2016): 153–159.

Kristensen, E., and E.B. Jakobsen. "Challenging the Myth of the Irrational Dairy Farmer; Understanding Decision-Making Related to Herd Health." *New Zealand Veterinary Journal*, 59, no. 1 (2011): 1–7.

Malik, Vasanti S., Walter C. Willett, and Frank B. Hu. "Global Obesity: Trends, Risk Factors and Policy Implications." *Nature Reviews Endocrinology*, 9, no. 1 (2013): 13.

McBride, Brooke B., C.A. Brewer, A.R. Berkowitz, and William T. Borrie. "Environmental Literacy, Ecological Literacy, Ecoliteracy: What Do We Mean and How Did We Get Here?" *Ecosphere*, 4, no. 5 (2013): 1–20.

McMichael, Anthony J. "Population, Environment, Disease, and Survival: Past Patterns, Uncertain Futures." *Lancet*, 359, no. 9312 (2002): 1145–1148.

Metelka, Julia, Colin Robertson, and Craig Stephen. "Japanese Encephalitis: Estimating Future Trends in Asia." *AIMS Public Health*, 2, no. 4 (2015): 601.

Poe, Melissa R., Karma C. Norman, and Phillip S. Levin. "Cultural Dimensions of Socioecological Systems: Key Connections and Guiding Principles for Conservation in Coastal Environments." *Conservation Letters*, 7, no. 3 (2014): 166–175.

Robertson, Colin, Dhan Kumar Pant, Durga Datt Joshi, Minu Sharma, Meena Dahal, and Craig Stephen. "Comparative Spatial Dynamics of Japanese Encephalitis and Acute Encephalitis Syndrome in Nepal." *PLoS One*, 8, no. 7 (2013): e66168.

Singh, M.D., and R. Kant. "Knowledge Management Barriers: An Interpretive Structural Modeling Approach." *International Journal of Management Science and Engineering Management*, 3, no. 2 (2008): 141–150.

Slovic, Paul, and Ellen Peters. "Risk Perception and Affect." *Current Directions in Psychological Science*, 15, no. 6 (2006): 322–325.

Stephen, Craig. "Rethinking Pandemic Preparedness in the Anthropocene." *Healthcare Management Forum* (Los Angeles: SAGE Publications, 2019), 0840470419867347.

Stephen, Craig, John Berezowski, and Vikram Misra. "Surprise Is a Neglected Aspect of Emerging Infectious Disease." *EcoHealth*, 12, no. 2 (2015): 208–211.

Stephen, Craig, and Colleen Duncan. "Can Wildlife Surveillance Contribute to Public Health Preparedness for Climate Change? A Canadian Perspective." *Climatic Change*, 141, no. 2 (2017): 259–271.

Stephen, Craig, Laura Griffith-Cochrane, and Joy Wade. "Addressing the Unique Challenges of Community-Based Capture-Hold-Release Aquariums Through a Facility Health Program." *Canadian Journal of Fisheries and Aquatic Sciences* 76, no. 9 (2019b): 1684–1687.

Stephen, Craig, Julie Wittrock, and Joy Wade. "Using a Harm Reduction Approach in an Environmental Case Study of Fish and Wildlife Health." *EcoHealth*, 15, no. 1 (2018): 4–7.

Thomson, Ann Marie, and James L. Perry. "Collaboration Processes: Inside the Black Box." *Public Administration Review*, 66 (2006): 20–32.

WAFWA. "Recommendations for Adaptive Management of Chronic Wasting Disease in the West." Western Association of Fish and Wildlife Agencies, Wildlife Health Committee and Mule Deer Working Group, Edmonton, Alberta, Canada, and Fort Collins, CO (2017): 20.

Weintraub, William S., Stephen R. Daniels, Lora E. Burke, Barry A. Franklin, David C. Goff, Jr, Laura L. Hayman, Donald Lloyd-Jones et al. "Value of Primordial and Primary Prevention for Cardiovascular Disease: A Policy Statement from the American Heart Association." *Circulation*, 124, no. 8 (2011): 967–990.

Wulff, Katharine, Darrin Donato, and Nicole Lurie. "What Is Health Resilience and How Can We Build It?" *Annual Review of Public Health*, 36 (2015): 361–374.

7 Building Health Surveillance for Decision Support at the Animal, Human, Environment Nexus

John Berezowski, Craig Stephen, and Luis Pedro Carmo

CONTENTS

The complex intersectoral and multi-level nature of the health of humans, animals, and environments means that it is not possible to prescribe an all-embracing policy that ensures decisions made to protect the health of one species, or generation, do not harm another. The world is a dynamic place where recent factors such as globalization, big data, climate change, and others have accelerated change from local to global levels. Changing conditions result in new problems and new opportunities with which societal organizations must deal. When new problems arise, existing information can be insufficient. New information will be needed to support decisions that enable optimal solutions. Information and knowledge are essential components of decision-making when uncertainty is present (Yim et al., 2004). Continuous production of new information and knowledge has become essential for organizations to be effective in this environment. Decision-makers need an expansive perspective so that decisions are less likely to have harmful unintended consequences, are more adaptable, and help reduce vulnerabilities to emerging and persistent harms. The nature of health under global change, and how it may be influenced by alternative decisions, is uncertain. Managing under these conditions requires new collaborations to gather information and perspectives to support integrated health management decisions.

Despite an urgency for interdisciplinary work in the Anthropocene, most surveillance programs remain bound to disciplinary conventions. During the last century, most population health programs were developed to deal with single diseases that occur in a single population of a single species. These single disease-species systems were effective for many contagious diseases that were important in the previous century. However, these systems were not completely successful and there is growing recognition that new approaches are needed for dealing with more complex problems, such as cross-species disease emergence, the effect of climate change on human, animal and ecosystem health, and antimicrobial resistance (Wilson et al., 2020). Since humans, animals, and other species live in an interconnected world that is constantly changing, we should not be surprised that health and disease constantly change and that new complex health problems constantly emerge. As organizations begin to operationalize new collaborative, cross-sectoral programs, they require a constant stream of new information. Surveillance at the interface of animals, health, and society needs to be reconceived as an information system made up of a chain of subsystems, each with its own knowledge of an aspect of the social and ecological drivers of health or harm. It must be conceptualized as a connected system of human knowledge and wisdom distributed throughout the socio-ecological chain of interactions that affect vulnerability and resilience.

HEALTH SURVEILLANCE AS AN INFORMATION SYSTEM

Decisions are built on a pyramid of data, information, knowledge, and wisdom (DIKW) (Figure 7.1). The DIKW hierarchy has been used to understand the role of information and knowledge in risk assessment in engineering (Aven, 2013), and we have adapted it to health surveillance.

Data are symbolic representations of observations of the real world. These can be observations of health outcomes, hazards, or other indices tracked by surveillance systems. Data alone have no intrinsic value. Information is data that are made useful, through processing and analyzing. Knowledge adds understanding (or knowing) of the importance of the information created. This entails examining information within the prevailing state of knowledge. Wisdom arises when knowledge is considered within its social and ecological reality. Wisdom helps us make the best decision based on the knowledge created, given the real-world opportunities, constraints, perspectives, and priorities.

DIKW systems can roughly be classed as static or dynamic. Static systems produce the largest volume and variety of DIKW used for health decision-making. It includes basic sciences – many applied disciplines such as pathology, epidemiology, sociology, human and veterinary medicine, engineering, animal science, experiential knowledge and wisdom held by community stakeholders, and more.

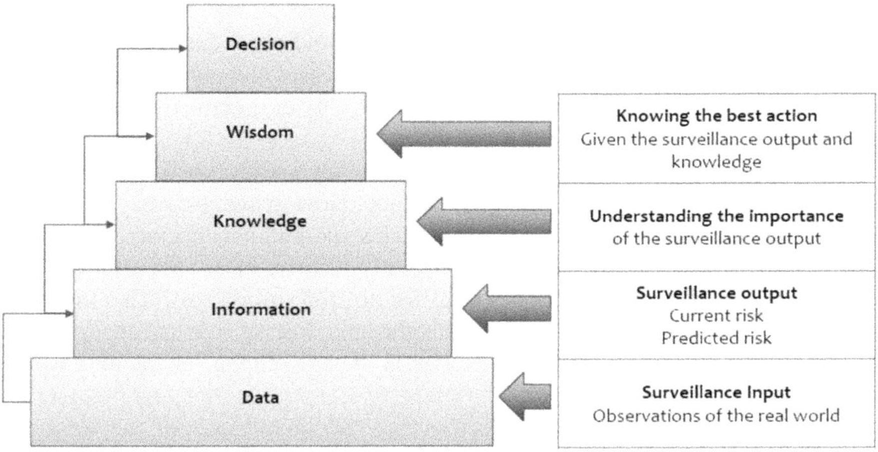

FIGURE 7.1 The data, information, knowledge, and wisdom (DIKW) hierarchy adapted to health surveillance.

These DIKWs are essential for understanding the complexity underlying health problems and solutions proposed to deal with them. The information and knowledge produced by static systems tend to change relatively slowly and, therefore, do not provide timely information about rapidly changing events.

Dynamic systems continuously generate DIKW through ongoing processes such as surveillance and monitoring. This type of information is often relatively simple but is essential for identifying changing health conditions or the emergence of new problems such as epidemics, as they occur. Population health surveillance, whether for humans, plants, or animals, can be defined as a continuous, dynamic process producing information that is used to inform decisions for dealing with health or disease-related problems (El Allaki et al., 2012). Surveillance systems create information that helps describe current health situations or helps forecast future health states. Since health surveillance is a continuous activity, it is particularly important for decisions that must be made in a constantly changing environment.

Health surveillance data are often collected from cases. What constitutes a case varies considerably, depending on the information output required. Cases can include laboratory-confirmed cases of a disease, health care provider (medical or veterinary) identification of suspect cases, patient self-reports of clinical symptoms, owner reports of clinical signs in animals, various surrogates for disease (for example counts of sales of diarrhoea medications as a surrogate for occurrences of diarrhoea in a population), antimicrobial-resistant bacterial or antimicrobial resistance genes, and many others (Antoine-Moussiaux et al., 2019). Other descriptive data are commonly combined with case data to provide contextual information. These include, at a minimum, the date and geographic location of case occurrences but can also include other data such as age, sex, breed, production type, and socio-ecological data such as socio-economic status of the case or the case owner, weather, nutrition, and other data. The additional data help to provide an epidemiological profile of cases and context for the cases that is useful for forecasting the risk of future events in specific times, places, or groups.

Surveillance data become information when they are processed and analyzed to answer specific questions. Examples include: How many cases have occurred in each region in the most recent period? Has the number of cases increased/decreased compared to previous time periods? Are there more cases among individuals with specific characteristics (e.g. age or production type)? This information is used to make inferences about whether the disease risk has changed and to make predictions about whether the risk is likely to change and among which regions or subpopulations.

Surveillance knowledge is understanding the importance of the information produced. It is required to answer the question: Does the most recent disease risk information warrant an action? Decision-makers must have access to additional knowledge to make decisions, most importantly an understanding of the potential harm that is likely to occur in a population for different disease risk estimates. Information and knowledge about the epidemiology of a disease, the expected spread of the disease, the vulnerability of the population, and the expected economic and other losses can enhance the value of surveillance information. One Health decision-makers who operate programs aimed at benefiting the health of humans, animals, and the environment require rich contextual knowledge. For example, they will need to understand the complex linkages between the health of many species and the environment if they are going to effectively estimate the potential harm that an increase in disease risk in one (or more) species may have on all species involved. Currently, population health is mostly domain specific, and decision-makers are educated mainly within the domain they work. One Health decision-makers will require skills and capacities to acquire, understand, and use the additional knowledge needed to make effective decisions.

Wisdom is being able to use surveillance information and knowledge in the most appropriate way, resulting in the least harm and greatest benefit to as much of the population and as many of the stakeholders as possible. Wise decision-makers must understand the significance of risk estimates and be able to design the most effective and appropriate actions. As we consider more species and populations in decision-making, the limitations of single species or single disease-focused surveillance quickly becomes apparent.

Health surveillance is typically a technical activity focused mainly on data creation and analysis leading to information creation. These activities are conducted by highly trained specialists who often perform their tasks with limited input from people outside their specialty. This specialist approach has elevated the importance of data and analysis, shifting emphasis away from knowledge and wisdom, and excluding many stakeholders who could provide valuable knowledge to enhance surveillance information.

The data information centric approach works well when health problems are well known, clearly defined, and the decision being made is unambiguous. For example, the U.S. Centers for Disease Control (CDC) defines an epidemic of listeriosis (caused by *Listeria monocytogenes*) as the occurrence of two or more related cases. The detection of two or more cases evokes an action (an outbreak investigation to find the source of food or drink contamination) (CDC, 2020). Behind this data information system is knowledge and wisdom gained from experiencing past epidemics such as knowing (i) that the occurrence of two related cases of listeriosis in the United States is abnormal and could indicate that an epidemic is ongoing in the population; and (ii) that listeriosis has a high case fatality ratio, making it important to respond immediately in order to minimize the number of people that become infected. The additional wisdom is understanding that the source of food or water contamination must be identified before the epidemic can be stopped, and therefore the best action is to conduct an outbreak investigation.

The importance of knowledge and wisdom, or the lack of it, is more evident in less well-defined situations. In the COVID-19 pandemic of 2020, the earliest piece of surveillance information was the identification of cases of a previously unseen, severe respiratory syndrome in Wuhan, China (Wang and Wang, 2020). At the time, knowledge of the importance of this information was uncertain, and as a result, some decision-makers decided not to immediately initiate a response (Liu et al., 2020). As the epidemic spread, it became more certain that the new disease was rapidly spreading and had a relatively high mortality rate. Decision-makers decided to respond based on wisdom gained from the first severe acute respiratory syndrome (SARS) epidemic in 2003 (Liu et al., 2020). Even after it became evident from the experience in China that the epidemic was a significant threat, decision-makers in different countries responded differently (Kluge, 2020). In some countries, the threat was immediately recognized and responses were rapidly implemented. In others, the threat was not recognized as early and disease control actions were delayed. The variation in rapidity and type of national responses could be attributed to many factors, but uncertainty in information and differences in the knowledge and wisdom of decision-makers in different countries are likely to have played a role.

Recognizing the importance of emerging disease signals identified by surveillance and implementing the most effective and efficient disease mitigation strategies requires decision-makers with considerable knowledge and wisdom. It is unlikely that any single individual will have the necessary knowledge and wisdom. Transdisciplinary teams that include experienced experts from a range of scientific and practical specialties as well as representatives of different parts of society are more likely to correctly identify signals that are important and develop interventions that are effective and efficient in terms of costs and negative effects on people, society, other species, and the environment. There are established methods to guide these types of evaluations such as rapid risk assessment methods to estimate the risk of a pathogen introduction and spread and the potential harm that it could cause (WHO, 2012).

ONE HEALTH SURVEILLANCE

The term One Health surveillance (OHS) has recently begun to appear in the scientific literature (Bordier et al., 2018). Collaborative OHS has been defined and characterized (Bordier et al., 2018; Berezowski et al., 2015; Stärk et al., 2015). OHS transgresses traditional boundaries between disciplines and sectors. It is the collaborative efforts between the human and animal health sectors working together to define health events, conduct systematic collection of event data, analysis of this data, and a timely dissemination of information and knowledge to guide interventions aimed at preventing or managing human and animal disease (Karimuribo et al., 2012; Stärk et al., 2015) defined OHS as "the systematic collection, validation, analysis, interpretation of data and dissemination of information collected on humans, animals and the environment to inform decisions for more effective, evidence and system-based health interventions."

Proposed benefits of OHS include a greater understanding of the complex dynamics of zoonotic diseases, saving resources by reducing duplication of infrastructure and processes, and ultimately reducing the burden of disease in people and animals (Hattendorf et al., 2017). These expected benefits are derived from the One Health principle that collaborating across sectors (human, animal, and environmental) is beneficial and that more is better. In other words, more or better information can be created by combining data from many sources. The value of combining data is made even more attractive by the promise of applying big data methods to health data (Mamlin and Tierney, 2016).

Methods and best practices for OHS have not been established. There is no consensus about what is or is not OHS, or what OHS can reasonably expected to accomplish. Methodological challenges are daunting, including how to integrate data that are collected, stored, and analyzed using different standards, definitions, scales, and techniques. Health surveillance practitioners have started thinking about the type of surveillance that will be needed to support decision-making for One Health programs (Berezowski et al., 2019). However, OHS is currently in its infancy.

AN ADAPTABLE PERSPECTIVE OF SURVEILLANCE STRUCTURE, OUTPUTS, AND OUTCOMES

All health surveillance systems are complex, involving different stakeholders, components, infrastructure, processes, policies, and regulations. They are often operationalized by government organizations that have jurisdiction over many people and animals. In this section, we describe common elements of health surveillance that exist across species. The intent is to focus on commonalities that might underlie the creation of surveillance systems using similar architectures and processes that ease data sharing and the creation of information and knowledge relevant to cross-sectoral decision-makers.

Public health surveillance is "the continuous, systematic collection, analysis and interpretation of health related data needed for the planning, implementation, and evaluation of public health practice" (WHO, 2020). Animal health surveillance has similarly been defined as "the systematic, continuous or repeated, measurement, collection, collation, analysis, interpretation and timely dissemination of animal health and welfare related data from defined populations." These data are then used to describe health hazard occurrence and to contribute to the planning, implementation, and evaluation of risk mitigation actions (Hoinville et al., 2013). A One Health surveillance system can be defined as "a system in which collaborative efforts exist between at least two sectors (among human health, animal health, plant health, food safety, wildlife and environmental health) at any stage of the surveillance process, to produce and disseminate information with the purpose of improving an aspect of human, animal or environmental health" (Bordier et al., 2018).

The definitions above share a common purpose of creating information for health decision-making. Differences between them relate to the number of species that should be monitored to create information, nature, number, and diversity of domains that should collaborate and the number of species that should benefit from the decisions informed by the surveillance system. They agree that health surveillance should be a systematic and ongoing process that creates information that is used for decision-making in health (El Allaki et al., 2012). Continuously creating information implies that these processes operate continuously or at least are repeated regularly. Information creation requires processes such as observation, and/or data collection, manipulation, cleaning, analysis, interpretation, and communication (El Allaki et al., 2012). Figure 7.2 is a visualization of a generic

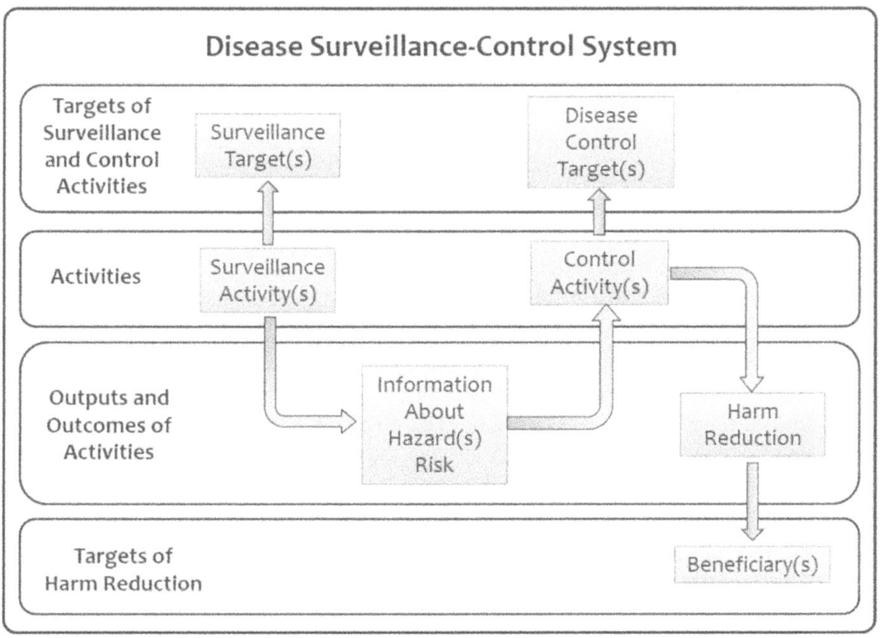

FIGURE 7.2 Attributes of a generic disease surveillance-control system.

TABLE 7.1
Attributes of a Generic Health Surveillance System

Attribute	Component	Description
Activities	Surveillance activities	Sampling and data collection that lead to the production of the required information; processing, cleaning, analysis, and interpretation to create information from observations and data and activities to disseminate the resulting information
	Disease control activities	Responses, interventions, and other harm-reduction activities that have been indicated from wisdom resulting from surveillance information
Tangible and measurable outputs and outcomes	Surveillance information output(s)	Estimates of the current and predicted hazard risk in the population under surveillance. These outputs are often very specific and are defined by the decision(s) that must be made to control the disease
	Harm reduction outcomes	Result from the actions that are set in motion because of a surveillance information output that reduce the harms of concern, such as mortality or economic or ecological impacts
Agents	Targets of surveillance	Agents upon which surveillance activities are directed to produce information
	Targets of control activities	Agents upon which the control actions are directed
	Targets of harm reduction	Agents that benefit from the disease control activity

disease surveillance-control system based on a general theory of surveillance. Table 7.1 defines the attributes of the model.

An example of an integrated disease surveillance-control program for West Nile Virus (WNV) in Italy will be used to illustrate the generic model (Paternoster et al., 2017; Calzolari et al., 2012; Angelini et al., 2010). WNV causes disease in humans, horses, birds, and other animals. It is transmitted by mosquitoes among birds, and from birds to horses and humans, who are dead-end hosts. The exception is that infected people can transmit the virus via blood and solid organ donations. The goal of this program was to mitigate the risk of WNV transmission to horses and humans and to reduce the risk of transmission between humans via blood transfusions and solid organ donations. The intended beneficiaries of the program (the agents of harm reduction in Table 7.1) were humans and horses living in the Emilia-Romagna region of Italy. WNV has a seasonal cycle in the region and the surveillance goal was early detection of increased risk of WNV circulation in the environment (surveillance outputs Table 7.1) that will trigger harm reduction actions. Surveillance activities that produced information included reporting and confirmation of cases of WNV in humans, horses, and birds, and screening of mosquito pools for the virus (Figure 7.2 surveillance targets). When increased risk was detected, specific preplanned actions were triggered, including informing the public about the risk so that people in the high risk area can reduce personal risk of mosquito bites and vaccinate horses; implementation of mosquito control activities; and testing of human blood donations for WNV (Figure 7.2 control activities). This was a continuous process that produced information for decision-making, involved multiple species, and was a collaborative effort involving human, animal, and environmental health. The program was reported to be very successful in reducing the health risk for the targets of harm reduction (people and horses) and demonstrated the benefit of a collaborative, multidisciplinary approach for dealing with a single-pathogen problem (Paternoster et al., 2017). From a One Health perspective, the scope of the program was relatively limited as it considers only two species (humans and horses), ignoring other species such as wild birds, whose populations have been shown to be adversely affected when WNV is introduced (Byas and Ebel, 2020).

Many of the reported OHS systems have been single hazard systems with hazard targets that include zoonotic diseases, antimicrobial resistance, food-borne pathogens, and one report of an environmental hazard (Bordier et al., 2018). Even though many are multisector collaborations, they were mainly designed with the goal of benefiting human populations. Ecosystem health and One Health have been promoted for dealing with health problems involving more than single hazards that benefit one population. Ecosystem health aims to achieve sustainable health for people, animals, and ecosystems (Buse et al., 2018). Although the practice of One Health may be narrower in scope, the goals are similar (AVMA, 2008). The scope of practice of One Health is increasing as researchers and practitioners learn more about the interconnectedness between humans and non-human species and as they respond to new threats to health resulting from climate change and other global changes. There is an opportunity to start thinking about new approaches for dealing with current and impending health risks, especially those associated with global changes.

The scope of health surveillance will need to be expanded to meet the information needs of emerging One Health and EcoHealth programs. This may be accomplished by increasing the number of species targeted for information production and broader consideration of the harms and benefits from interventions. For example, control programs for mosquito-transmitted diseases can include spraying pesticides around homes and in cities to eliminate mosquitoes. This intervention can have harmful effects on people and the other species with which they live (Zikankuba et al., 2019). To represent these species and the ecosystem as a whole, the processes for informing disease control and health promotion interventions requires knowledge and wisdom from a wide range of people. They could include discipline specialists who have practical knowledge and wisdom relating to the undesired or unintended effects of interventions; citizens and traditional knowledge holders such as farmers and gardeners who have local information on critical relationships that can influence the breadth of intervention impacts; and local residents who can inform decision-makers about prevailing values that can influence risk versus benefit determinations for intervention implementation decisions. In the preceding WNV example, surveillance targets and information production could be expanded to better inform current and future decisions about which type of control program would be most beneficial or desirable by the community. For example, surveillance could be expanded to produce continuous information about local pollinator insects and other species, as well as cases of respiratory or other syndromes in people that could be the result of pesticide spraying to control vector mosquitoes. This would allow the production of information about the non-target effects of WNV control programs and promote a more comprehensive view of the effects of interventions on the local healthscape (see Chapter 8 for more information on healthscapes).

Making It Happen

There are many governmental and non-governmental sources of surveillance and monitoring data, and information that could be used to create an integrated, real-time picture of the changing patterns of the target disease and the effects of the control measures on humans, animals, and the environment. Even though these data are available, there are challenges to combining or collating them (Wendt et al., 2014). Close collaboration will be required between many of the people involved in surveillance at all levels from data collectors to information technology technicians, to analysts, to epidemiologists, to policy and decision-makers (Bordier et al., 2019; Houe et al., 2019; Johnson et al., 2018; Bordier et al., 2018; Wendt et al., 2014). One Health leaders will need to support these initiatives with policies, funding, and especially leadership that models collaborative behaviours (Johnson et al., 2018).

Many surveillance roles will need to change to meet this vision of integrated, cross-sectoral surveillance. For example, data collected by one surveillance initiative in one domain may need to be expanded to capture additional data needed to meet the information needs of the overall surveillance-control program. Data cleaners and processors will need to create datasets that are standardized, allowing collation of multiple datasets (Wendt et al., 2016). Disease control and response interventions across multiple species will need to be coordinated by a multi-domain, collaborative, overarching organization (Paternoster et al., 2017). These changes will not be possible without changes in the attitudes of people working in disease surveillance and control in government departments (Antoine-Moussiaux et al., 2019; Johnson et al., 2018). Rethinking disease surveillance-control in these terms is not insurmountable. Changes are occurring, suggesting a shift in attitudes may be taking place. For example, some governments have established One Health offices and departments with the aim of promoting collaborative approaches within and outside of government (BLV/Food Safety and Veterinary Office, 2020; USDA-APHIS, 2020). There are examples of surveillance systems that integrate data from multiple sectors (Dente et al.,

2019; Hutchison et al., 2019; Bordier et al., 2018), and tools have been developed to facilitate government collaborations (Errecaborde et al., 2017).

HEALTH SURVEILLANCE IN COMPLEX DYNAMIC SITUATIONS

The generic model of health surveillance presented above is insufficient for capturing the complexity that is present in most operational disease surveillance-control systems. The model is not well suited for capturing the characteristics and contextually of the relationships and collaborations that can be present in One Health programs, or the different types of decisions being made in a One Health context. It also does not capture some of the other uses for health surveillance information in disease control, and, of particular importance, it does not capture the continuous nature of surveillance information production.

At a population level, the occurrence of new cases of most diseases is a continuous process. The incidence of new cases can vary considerably depending on the changing characteristics of the hazards and dynamic socio-ecological determinants of disease transmission or occurrence (in the case of non-infectious disease) (Da Costa et al., 2019; Homer and Hirsch, 2006). To be effective, health surveillance must create information about changing disease conditions at a rate that is appropriate for the disease and the decisions that need to be made to deal with it. Making decisions for slowly changing processes like the obesity epidemic will require information at a much slower rate than fast moving processes such as the recent emergence of a novel coronavirus in China and its rapid intercontinental spread (Munster et al., 2020).

The rate of occurrence of new cases of disease in a population of any species is the result of complex processes made up of many interactions between many agents (humans, animals, plants, microbes, organizations, societies, governments, etc.) and the environment. These collections of agents and the environment are called complex adaptive systems (CAS) (see Chapter 13 for more information on complexity in health). CAS are ubiquitous. They are self-organized hierarchical structures of nested systems. For example, metabolic pathways are CAS nested within cells, which are CAS nested within organ systems, which are CAS nested within animals, which are CAS nested within farms, which are CAS nested within local ecosystems, and nested also within production systems, and so on. There are many interactions between individual actors within levels of the hierarchy and up and down the hierarchy. The constant interaction within and between the levels results in the system being in a constant state of change at all levels of the hierarchy. These systems are not isolated from their surroundings. For example, there is no physical structure that bounds the local ecosystem within which a farm is located. There are external inputs into the farm, such as changing weather conditions, and there is communication and exchange of materials over short and long distances. Farmers purchase animals, feed, and other commodities and receive visitors from locations that may be many kilometres distant from the farm. All of these are the channels through which information and hazards can be introduced onto the farm. Surveillance systems for the Anthropocene will need to be adaptable to these types of complex, dynamic situations.

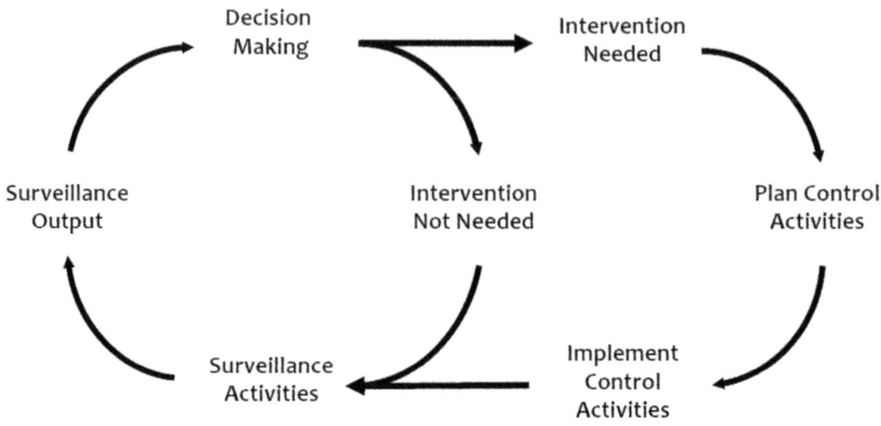

FIGURE 7.3 An adaptive management system for health surveillance.

Adaptive management provides frameworks that can be tailored to health surveillance. Adaptive management is a continuous, cyclic process that has been extensively used to model continuous decision-making processes in business (Landström et al., 2018), ecology (Jokinen et al., 2018), and it has recently been proposed for surveillance and disease control (Miller and Pepin, 2019; Stärk et al., 2018). Adaptive management cycles begin with a situation requiring improvement. Decision-makers wanting to improve the situation design and implement policies and actions to change the situation. The situation is then monitored to evaluate the effect of the policies and actions. This is followed by a critical re-evaluation of the situation to determine whether the desired improvement has been achieved (Jokinen et al., 2018).

Figure 7.3 blends adaptive management and health surveillance concepts with decision-making for health management and disease control. We explain the model by considering a common situation such as when an endemic disease is present in a population at an acceptable level. If the disease increases to a certain level, control actions would be considered to reduce the level of disease in the population. We begin the description of the adaptive management cycle when surveillance activities are underway to assess the status of the disease (Surveillance Activities in Figure 7.3). As in the generic surveillance model (Figure 7.2), surveillance outputs are estimates of the current disease risk. The decision of importance (decision-making in Figure 7.3) is whether the current estimated disease risk (Surveillance Output in Figure 7.3) is high enough to warrant an intervention or not. If it remains at an acceptable level, no intervention is needed, and surveillance activities continue to generate risk estimates that are continuously evaluated by decision-makers. However, if the surveillance risk estimate exceeds an acceptable level, an intervention will be warranted, and control activities will be planned and implemented. In many cases, control activities are preplanned as in responses to known emergencies, where pre-established plans are activated when significant risks are identified by surveillance.

An example of this adaptive management model for health surveillance is the Kyasanur Forest Disease (KFD) control program implemented by Western Indian State Health Departments in the regions endemic for the disease (Shah et al., 2018). KFD virus is a tick-borne virus that causes severe, sometimes fatal, disease in humans (case fatality ranges from 2% to 10%) and monkeys in Western India (Kasabi et al., 2013). There is currently no definitive treatment for the disease (Oliveira et al., 2019). Virus reservoirs include small forest-dwelling rodents, shrews, bats, and birds (Oliveira et al., 2019). Epidemics in humans and monkeys can occur when local environmental conditions support the proliferation of ticks. A surveillance system maintained by state health departments operates continually during the tick season. It targets humans (reports of suspect and confirmed clinical cases), monkeys (reports of dead monkeys by villagers), and ticks (population density estimates), outputting information about the human risk of infection. When the surveillance system detects an increase in risk in a geographic region (surveillance output), a decision is made whether to implement preplanned disease control activities or not. These can include vaccination, tick control, public awareness campaigns informing the public to take personal protective measures against tick bites, and avoiding dead monkeys (Gupta, 2020).

This cyclical adaptive management model is very simple; however, it can be generalized to most disease surveillance-control situations that have been reported (Häsler et al., 2011):

1. A known pathogen of importance that is not currently present in the population and a response is required if the pathogen is introduced.

    Surveillance activities will be targeted towards identifying a confirmed case as quickly as possible (surveillance output), and the decision will be to respond or not, to eliminate the pathogen from the population.

2. The emergence of a previously unknown pathogen that causes significant harm and requires an intervention.

    Surveillance activities must be quite broad to detect the emergence of any new disease as quickly after the occurrence of the emergence as possible (surveillance output). The decision about whether to respond may be quite difficult, especially in the early stages of the disease emergence, as little will be known about the potential harm that may be caused by the new pathogen.

3. The occurrence of an epidemic of an important disease that escapes the current control activities.

    The surveillance system produces a variety of information (incidence, geographic distribution, production types affected) that is used to develop a new more effective control strategy and decide about the type of intervention.

4. During the time when disease control activities are ongoing.

    Surveillance produces information about the changing amount of disease in the population. This information is used to evaluate the effectiveness of control activities. In this case, the decision is whether to continue with the current control activities or to consider new ones.

THE IMPLICATIONS OF COMPLEXITY FOR ONE HEALTH SURVEILLANCE

One of the first models aimed at understanding the complex nature of disease causality was the epidemiological triad which modelled disease occurrence as the result of the interactions between characteristics of pathogens, hosts, and the environment. These models have been expanded in recognition that infectious and non-infectious disease occurrence is the result of the complex interactions between many dynamic socio-economic, behavioural, ecological, environmental, pathogen and other factors (by factors we mean any characteristics or descriptors that can be measured or counted). Identifying these factors, often called risk factors or determinants (here we use the terms interchangeably), has been a primary focus of epidemiology since its emergence (Galea et al., 2010). Risk factors are most commonly identified using studies that measure the association (positive or negative) between potential risk factors and the occurrence of cases of disease (Joffe et al., 2012). The motivation for identifying these risk factors is to use them to develop strategies and interventions aimed at nullifying (where the effect is negative) or promoting their effect (where the effective is protective or health promoting), thereby improving the health of individuals, groups (e.g. farms), or populations. Risk factor models have been integrated into the practice of human and veterinary medicine and can be found in the epidemiology sections of medical and veterinary textbooks.

Since disease occurrences are known to be the result of the interaction between many factors, it follows that there will be some factors that increase the occurrence of disease, and others that decrease it. Both positive and negative risk factors are in constant operation at varying degrees and time scales. The constant change in risk factors explains much of the random variation that is seen in time series of counts of cases of endemic disease that is common to all surveillance data. Large shifts in the rate of production of endemic disease can result from changes in any single risk factor or a combination of multiple risk factors. Figure 7.4 is an oversimplified illustration of the way in which risk factors influence the occurrence of disease cases that are counted in a surveillance system.

The surveillance models presented up to this point are based on the currently accepted paradigm that the information created by surveillance activities is based on cases of disease or infection, in humans and other species. Case-based surveillance is limited in its timeliness to creating information only after the onset of a disease event in a population. There is a strong motivation for producing information about changing disease risks earlier in the chain of events leading up to a

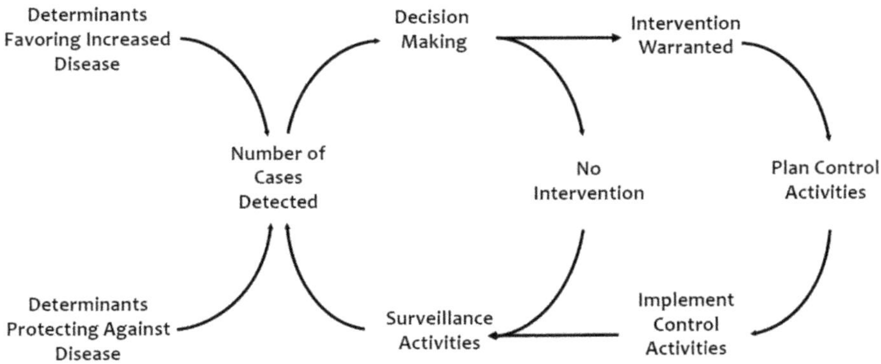

FIGURE 7.4 The effect of determinants of disease case occurrence as seen via a disease surveillance system.

significant health event such as rapidly spreading epidemic. Basing decisions on out-of-date information can result in ineffective decisions that can have devastating consequences on humans and animals (Kitching et al., 2006). Earlier information about an impending change in the disease status of a population would be beneficial, as it would provide more time to design interventions to prevent or mitigate the harm caused by the disease. Overcoming the time constraint will require rethinking the way we do surveillance. In addition to producing information about current disease risks, surveillance could also produce information predicting impending change in disease risks or population vulnerability. For example, a well-known risk factor for an epidemic is the proportion of the population that has resistance to infection with the pathogen causing the epidemic. When the proportion drops below the level of herd immunity in the population, the introduction of the pathogen into the population can result in a sustained epidemic. The recent measles outbreaks in North America and Europe occurred because communities of parents decided to withhold vaccination of their children, to the point where community-based epidemics were sustained. The risk factor "reduced proportion of resistant children in the community" is commonly used to model epidemic progression. It is very simple and easy to model but fails to convey the complexity of the situation. A large number of interconnected risk factors for parents failing to vaccinate their children have been identified and grouped into the 5A's taxonomy (Access, Affordability, Awareness, Acceptance, and Activation) (Bell et al., 2020). Understanding the complexity of these interconnected risk factors is difficult. However, a surveillance system that recorded the number of people refusing measles vaccination for their children would have provided some warning of which communities were reaching a point where an epidemic was imminent. This information would provide additional time for the development of interventions such as campaigns to inform people of the risks of not vaccinating their children.

Predicting increased disease risk for known endemic diseases is a relatively common practice in humans, animals, and plant health, and ecological sciences

(Funk et al., 2019; Kleczkowski et al., 2019; Thompson and Brooks-Pollock, 2019; Lee et al., 2017). It has been especially promoted for predicting changing vector-born disease risk because vector populations are affected by weather (temperature, precipitation, and humidity) (Morin et al., 2018; Lee et al., 2017). It has also been reported for predicting changing risk for diseases (e.g. Ebola or obesity) that are affected by changing social behaviours (Funk et al., 2019; Scarpino and Petri, 2019; Galea et al., 2010). Epidemic prediction has been likened to weather forecasting (Morin et al., 2018; Moran et al., 2016). We can speculate that epidemic prediction will improve over time as methods and the amount and quality of data improve, similar to the way that weather forecasting has improved over time. Because of the current state of the art of epidemic prediction methods and the associated uncertainties with epidemic prediction, this type of surveillance is not likely to replace conventional surveillance in the immediate future. However, it could be used to provide additional information to supplement conventional surveillance information.

Risk factor surveillance has been proposed for animal health surveillance (Rich et al., 2013), but is not widely practiced. It is more common in public health surveillance. The Behavioural Risk Factor Surveillance System has been used since 1981 in the United States to monitor a variety of risk factors for human disease, including immunization, preventative testing, physical activity, chronic conditions, mental health, obesity, tobacco use, alcohol and substance abuse, and health risks associated with sexual behaviour, injury, and violence (Pierannunzi et al., 2013). It has successfully predicted epidemics (for example diabetes) and has been used by most of the U.S. states to design and implement disease control interventions (including legislation) aimed at reducing disease risk predicted by the surveillance system (Mokdad, 2009).

Both case-based and risk factor surveillance can be practiced at the same time. Surveillance targeting disease cases could be continued as usual, producing information that is used to make decisions about the current situation in the population. Risk factor surveillance could be added and used to produce information predicting future changes in disease risk (Figure 7.5). This information would be used to make decisions about implementing prevention activities aimed at reducing the future risk of disease.

Monitoring several individual risk factors as independent variables is not likely to provide a complete understanding of the processes affecting disease production (Joffe et al., 2012). There are at least two reasons for this. The first is that many risk factors do not act independently. They interact with each other and with other risk factors in ways that make it difficult to understand their behaviour without understanding the behaviour of the complex system in which the species under surveillance resides (Joffe et al., 2012; Galea et al., 2010). The second is that monitoring risk factors independently ignores interconnectedness of the health of humans, animals, and the environment. Understanding the connections between many species and the environment they reside in, as a system, will be essential for understanding how changes in risk factors for disease in one species may affect the health of another species, or how interventions

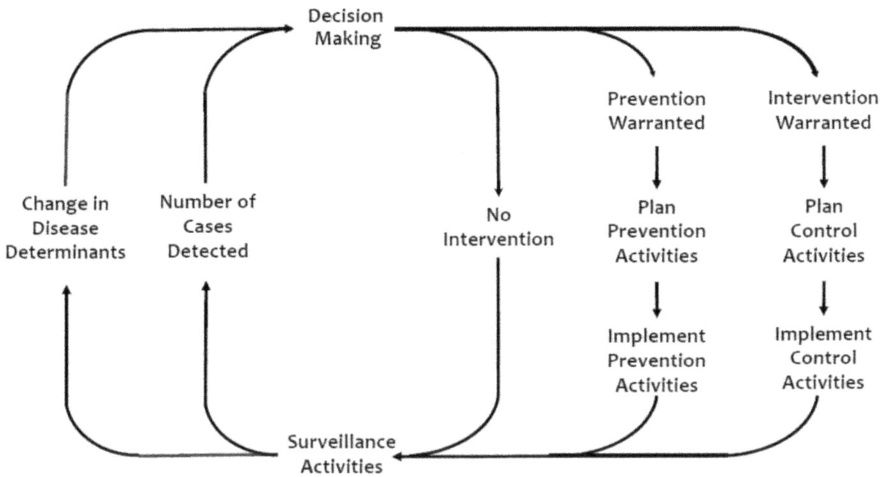

FIGURE 7.5 Adaptive management model of surveillance that monitors disease cases and determinants of disease.

aimed at reducing disease in one species may have beneficial or harmful effects on another.

Even though there is a toolbox of methods that have been developed in the field of complexity science that could be used in One Health, the use of complex system methods has not been widely reported in One Health research (see Chapter 13 for more on complexity in On Health). There are studies that have identified the relationships between risk factors and disease in single species using complexity science methods, and complexity methods in combination with participatory approaches have been used to select the most appropriate disease control activities and to develop health policy (Mumba et al., 2017). At the time of writing, we found no studies that used complexity science methods to select targets for One Health surveillance. The reasons for this are unknown. It may be that complexity science is not well known among One Health researchers and practitioners, or it may be that additional basic research is needed to support the use of complexity science in One Health. It may also relate to the relatively narrow scope of One Health that is currently being implemented. Most reported applications of One Health focus on zoonotic diseases. Relationships between species are included only for those species that act as reservoirs or vectors for human disease, and only risk factors that affect disease in people are considered. This narrow view is a one-way causal model pointing towards people. These are not full models, because they ignore the effects of interventions to protect human health on the health of non-human species, ecosystems, and the environment. Furthermore, they do not consider how the health of animals and the environment effects the health of humans. Complex system approaches have the potential to aid our understanding of all these interactions and to help to develop interventions that are more likely to be considerate of the health of non-human species and the environment. However, complex system approaches will not likely be seen as necessary until organizations responsible for health fully embrace One Health and develop programs that accept responsibility for the health of all species and the environment.

SUMMARY

Surveillance is a tool for creating information for decision-making. The activities and information created by health surveillance are therefore limited by the decisions it supports. Since most One Health programs target single diseases and aim to benefit only humans, OHS is currently limited to a small number of activities that produce a limited amount of information. If the decisions in One Health remain univariate in nature (have one outcome), there will be little or no need for change in the type of surveillance information produced. However, considering the impact of humanity on all species on this planet (and vice versa), it is imperative that health decision-makers begin to think about policy that aims to concurrently benefit people, non-human species, and the environment. This will require broadening of the scope of their responsibility and jurisdiction to include the health of many species. Changing health legislation and organizational structures is not an easy process. It will likely require a shift in societal values to the point where all species, ecosystems, and the environment are highly valued.

One Health practice, on the other hand, aims to implement solutions to health-related problems that provide multiple species benefits. It should be collaborative in nature, engaging a wide range of stakeholders, including those affected by the health problem in question, and those affected by proposed solutions. Solutions developed in this manner are more likely to be effective and sustainable than top-down approaches where health ministries develop policies with little or no engagement with stakeholders or affected communities. Because of the complex, multi-stakeholder nature of One Health practice, decision-making is likely to require more information and knowledge than is needed for single disease-species approaches.

REFERENCES

Allaki, Farouk El, Michel Bigras-Poulin, Pascal Michel, and André Ravel. "A Population Health Surveillance Theory." *Epidemiology and Health*, 34, (2012): e2012007.

Angelini, P., M. Tamba, A.C. Finarelli, R. Bellini, A. Albieri, P. Bonilauri, F. Cavrini et al. "West Nile Virus Circulation in Emilia-Romagna, Italy: The Integrated Surveillance System 2009." *Eurosurveillance*, 15, no. 16 (2010): 11–15.

Antoine-Moussiaux, Nicolas, Luc Janssens de Bisthoven, Stéphane Leyens, Timo Assmuth, Hans Keune, Zinsstag Jakob, Jean Hugé et al. "The Good, the Bad and the Ugly: Framing Debates on Nature in a One Health Community." *Sustainability Science*, 14, no. 6 (2019): 1729–1738.

Aven, Terje. "A Conceptual Framework for Linking Risk and the Elements of the Data-Information-Knowledge-Wisdom (DIKW) Hierarchy." *Reliability Engineering and System Safety*, 111, (2013): 30–36.

AVMA. "One Health: A New Professional Imperative." One Health Initiative Task Force Final Report, American Veterinary Medical Association, 2008.

Bell, Sadie, Vanessa Saliba, Mary Ramsay, and Sandra Mounier-Jack. 2020. "What Have We Learnt from Measles Outbreaks in 3 English Cities? A Qualitative Exploration of Factors Influencing Vaccination Uptake in Romanian and Roma Romanian Communities." *BMC Public Health*, 20, no. 1 (2020): 1–10.

Berezowski, John, Judy Akkina, Victor Del Rio, Katrina DeVore, Fernanda Dorea, Celine Dupuy, Melody Maxwell et al. "Do We Need One Health Surveillance." *One Health Newsletter*, 8, no. 1 (2015): 5–7.

Berezowski, John, Simon R. Rüegg, and Céline Faverjon. "Complex System Approaches for Animal Health Surveillance." *Frontiers in Veterinary Science*, 6, (2019): 1–11.

BLV/Food Safety and Veterinary Office. "Sub-Organ One Health." Swiss Government, 2020. https://www.blv.admin.ch/blv/de/home/das-blv/organisation/kommissionen/unterorgan-one-health.html (accessed May 2020).

Bordier, Marion, Camille Delavenne, Dung Thuy Thi Nguyen, Flavie Luce Goutard, and Pascal Hendrikx. "One Health Surveillance: A Matrix to Evaluate Multisectoral Collaboration." *Frontiers in Veterinary Science*, 6, (2019): 1–12.

Bordier, Marion, Theethawat Uea-anuwong, Aurélie Binot, Pascal Hendrikx, and Flavie L Goutard. "Characteristics of One Health Surveillance Systems: A Systematic Literature Review." *Preventive Veterinary Medicine*, S0167-5877, no. 18 (2018): 30365–30369.

Buse, Chris G., Jordan Sky Oestreicher, Neville R. Ellis, Rebecca Patrick, Ben Brisbois, Aaron P. Jenkins, Kaileah McKellar et al. "Public Health Guide to Field Developments Linking Ecosystems, Environments and Health in the Anthropocene." *Journal of Epidemiology and Community Health*, 72, no. 5 (2018): 420–425.

Byas, Alex D., and Gregory D. Ebel. "Comparative Pathology of West Nile Virus in Humans and Non-Human Animals." *Pathogens*, 9, no. 1 (2020): 48.

Calzolari, Mattia, Paolo Gaibani, Romeo Bellini, Francesco Defilippo, Anna Pierro, Alessandro Albieri, Giulia Maioli et al. "Mosquito, Bird and Human Surveillance of West Nile and Usutu Viruses in Emilia-Romagna Region (Italy) in 2010." *PLoS One*, 7, no. 5 (2012): e38058.

CDC. "Listeria Outbreaks." Centers for Disease Control and Prevention, 2020. https://www.cdc.gov/listeria/outbreaks/index.html (accessed May 2020).

Costa, Anderson Luiz Pena Da, Orlando Alves Rodrigues Neto, and Antonio Carlos Souza Silva-Júnior. "Conditioners of the Infectious Diseases Dynamics." *Estação Científica (UNIFAP)*, 8, (2019): 09. https://doi.org/10.18468/estcien.2018v8n3.p09-23.

Dente, Maria Grazia, Flavia Riccardo, Francesco Bolici, Nello Augusto Colella, Verica Jovanovic, Mitra Drakulovic, Milena Vasic et al. "Implementation of the One Health Approach to Fight Arbovirus Infections in the Mediterranean and Black Sea Region: Assessing Integrated Surveillance in Serbia, Tunisia and Georgia." *Zoonoses and Public Health*, 66, no. 3 (2019): 276–287.

Errecaborde, Kaylee Myhre, Katharine M. Pelican, Heidi Kassenborg, Ong Orn Prasarnphanich, Linda Valeri, Erinaldi Yuuzar, Rama Prima Syahti Fauzi et al. "Piloting the One Health Systems Mapping and Analysis Resource Toolkit in Indonesia." *EcoHealth*, 14, no. 1 (2017): 178–181.

Funk, Sebastian, Anton Camacho, Adam J. Kucharski, Rachel Lowe, Rosalind M. Eggo, and W. John Edmunds. "Assessing the Performance of Real-Time Epidemic Forecasts: A Case Study of Ebola in the Western Area Region of Sierra Leone, 2014–15." *PLoS Computational Biology*, 15, no. 2 (2019): 1–17.

Galea, Sandro, Matthew Riddle, and George A. Kaplan. "Causal Thinking and Complex System Approaches in Epidemiology." *International Journal of Epidemiology*, 39, no. 1 (2010): 97–106.

Gupta, Promila. "Kyasanur Forest Disease: A Public Health Concern." CD Alert, National Centre for Disease Control, 2020. https://idsp.nic.in/showfile.php?lid=3946 (accessed May 2020).

Häsler, Barbara, Keith S. Howe, and Katharina D.C. Stärk. "Conceptualising the Technical Relationship of Animal Disease Surveillance to Intervention and Mitigation as a Basis for Economic Analysis." *BMC Health Services Research*, 11, no. 1 (2011): 225.

Hattendorf, Jan, Kevin Louis Bardosh, and Jakob Zinsstag. "One Health and Its Practical Implications for Surveillance of Endemic Zoonotic Diseases in Resource Limited Settings." *Acta Tropica*, 165, (2017): 268–273.

Hoinville, L.J., L. Alban, J.A. Drewe, J.C. Gibbens, L. Gustafson, B. Häsler, C. Saegerman et al. "Proposed Terms and Concepts for Describing and Evaluating Animal-Health Surveillance Systems." *Preventive Veterinary Medicine*, 112, no. 1–2: 1–12.

Homer, Jack B., and Gary B. Hirsch. "System Dynamics Modeling for Public Health: Background and Opportunities." *American Journal of Public Health*, 96, no. 3 (2006): 452–458.

Houe, Hans, Søren Saxmose Nielsen, Liza Rosenbaum Nielsen, Steen Ethelberg, and Kåre Mølbak. "Opportunities for Improved Disease Surveillance and Control by Use of Integrated Data on Animal and Human Health." *Frontiers in Veterinary Science*, 6, (2019): 1–8.

Hutchison, Jennifer, Catriona Mackenzie, Ben Madin, Jonathan Happold, Edwina Leslie, Emma Zalcman, Anne Meyer et al. "New Approaches to Aquatic and Terrestrial Animal Surveillance: The Potential for People and Technology to Transform Epidemiology." *Preventive Veterinary Medicine*, 167, (2019): 169–173.

Joffe, Michael, Manoj Gambhir, Marc Chadeau-Hyam, and Paolo Vineis. "Causal Diagrams in Systems Epidemiology." *Emerging Themes in Epidemiology*, 9, no. 1 (2012): 1.

Johnson, I., A. Hansen, and P. Bi. "The Challenges of Implementing an Integrated One Health Surveillance System in Australia." *Zoonoses and Public Health*, 65, no. 1 (2018): e229– e236.

Jokinen, Pekka, Malgorzata Blicharska, Eeva Primmer, Ann Van Herzele, Leena Kopperoinen, and Outi Ratamäki. "How Does Biodiversity Conservation Argumentation Generate Effects in Policy Cycles?" *Biodiversity and Conservation*, 27, no. 7 (2018): 1725–1740.

Karimuribo, Esron D., Kuya Sayalel, Eric Beda, Nick Short, Philemon Wambura, Leonard G. Mboera, Lughano J.M. Kusiluka et al. "Towards One Health Disease Surveillance: The Southern African Centre for Infectious Disease Surveillance Approach." *Onderstepoort Journal of Veterinary Research*, 79, no. 2 (2012): 1–7.

Kasabi, Gudadappa S., Manoj V. Murhekar, Vijay K. Sandhya, Ramappa Raghunandan, Shivani K. Kiran, Gowdra H. Channabasappa, and Sanjay M. Mehendale. "Coverage and Effectiveness of Kyasanur Forest Disease (KFD) Vaccine in Karnataka, South India, 2005–10." *PLoS Neglected Tropical Diseases*, 7, no. 1 (2013): 13–16.

Kitching, R. P., M. V. Thrusfield, and N. M. Taylor. "Use and Abuse of Mathematical Models: An Illustration from the 2001 Foot and Mouth Disease Epidemic in the United Kingdom." *OIE Revue Scientifique et Technique*, 25, no. 1 (2006): 293–311.

Kleczkowski, Adam, Andy Hoyle, and Paul McMenemy. "One Model to Rule Them All? Modelling Approaches Across One Health for Human, Animal and Plant Epidemics." *Philosophical Transactions of the Royal Society B: Biological Sciences*, 374, no. 1775 (2019): 20180255.

Kluge, Hans Henri P. "Statement – Where Do We Stand Today on COVID-19, and What Have We Learned?" World Health Organization, 2020. http://www.euro.who.int/en/health-topics/health-emergencies/coronavirus-covid-19/statements/statement-where-do-we-stand-today-on-covid-19,-and-what-have-we-learned (accessed April 2020).

Landström, Anna, Peter Almström, Mats Winroth, Carin Andersson, Anna Ericson Öberg, Martin Kurdve, Sasha Shahbazi et al. "A Life Cycle Approach to Business Performance Measurement Systems." *Procedia Manufacturing*, 25 (2018): 126–133.

Lee, Jung Seok, Mabel Carabali, Jacqueline K. Lim, Victor M. Herrera, Il Yeon Park, Luis Villar, and Andrew Farlow. "Early Warning Signal for Dengue Outbreaks and Identification of High Risk Areas for Dengue Fever in Colombia Using Climate and Non-Climate Datasets." *BMC Infectious Diseases*, 17, no. 1 (2017): 1–11.

Liu, Wei, Xiao-Guang Yue, and Paul B. Tchounwou. "Response to the COVID-19 Epidemic: The Chinese Experience and Implications for Other Countries." *International Journal of Environmental Research and Public Health*, 17, no. 7 (2020): 2304.

Mamlin, Burke W., and William M. Tierney. "The Promise of Information and Communication Technology in Healthcare: Extracting Value from the Chaos." *The American Journal of the Medical Sciences*, 351, no. 1 (2016): 59–68.

Miller, Ryan S., and Kim M. Pepin. "Board Invited Review: Prospects for Improving Management of Animal Disease Introductions Using Disease-Dynamic Models." *Journal of Animal Science*, 97, no. 6 (2019): 2291–2307.

Mokdad, Ali H. "The Behavioral Risk Factors Surveillance System: Past, Present, and Future." *Annual Review of Public Health*, 30, (2009): 43–54.

Moran, Kelly R., Geoffrey Fairchild, Nicholas Generous, Kyle Hickmann, Dave Osthus, Reid Priedhorsky, James Hyman et al. "Epidemic Forecasting Is Messier than Weather Forecasting: The Role of Human Behavior and Internet Data Streams in Epidemic Forecast." *Journal of Infectious Diseases*, 214 (2016): S404–S408.

Morin, Cory W., Jan C. Semenza, Juli M. Trtanj, Gregory E. Glass, Christopher Boyer, and Kristie L. Ebi. "Unexplored Opportunities: Use of Climate- and Weather-Driven Early Warning Systems to Reduce the Burden of Infectious Diseases." *Current Environmental Health Reports*, 5, no. 4 (2018): 430–438.

Mumba, Chisoni, Eystein Skjerve, Magda Rich, and Karl M. Rich. "Application of System Dynamics and Participatory Spatial Group Model Building in Animal Health: A Case Study of East Coast Fever Interventions in Lundazi and Monze Districts of Zambia." *PLoS One*, 12, no. 12 (2017): 1–21.

Munster, Vincent J., Marion Koopmans, Neeltje van Doremalen, Debby van Riel, and Emmie de Wit. "A Novel Coronavirus Emerging in China- Key Questions for Impact Assessment." *New England Journal of Medicine*, 382, no. 8 (2020): 1–3.

Oliveira, Annet, Kalaiselvi Selvaraj, Jaya Prasad Tripathy, Utkarsh Betodkar, Jagadish Cacodcar, and Abhijit Wadkar. "Kyasanur Forest Disease Vaccination Coverage and Its Perceived Barriers in Goa, India: A Mixed Methods Operational Research." *PLoS One*, 14, no. 12 (2019): 1–13.

Paternoster, Giulia, Sara Babo Martins, Andrea Mattivi, Roberto Cagarelli, Paola Angelini, Romeo Bellini, Annalisa Santi et al. "Economics of One Health: Costs and Benefits of Integrated West Nile Virus Surveillance in Emilia-Romagna." *PLoS One*, 12, no. 11 (2017): 1–16.

Pierannunzi, Carol, Shaohua Sean Hu, and Lina Balluz. "A Systematic Review of Publications Assessing Reliability and Validity of the Behavioral Risk Factor Surveillance System (BRFSS), 2004–2011." *BMC Medical Research Methodology*, 13, no. 1 (2013): 1–14.

Rich, Karl M, Matthew J. Denwood, Alistair W. Stott, Dominic J. Mellor, Stuart W.J. Reid, and George J. Gunn. "Systems Approaches to Animal Disease Surveillance and Resource Allocation: Methodological Frameworks for Behavioral Analysis." *PLoS One*, 8, no. 11 (2013): e82019.

Scarpino, Samuel V., and Giovanni Petri. "On the Predictability of Infectious Disease Outbreaks." *Nature Communications*, 10, no. 1 (2019): 1–8.

Shah, Syed Z., Basit Jabbar, Nadeem Ahmed, Anum Rehman, Hira Nasir, Sarooj Nadeem, Iqra Jabbar et al. "Epidemiology, Pathogenesis, and Control of a Tick-Borne Disease-Kyasanur Forest Disease: Current Status and Future Directions." *Frontiers in Cellular and Infection Microbiology*, 8, (2018): 149.

Stärk, Katharina D.C., Montserrat Arroyo Kuribreña, Gwenaelle Dauphin, Sandra Vokaty, Michael P. Ward, Barbara Wieland, and Ann Lindberg. "One Health Surveillance: More than a Buzz Word?" *Preventive Veterinary Medicine*, 120, no. 1 (2015): 124–130.

Stärk, K.D.C., S. Babo Martins, J. Rushton, and B. Häsler. "Surveillance Without Intervention: Is There a Value?" in *15th Veterinary Epidemiology and Economics, ISVEE 2018* (Thailand: Chiang Mai, 2018). https://www.researchgate.net/conference-event/ISVEE_Veterinary-Epidemiology-and-Economics_2018/132719.

Thompson, Robin N., and Ellen Brooks-Pollock. "Detection, Forecasting and Control of Infectious Disease Epidemics: Modelling Outbreaks in Humans, Animals and Plants." *Philosophical Transactions of the Royal Society B: Biological Sciences*, 374, no. 1775 (2019) 20190038.

USDA-APHIS. "USDA APHIS Veterinara Services One Health Coordination Center." U.S. Department of Agriculture, Animal and Plant Health Inspection Service, 2020. https://www.aphis.usda.gov/animal_health/One_Health/downloads/One_Health_coordination_office.pdf (accessed May 2020).

Wang, Jia, and Zhifeng Wang. "Strengths, Weaknesses, Opportunities and Threats (SWOT) Analysis of China's Prevention and Control Strategy for the COVID-19 Epidemic." *International Journal of Environmental Research and Public Health*, 17, no. 7 (2020): 2235.

Wendt, A., L. Kreienbrock, and A. Campe. "Zoonotic Disease Surveillance: Inventory of Systems Integrating Human and Animal Disease Information." *Zoonoses and Public Health*, 62, no. 1 (2014): 61–74.

Wendt, A., L. Kreienbrock, and A. Campe. "Joint Use of Disparate Data for the Surveillance of Zoonoses: A Feasibility Study for a One Health Approach in Germany." *Zoonoses and Public Health*, 63, no. 7 (2016): 503–514.

WHO. "Rapid Risk Assessment of Acute Public Health Events." World Health Organization, 2012. https://apps.who.int/iris/bitstream/handle/10665/70810/WHO_HSE_GAR_ARO_2012.1_eng.pdf;jsessionid=8E5E9D4C4FC07A3CE521793777D486B3?sequence=1 (accessed May 2020).

WHO. "Public Health Surveillance." World Health Organization, 2020. https://www.who.int/immunization/monitoring_surveillance/burden/vpd/en/ (accessed May 2020).

Wilson, J.B., M. Salman, E. Janzen, O. Sparagano, N. Speer, L. Pantaleon, C. La Jeunesse et al. "Community Network Integration: An Approach to Alignment of One Health Partners for Solutions to 'Wicked' Problems of Antimicrobial Resistance." *Preventive Veterinary Medicine*, 175 (2020): 104870.

Yim, Nam Hong, Soung Hie Kim, Hee Woong Kim, and Kee Young Kwahk. "Knowledge Based Decision Making on Higher Level Strategic Concerns: System Dynamics Approach." *Expert Systems with Applications*, 27, no. 1 (2004): 143–158.

Zikankuba, Vumilia Lwoga, Gaspary Mwanyika, Julius Edward Ntwenya, and Armachius James. "Pesticide Regulations and Their Malpractice Implications on Food and Environment Safety." *Cogent Food and Agriculture*, 5, no. 1 (2019): 1–15.

8 Traversing the Eco-Healthscape: The Final Frontier in Understanding Shared Determinants of Health at the Animal-Society Interface

Colin Robertson

CONTENTS

INTRODUCTION

The World Health Organization's Ottawa Charter for Health Promotion (WHO, 1986) states: "Health is created and lived by people within the settings of their everyday life; where they learn, work, play and love." This has been a cornerstone in the development of a healthy settings approach to health promotion and public health (Kickbusch, 2003). This approach recognizes the need to develop strategies, programmes, and policies that contribute to total population health in a given setting, rather than targeting specific at-risk groups. This chapter examines the healthy settings approach in a One Health world by considering how geographical contexts have been framed in research relevant to animal, the environment, and human health.

Over the last two decades, the role of geography in understanding, predicting, and shaping health and disease has grown, through the expansion of health geography and spatial epidemiology (Meade, 2014; Kearns and Moon, 2002). Locations linked with health surveillance data, for example, can help reveal patterns of disease relative to the underlying population. Spatial patterns can reveal clues about aetiology and risk factors. In health surveillance, this sort of analysis can lead to early detection of emergence, spillover, geographic spread, or evidence of co-infection and interaction among pathogens, hosts, and vectors (Mayer, 2000; Robertson et al., 2010). Disease data linked to other environmental datasets through location references can help identify spatial risk factors. This approach is widely used in the study of distribution of both chronic and infectious diseases in humans and increasingly in animals.

Despite these successes in "spatial" approaches to health and disease, key challenges remain. Epidemiologically significant events can be (and often are) the outcome of unpredictable and seemingly random interactions. The challenges of making useful spatial associations begin to compound for diseases that are shared between species, where early risk factors interact in unpredictable ways in extremely complex processes. Animal and human data are collected for widely varying purposes, with different methods, sampling approaches, and data formats, which complicates developing a data acquisition and analysis strategy (Robertson et al., 2016). Needing to monitor interactions between multiple species, their environment, and human populations (all of which have different restrictions across political boundaries, and different surveillance methods and approaches) creates difficulties for spatial risk factor surveillance for emerging zoonoses. While spatially explicit methods are commonly used in understanding wildlife, spatial information alone rarely tells us much about population health overall. Population declines or collapses are rarely explained only by where wildlife are located. Integration of spatially explicit health information, encompassing wildlife, domestic animals, humans, and the environment, remains elusive. A shift in thinking is now promoting One Health through a health promotion lens: to monitor and understand those shared dimensions that contribute positively to health – across generations, places, and species boundaries (see Chapter 2 for more on the evolution of One Health).

We propose the idea of an "Eco-Healthscape," which combines what we might traditionally consider as EcoHealth approaches to health assessment, with the notion of landscape approaches to health. A "landscape," in a geographical sense, is defined relationally rather than as a particular spatial scale. A landscape is a geographical unit, which may change and fluctuate over time that incorporates some heterogeneity with respect to the habitat experienced by an individual species. A landscape definition is linked to habitat area, dispersal characteristics, habitat quality/disturbances, and species interactions. Landscapes are in constant interaction with the ecosystems and social systems within which they are embedded. An Eco-Healthscape emerges from the complex interactions among constituent species and their landscapes. Moving beyond identifying threats to one species to a more holistic multi-species assemblage perspective means thinking about ecosystem functions, actors, and populations on their own intrinsic terms rather than in relation to the services they provide to humans.

In this chapter, we explore how to identify, shape, and create a healthy Eco-Healthscape. We review spatial approaches to health and disease, followed by considering how recent ideas from health geography can inform the conceptualization and operationalization of Eco-Healthscape approaches to One Health. A caribou case study illustrates some of the ideas in practice, followed by a discussion of preliminary conclusions regarding an Eco-Healthscape approach to promoting health of animals, humans, and our shared environment.

Prioritizing Space in Ecological Research

The idea that ecological systems can be understood without considering the geographic contexts and their effects within which those systems operate was called into question by Peter Kareiva in the journal *Ecology* in 1994. Further papers (e.g. Holmes et al., 1994) identified that spatially explicit processes were vital to specific ecological systems, and modelling their spatial patterns and dynamics was needed to understand them. Incorporating spatial dynamics into ecological theory helped make better predictions, create sound management strategies, and understand ecological responses to disturbances. The spatial ecology field has grown to encompass a variety of frameworks, methods, and theoretical developments. Some of these were made possible by new technologies for monitoring landscapes, tracking animals through space and time, and mapping and monitoring disturbances, their impacts, and responses. However, by the mid-2000s – despite hundreds of papers describing spatial effects in ecological systems – it became clear that spatial ecological theory, techniques, and applications were not enough to aid in the management and conservation of declining and threatened wildlife populations (e.g. Mladenoff, 2005).

Most ecological theory and the knowledge it generates relate to systems in the absence of humans. Humans, however, are very much a major part of ecological development, maintenance, and resilience (Burke and Mitchell, 2007). Incorporating humans into conservation and ecosystem restoration is now considered a vital aspect of successful restoration and conservation efforts. The field of restoration ecology grew to try to fill the gaps in applying ecological theory to restoring real-world systems.

This tension between "space" (i.e. spatial properties and relationships as an important object of study) and "place" (i.e. spatial properties and relationships as perceived and valued by people) has been going on in the geographical literature since at least the 1960s. On the one hand, spatial effects and properties can increasingly be mapped and monitored and used to create complex spatially explicit models of disease risk or species niches. On the other hand, understanding how human values, attitudes, and behaviours coincide to shape priorities and approaches to understanding animal and environmental health remains more sporadic and disjointed. A healthy-settings approach that leads to healthy lives for humans and animals needs to consider both spatial and *platial* factors in its definition of "settings."

"LANDSCAPES" AS "SETTINGS" FOR ONE HEALTH PROMOTION

A landscape is an ecological research unit of analysis which defines a region containing some spatial differentiation in habitat quality or characteristics: often described as spatial heterogeneity (Turner, 1989; Picket and Cadenasso, 1995). Studies of environmental relationships to human, vector-borne, zoonotic, and animal diseases through the lens of spatial and/or landscape epidemiology (so-called "landscape approaches") have led to new frameworks, methodologies, and thousands of empirical studies linking disease and disease risks to features of the environment (Robertson, 2017). Ten key propositions of landscape approaches to health (based on a review of several studies) are outlined in Table 8.1 (Lambin et al., 2010). We have examined these propositions to consider how they can inform the characteristics of a healthy settings approach (Table 8.1).

Almost all the strictly "spatial" dimensions of these propositions focus on static aspects of habitats: the presence or absence of landscape features; how they are configured in the landscape; and how their configuration influences connections and movement between patches. Other characteristics of the propositions relate to human-mediated processes within landscapes: how aspects of land use change, land ownership, and human movement and behaviour define and shape key aspects of infection risk.

The propositions identified by Lambin et al. (2010), while important for identifying many spatial components of disease emergence and distribution processes, are inadequate for defining the Eco-Healthscape. Landscape approaches typically focus on associations between environmental features and disease incidence to isolate areas of special concern (e.g. clusters or high-risk areas). Analyses frequently stop there, rather than identifying what actions should follow in these demarcated areas (e.g. see Rothman, 1990). Identifying the high-risk areas on a map is in many ways a stark contrast to the "total population" approach articulated in healthy-settings literatures. Place-based and more recent health geography ideas have transformed how health research conceives space and geographic context. These ideas can help us understand how to use a broader articulation of landscapes suitable for One Health.

In geographical studies of health and disease, the turn from a disease-ecology focus to a health-promotion focus has been occurring for over 25 years. Kearns (1993) identified place, health services, and community resources as central to a new "health geography." This shift in emphasis can be seen in a methodological sense in the shift from space-centric to place-centric studies. Kearns and Moon (2002) describe this as changes in how geographic context is represented, from treating space as "an unproblematized activity container" to that of "alternative perspectives emphasizing constructed meanings and the experiential aspects of place."

TABLE 8.1
Review of the Ten General Principles Governing Landscape Epidemiology Described by Lambin et al. (2010) Focusing on Vector-Borne Zoonoses

Landscape Epidemiology Propositions	Healthscape Interpretation	Spatial or Platial?
1. Landscape attributes may influence the level of transmission of an infection	Landscape features, such as flooded banks, can be mapped from high-resolution imagery and examined for association with disease variables	Highly spatial
2. Spatial variations in disease risk depend not only on the presence and area of critical habitats but also on their spatial configuration	Spatial measures of habitat configuration (e.g. edge density) can be important indicators of disease risk for both animals and people	Highly spatial
3. Disease risk depends on the connectivity of habitats for vectors and hosts	Landscapes as a site of heterogeneity imply different levels of connectivity between habitat patches. Processes driving habitat fragmentation and reducing connectivity can predict changes in disease risk	Highly spatial, moderately platial
4. The landscape is a proxy for specific associations of reservoir hosts and vectors linked with the emergence of multi-host diseases	Landscape-level associations that can be easily mapped and modelled are frequently proxies for specific interactions and processes that cannot easily be measured Understanding specific associations is vital if interventions are to be designed to mitigate risks. Place factors typically operate at the level of specific associations, not the landscape	Moderately spatial Moderately platial
5. To understand ecological factors influencing spatial variations of disease risk, one needs to take into account the pathways of pathogen transmission between vectors, hosts, and the physical environment	Pathways of transmission are defined to be social interactions and processes, whether between animals or between animals and humans. Mapping their locations is insufficient to understand the causal mechanisms driving the interactions	Weakly spatial, highly platial
6. The emergence and distribution of infection through time and space is controlled by different factors acting at multiple scales	Scale is a key consideration of both spatial and platial factors that influence disease emergence and distribution	Highly spatial, highly platial
7. Landscape and meteorological factors control not just the emergence but also the spatial concentration and spatial diffusion of infection risk	Healthscapes are in part defined by abiotic conditions that correlate with life history requirements for survival. This applies to both pathogens and hosts	Highly spatial

(Continued)

TABLE 8.1 (*Continued*)
Review of the Ten General Principles Governing Landscape Epidemiology Described by Lambin et al. (2010) Focusing on Vector-Borne Zoonoses

Landscape Epidemiology Propositions	Healthscape Interpretation	Spatial or Platial?
8. Spatial variation in disease risk depends not only on land cover but also on land use, via the probability of contact between, on one hand, human hosts and, on the other hand, infectious vectors, animal hosts, or their infected habitat	Land use is governed by an assemblage of human and place-based factors, including cultural, economic, political, and social practices. Land use change is typically cited as disease emergence risk factor. Land use continuity may play an important role in conservation of healthy populations	Highly spatial highly platial
9. The relationship between land use and the probability of contact between vectors and animal hosts and human hosts is influenced by land ownership	Land ownership controls access to land and helps to define flows of people and animals on the landscape. As such, this can determine infection contact rates. Different forms of land ownership arrangements may contribute to healthful populations	Moderately spatial Highly platial
Human behaviour is a crucial controlling factor of vector-human contacts, and of infection	Whether people interact with pathogen vectors or animal reservoirs is an outcome of human behaviour. Animal health is also largely an outcome of human behaviour as it relates to change, contaminants, and other sources of animal morbidity/mortality	Weakly spatial Highly platial

Kearns (1993) called for the geography of health to consider the "dynamic relationship between health and place and the impacts of both health services and the health of population groups on the vitality of places." This latter consideration how subpopulations' health relates to the overall "vitality" of places has important connotations for a healthy settings perspective. Settings are the environments of everyday life. Much health policy and promotion literature focuses on distinct and various types of settings, such as schools and hospitals, to more variable entities such as homes, communities, and cities. Settings can equally be seen as population-level activity spaces, where people spend significant amounts of time engaged in activities of life, undergoing a variety of exposures at a multitude of scales. A settings framework from health policy and programming addresses health issues in a holistic nature, focusing on provision of resources that address health priorities.

The question of how to apply concepts from the "new" health geography to the shared health of humans and animals remains to be fully explored. Neely and Nading (2017) offered critical insights in their critique of global health research and practice from the perspective of place-based research. They considered the social construction of place through a unique set of social relations, understanding, and experiences (King, 2010) within an underlying biophysical environment. Biota of all types (i.e. so-called "non-human elements") are equally considered in this construction of place. Yet precisely how to operationalize these concepts remains challenging. Neely and Nading (2017) reviewed the case of an elderly Zulu woman who contracted a rare infectious disease in rural South Africa. The factors that convened to determine her diagnoses and treatment included internationally funded global health HIV programmes (which initially refused to treat her and presumed she was HIV positive), which stipulated treatment options for HIV patients, individual factors such as her age and socio-economic status, as well as the local geographic context characterized by "high numbers of immunocompromised people, rampant tuberculosis, crowding, and poor ventilation" within which she was embedded. Although used in Neely and Nading (2017) to critique the place-deaf specification of global health programme protocols, we see here how local and global forces intermix with local environment and individual factors to determine both the health and the health care of this person.

The Intergovernmental Platform on Biodiversity and Ecosystem Services (IPBES) addressed nature/society linkages in its own conceptual framework (Figure 8.1). Key here is the framing of benefits of the natural world to humans through the lens of ecosystem services. This framing, while addressing the role of governance and institutions in structuring natural drivers of change, narrowly conceives of health in human terms. Work on animal geographies and health has begun to reconsider these anthropocentric concerns, though empirical work in this vein is challenging.

In a compelling example, Enticott (2008) used sociological concepts of health to construct understandings of bovine tuberculosis biosecurity initiatives in the United Kingdom. They noted that "the population approach to animal health attempts an important transformation within farming styles," albeit without considering the highly localized nature of farming cultures and understandings. Here, farmers' own understandings of bovine tuberculosis risks and outcomes – termed lay epidemiology – which include some officially communicated risk factors, as well as luck and circumstance, were validated by (geographically) generic risk messaging from animal health authorities, because there are just as many other unknown or unmeasurable factors that relate to transmission. They note that such an ecological paradox emerges when there is a

> fundamental gap which exists between a person's experience of a given reality and science's explanation of that same reality (Gifford, 1986). This gap can be traced to a clash of spatial logics inherent within the population approach and farmers' understandings of animal health.

Thus, farmers' own cultural understandings of disease, outcomes, risky and safe behaviours, and their concern for the welfare of their animals, which are locally

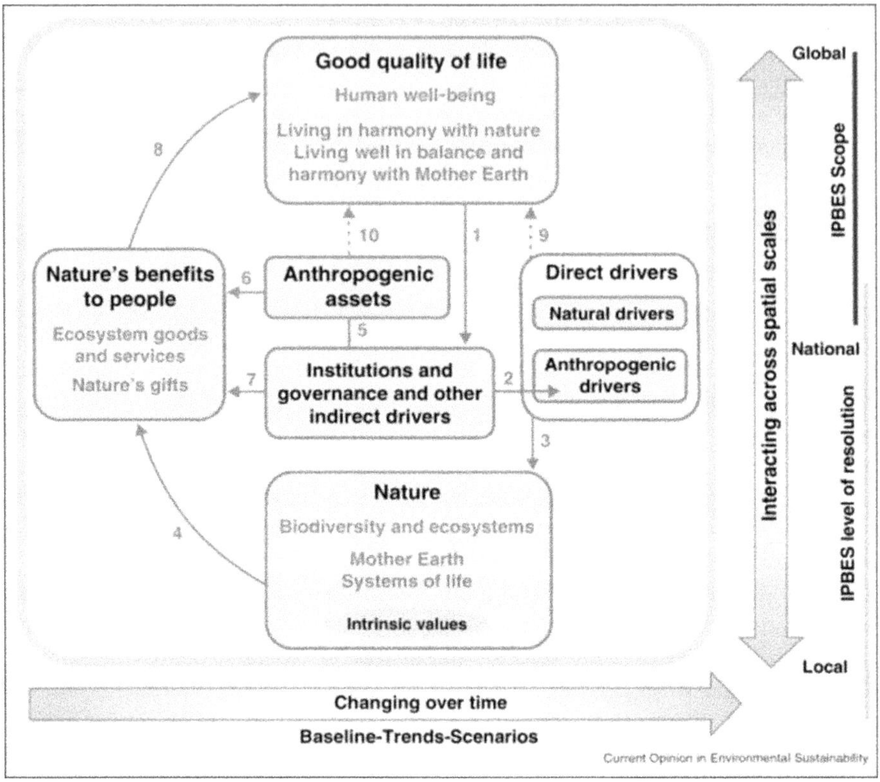

FIGURE 8.1 Intergovernmental Platform on Biodiversity and Ecosystem Services (IPBES) conceptual framework. (From Díaz et al., 2015.)

rooted in cultural geographies, needed deeper consideration and engagement in creating appropriate health promotion tools. These are critical variables of the Eco-Healthscape within which health is actively created, maintained, and sometimes reduced.

CONCEPTS FOR TRAVERSING ECO-HEALTHSCAPES

Figure 8.2 provides a conceptual model that can be adopted and specified for humans and animals. The factors in Figure 8.2 are either influences external to the individual or population or influences internal to the individual or population. For example, all the factors and processes outlined in Lambin et al. (2010) can be considered external. However, internal factors (personal connections to specific areas, therapeutic benefits derived from specific place assemblages, etc.) play an important role in health, and in creating resources for healthy environments. We now turn to a brief review of some key internal factors and how they might inform an Eco-Healthscape approach.

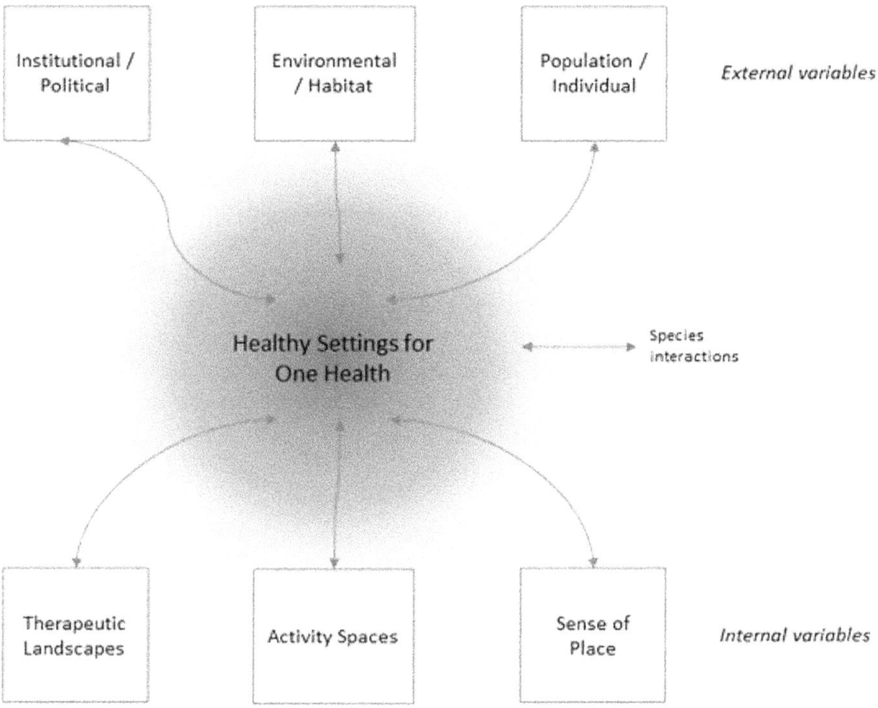

FIGURE 8.2 Proposed conceptual model for integrating place and health for shared health at human-animal interface.

Therapeutic Landscapes

One of the most important and widely adopted frameworks in understanding the intersection of place and health is that of therapeutic landscapes. The concept was introduced to describe how and why certain environments can provide healing (Gesler, 1992). Therapeutic landscapes research tends to come out of the humanist tradition in geography, which emphasizes the lived experiences of individuals above identifying general mechanisms or associations. Williams (1998) described therapeutic landscapes as "those changing places, settings, situations, locales, and milieus that encompass both the physical and psychological environments associated with treatment or healing." The concept, while initially focused on extraordinary spaces such as sites of ceremonial significance, has since been expanded to consider therapeutic benefits and restorative nature of everyday spaces such as farms (Kaley et al., 2019) and forests (Morita et al., 2007). Work considering animals in therapeutic landscapes has only recently begun, generally within the context of health-benefits derived from human interaction with animals – such as "care farms" (Kaley et al., 2019) or companion animals (Fletcher and Platt, 2018). The therapeutic benefits derived from place are not intrinsic qualities of the spaces in and of themselves

(Gorman, 2017). Thus, the therapeutic nature of movements/mobilities has also recently been explored (Gatrell, 2013).

Health benefits (i.e. not just risks) can be relationally derived and experienced by individuals through their interactions with specific locales. These relations are defined and structured by a complex assemblage of factors. How animals perceive and experience space remains speculative, but trying to understand the intersection of animal behaviour(s) and specific environments, as opposed to habitat categories, might shed some light on this question, especially as it relates to understanding animal-human relations. Typical ecological approaches to understanding habitat requirements and quality resort to enumerating features of the environment at sites where they are present (and comparing to where they are absent). More consideration could be given to, for example, what animals are doing at a specific locale, to inform that site's importance (functionally and ecologically) to that animal.

Sense of Place

One's personal connection to a given setting is frequently theorized in geography through the construct of their so-called sense of place; that is, the sum of their emotional and personal feelings towards a specific locale (Tuan, 1975). Sense of place is a key resource at the intersection of geography and health. As people spend time and interact with a place, they develop attachments to these places in complex ways. Knowing, interacting, and living within natural areas has been positively linked to sense of place development and positive health benefits (Russell et al., 2013). Sense of place is also linked to the concept of ecological identity (Thomashow, 1996), which relates to conservation values, attitudes, and behaviours. Personal values can be a critical driver of political and institutional actions (Díaz et al., 2015). There has been increased research interest on how sense of place drives conservation and biodiversity protections (Russell et al., 2013). Although rooted in humanist and phenomenological traditions, sense of place has also been frequently used and examined in quantitative settings, often proxied with a variable representing "time of residence" or in some way quantifying duration of time spent in a location which tends to be positively correlated with sense of place sentiments (Lewicka, 2011).

Neuroscience has also examined dimensions of place and sense of place in humans and animals. This research has identified "place-cells" that encode spatial information and are instrumental in navigation and spatial memory (Moser et al., 2008). O'Keefe and Conway (1978) identified the hippocampus as a sort of cognitive map responsible for spatial cognition and memory functions. Experimental evidence in rats (Save et al., 1998), monkeys (Rolls, 1999), and humans (Ekstrom et al., 2003) has confirmed the importance of hippocampal pyramidal cells in spatial cognition and memory functions. Multiple types of environmental sensory input (visual, auditory, tactile) influence firing of place-cells (Best and Thompson, 1989). It is, therefore, plausible that changes or characteristics of environment may impact spatial memory. Jeansok et al. (2007) showed that stressful environments impaired the stability of firing rates of place-cells in rats. Neuroscience research has concluded that there is convincing evidence that place constitutes a distinct dimension in neural processing. It is, therefore, plausible that the connections found to be vital between humans and their environments for health may also be at play in some animals.

Activity Spaces

The spatial dimensions of place are equally challenging as the psychological and neurological dimensions but are fundamental in defining an Eco-Healthscape across species boundaries. In classic health/medical geography studies, the home address was often used as the unit of analysis, usually summarized at the block or neighbourhood levels. Health outcomes (usually expressed as rates) over these so-called "small areas" could then be investigated in relation to other site-specific spatial variables for statistical associations. Statistically significant associations were deemed indicative of causal relations. The implicit connection was that the home address was a useful proxy for the exposures experienced by residents, and that if these exposures had a causal link with the health condition under study, these would exhibit a consistent pattern across the larger study area. A surge in research in the 2000s investigating space/place/health sought to identify "contextual" effects at the neighbourhood level through multi-level modelling. For example, some variables such as income inequality or social capital might only relate to health at an aggregate scale. These effects could then be embedded within a spatial hierarchy of factors that relate individual, neighbourhood, and contextual factors to the health outcome of interest. However, in recent years, several problems have been identified with using home addresses and residential neighbourhoods as proxies for spatial exposure.

Firstly, people move around at different rates to different locations, thereby creating different exposure profiles for people living in the same areas. This is exacerbated by shifts in commuting patterns and working arrangements to more variable and complex flows than suburb-central core commuting characteristic of the post-war era. The notion of spatial polygamy has been used to describe the "simultaneous belonging or exposure to multiple nested and non-nested; social and geographic; real, virtual, and fictional; and past and present contexts" (Matthews, 2011). These spaces vary individually but also along income, racial, and gendered lines (Matthews and Yang, 2013). They also interact differentially with health. Vallée et al. (2011), for example, found that people in deprived neighbourhoods that had more localized activity spaces were linked to negative mental health outcomes, whereas localized activity space footprints in advantaged neighbourhoods actually had a protective effect (Vallée et al., 2011).

A second concern is that people may interact with features of their environment at vastly different rates. This becomes a concern for approaches that equate presence of some environmental resource with a constant rate of use of that resource. For example, many studies have investigated neighbourhood-level green space and indicators of stress, cardiovascular disease, and overall self-reported well-being. These studies often fail to account for the differential use of greenspace among residents in the same community, which may, for example, vary systematically with age or income. Living near a park can mean you are more likely to be physically active, or more people that are predisposed to active lifestyles choose to buy homes near parks (Cummins and Fagg, 2012); these two processes cannot be untangled from observational spatial associations (Robertson and Feick, 2018). Tools for understanding animal use of space (i.e. their geographic context) are well developed for home range mapping via a sophisticated array of tracking technologies and analytical methods to identify locations of significance for individuals and populations (see Long and Nelson, 2013). Some of these are incorporating aspects of spatial memory into their conception of home ranges.

The uncertainty in one's true geographic context is termed the *uncertain geographic context problem*: it is not totally clear what spatial contexts apply to what individuals (Kwan, 2012). Ignoring this problem risks incorrect inferences about associations between health outcomes and environmental factors. These sorts of problems are compounded for more complex health issues such as vector-borne diseases, where the number of mobile actors in a transmission chain can be two, three, or more. Added to that, we typically have far less information about animals' use of space and their interactions and differential experiences; these issues can limit the value of spatial associations in these systems (Robertson et al., 2017). Understanding how individuals' activity space (physically and perceptively/psychologically) intersects with those both within and across species divides may shed light on positive and negative effects on shared health outcomes.

CASE STUDY: BATHURST CARIBOU HERD POPULATION DECLINE IN NORTHERN CANADA

Major changes in barren-ground caribou (*Rangifer tarandus groenlandicus*) habitat around the circumpolar north are leading to significant population declines (Vors and Boyce, 2009). There is no singular global driver of these declines. Herds are experiencing a mix of local, regional, and hemispheric anthropogenic and climate-driven changes that can alter caribou ecology (Post and Forchhammer, 2008; Vors and Boyce, 2009; Rickbeil et al., 2017; Zamin et al., 2017). In Canada's north, barren-ground, caribou habitat has recently been subject to two principal drivers of change: intensification and expansion of industrial land uses; and climate warming and resultant changes to eco-hydrological processes (Joly et al., 2011).

The Bathurst caribou herd (BCH) traverses a range from their calving grounds on Bathurst Inlet south to their winter range north of Great Slave Lake. The BCH is an integral part of the cultural and spiritual lives of the Tłı̨chǫ peoples who have lived with the herd for centuries. Population surveys have estimated a decline of over 95%, from over 470,000 animals in the 1980s to less than 10,000 in the most recent estimates (ENR, 2020).

Without a single known cause for the decline, management "solutions" must be locally embedded within the cultural communities that interact with the herd and its range. The Government of Northwest Territories (GNWT) has

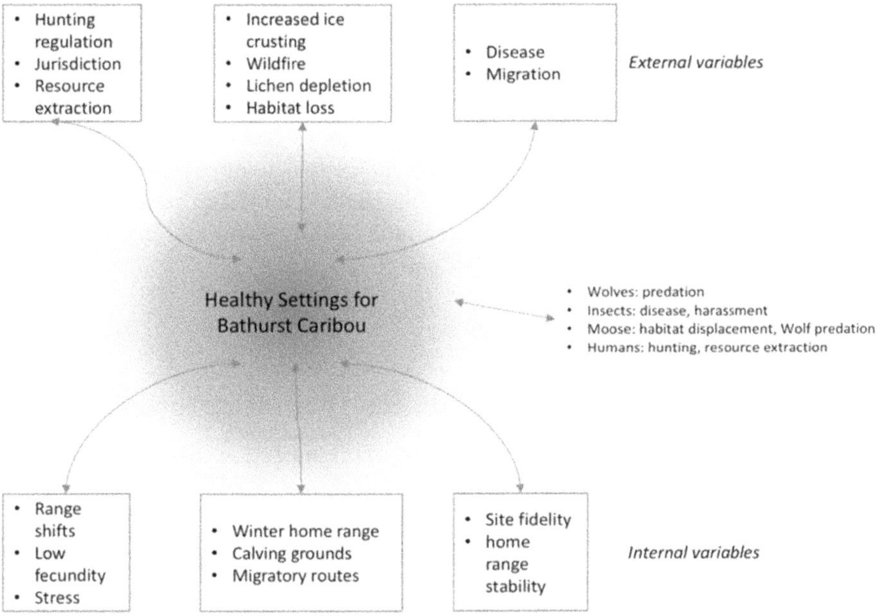

FIGURE 8.3 Proposed conceptual model applied to Bathurst caribou herd population decline in Northwest Territories, Canada.

regulatory responsibility for the herd. Natural resource management decisions of GWNT must recognize traditional knowledge as a "valid and essential source of information about the natural environment and its resources" (GNWT Traditional Knowledge Policy, 2005). In our Eco-Healthscape model for the herd (Figure 8.3), it is essential to consider which actors are involved in each variable and how they might be engaged to achieve healthier outcomes. For instance, Tłıchǫ communities in and around the winter range have long hunted and monitored the herd. Yet, with declines, modernization, loss of traditional ways of living, these "eyes on the land" need bolstering through targeted investments and resources to get local people involved in monitoring the herd. Such culturally integrated monitoring might then lead to insights that direct resource extraction and hunting regulations. Indigenous governments are indeed taking these steps. The Ekwǫ̀ Nàxoède K'è: Boots on the Ground Programme developed by the Tłı̨chǫ government, for example, focuses on field monitoring in the BCH summer range and "everything that relates to them" – focusing on four broad indicators: (1) habitat, (2) caribou, (3) predators, and (4) industrial development (Tłı̨chǫ Research and Training Institute, 2019). Such Indigenous-led stewardship is a viable means for conservation of ecosystem integrity and individual populations (Hunter, 2008; Artelle et al., 2019).

Scientific management and monitoring, including GWNT ungulate biologists, data analysts, academic researchers, consultants, etc., remain essential components of research and monitoring herd size, habitat quality, and developing actionable insights into BCH management and recovery. The GNWT traditional knowledge implementation programme is helping to build a community of conservation with indigenous communities and scientific staff. Thus, when controversial control measures such as a total hunting ban are proposed, as was the case in 2016, they are more acceptable by both scientific and indigenous communities than would otherwise be the case. These proactive approaches, rooted in co-management, relationship building, and culturally integrated health promotion, reveal promising tools for an Eco-Healthscape approach.

CONCLUSIONS

Practical Considerations in Taking an Eco-Healthscape Approach

The complexity involved in understanding and taking action that changes health outcomes in favour of a group of species is staggering. There will be no statistical recipe or data analytic approach that will work in every context. We have, however, identified several key practical considerations when considering how to employ space and place thinking across species boundaries.

1. *Getting the scale(s) right*
   The geographic scale(s) of health promotion activities is critically important. When attempting to translate a finding about, for example, features of the landscape (e.g. edge habitat) impacting disease risk, one must consider the magnitude of this feature in a given locale, and how people already interact with such features and how people's experiences and activities relate to the landscape configuration. If human-modified aspects of the landscape relate to the population health, consider how to engage with local cultural communities shaping those places rather than through generic messaging.
2. *Isolating key variables without reducing complexity*
   While any given Eco-Healthscape context necessarily includes a multitude of species and their internal and external factors, it becomes important to isolate key variables that create conditions for health. For example, while "climate change" may be a driver of ecosystem change and community response(s), tracking changes in temperature may not be the most effective means to understand or impact future health states. What data can be collected or analysis performed that maximizes coverage of key processes? In the case study above, while Indigenous-led stewardship may not reduce threats, they will inform on key internal variables, and do so in a way that can incentivize future actions.
3. *Developing generic data synthesis tools*
   The most important data analytic task that will facilitate Eco-Healthscape research is developing tools for data integration. Given that data are frequently repurposed to look at spatial and platial associations with health, bringing data into a common representation is critical. Often, data

integration is akin to mapping spatial data onto specific units of analysis, for example census units or ecozones. Generic data representations, such as pixel arrays, or functions of space such as Delaunay triangulations, can provide consistent, reproducible tools for bringing disparate datasets into a common spatial fabric. Discrete global grid system is an emerging geographic data standard that can also serve as an intermediate data integration node (Robertson et al., 2020). Given the need to mobilize analytical results across different user communities, having clearly documented data collection, integration, and analysis procedures is important.

REFERENCES

Artelle, Kyle A., Melanie Zurba, Jonaki Bhattacharrya, Diana E. Chan, Kelly Brown, Jess Housty, and Faisal Moola. "Supporting Resurgent Indigenous-Led Governance: A Nascent Mechanism for Just and Effective Conservation." *Biological Conservation*, 240, (2019): 108284.

Best, Phillip J., and Lucien T. Thompson. "Persistence, Reticence, and Opportunism of Place-Field Activity in Hippocampal Neurons." *Psychobiology*, 17, no. 3 (1989): 236–246.

Burke, Sarah M., and Neil Mitchell. "People as Ecological Participants in Ecological Restoration." *Restoration Ecology*, 15, no. 2 (2007): 348–350.

Conradson, David. "Landscape, Care and the Relational Self: Therapeutic Encounters in Rural England." *Health and Place*, 11, no. 4 (2005): 337–348.

Cummins, S., and J. Fagg. "Does Greener Mean Thinner? Associations Between Neighbourhood Greenspace and Weight Status Among Adults in England." *International Journal of Obesity*, 36, no. 8 (2012): 1108–1113.

Díaz, Sandra, Sebsebe Demissew, Julia Carabias, Carlos Joly, Mark Lonsdale, Neville Ash, Anne Larigauderie et al. "The IPBES Conceptual Framework: Connecting Nature and People." *Current Opinion in Environmental Sustainability*, 14 (2015): 1–16.

Ekstrom, Arne D., Michael J. Kahana, Jeremy B. Caplan, Tony A. Fields, Eve A. Isham, Ehren L. Newman, and Itzhak Fried. "Cellular Networks Underlying Human Spatial Navigation." *Nature*, 425, no. 6954 (2003): 184–188.

Ekwǫ̀ Nàxoède K'è: Boots on the Ground | Tlicho Research and Training Institute. (2019). Retrieved August 31, 2020, from https://research.tlicho.ca/research/bootsontheground

Enticott, Gareth. "The Ecological Paradox: Social and Natural Consequences of the Geographies of Animal Health Promotion." *Transactions of the Institute of British Geographers*, 33, no. 4 (2008): 433–446.

Environment and Natural Resources. Government of the Northwest Territories. (2020.). *Bathurst Herd* [Information]. Retrieved August 31, 2020, from https://www.enr.gov.nt.ca/en/node/670 https://www.enr.gov.nt.ca/en/services/barren-ground-caribou/bathurst-herd

Fletcher, Thomas, and Louise Platt. "(Just) a Walk with the Dog? Animal Geographies and Negotiating Walking Spaces." *Social and Cultural Geography*, 19, no. 2 (2018): 211–229.

Gatrell, Anthony C. "Therapeutic Mobilities: Walking and 'Steps' to Wellbeing and Health." *Health and Place*, 22 (2013): 98–106.

Gesler, Wilbert M. "Therapeutic Landscapes: Medical Issues in Light of the New Cultural Geography." *Social Science and Medicine*, 34, no. 7 (1992): 735–746.

Gesler, Wilbert M., Matthew Hayes, Thomas A. Arcury, Anne H. Skelly, Sally Nash, and April C.M. Soward. "Use of Mapping Technology in Health Intervention Research." *Nursing Outlook*, 52, no. 3 (2004): 142–146.

Gifford SM. "The Meaning of Lumps: A Case Study of the Ambiguities of Risk." In: Janes CR, Stall R, Gifford SM (eds). *Anthropology and Epidemiology*. Dordrecht: D. Reidel Publishing (1986): 213–46.

GNWT Traditional Knowledge Policy, 2005 https://www.eia.gov.nt.ca/sites/eia/files/content/53.03-traditional-knowledge.pdf. (accessed October 2019).

Gorman, Richard. "Therapeutic Landscapes and Non-Human Animals: The Roles and Contested Positions of Animals within Care Farming Assemblages." *Social and Cultural Geography*, 18, no. 3 (2017): 315–335.

Holmes, Elizabeth E., Mark A. Lewis, J.E. Banks, and R.R. Veit. "Partial Differential Equations in Ecology: Spatial Interactions and Population Dynamics." *Ecology*, 75, no. 1 (1994): 17–29.

Hunter, Troy. "Aboriginal Stewardship: A Better Way to Save the Mountain Caribou." *LawNow*, 33, (2008): 51.

Jeansok, Kim J., Hongjoo J. Lee, Adam C. Welday, EunYoung Song, Jeiwon Cho, Patricia E. Sharp, Min W. Jung, et al. "Stress-Induced Alterations in Hippocampal Plasticity, Place Cells, and Spatial Memory." *Proceedings of the National Academy of Sciences of the United States of America*, 104, no. 46 (2007): 18297–18302.

Joly, Kyle, David R. Klein, David L. Verbyla, T. Scott Rupp, and F. Stuart Chapin, III. "Linkages Between Large-Scale Climate Patterns and the Dynamics of Arctic Caribou Populations." *Ecography*, 34, no. 2 (2011): 345–352.

Kaley, Alexandra, Chris Hatton, and Christine Milligan. "Therapeutic Spaces of Care Farming: Transformative or Ameliorating?" *Social Science and Medicine*, 227 (2019): 10–20.

Kearns, Robin A. "Place and Health: Towards a Reformed Medical Geography." *The Professional Geographer*, 45, no. 2 (1993): 139–147.

Kearns, Robin, and Graham Moon. "From Medical to Health Geography: Novelty, Place and Theory After a Decade of Change." *Progress in Human Geography*, 26, no. 5 (2002): 605–625.

Kickbusch, Ilona. "The Contribution of the World Health Organization to a New Public Health and Health Promotion." *American Journal of Public Health*, 93, no. 3 (2003): 383–388.

King, B. 2010. Political Ecologies of Health. *Progress in Human Geography* 34 (1):38–55.

Kwan, Mei-Po. "The Uncertain Geographic Context Problem." *Annals of the Association of American Geographers*, 102, no. 5 (2012): 958–968.

Lambin, Eric F., Annelise Tran, Sophie O. Vanwambeke, Catherine Linard, and Valérie Soti. "Pathogenic Landscapes: Interactions Between Land, People, Disease Vectors, and Their Animal Hosts." *International Journal of Health Geographics*, 9, no. 1 (2010): 54.

Lewicka, Maria. "Place Attachment: How Far Have We Come in the Last 40 Years?" *Journal of Environmental Psychology*, 31, no. 3 (2011): 207–230.

Long, Jed A., and Trisalyn A. Nelson. "A Review of Quantitative Methods for Movement Data." *International Journal of Geographical Information Science*, 27, no. 2 (2013): 292–318.

Matthews, Stephen A. "Spatial Polygamy and the Heterogeneity of Place: Studying People and Place via Egocentric Methods," in *Communities, Neighborhoods, and Health* (New York: Springer, 2011), 35–55.

Matthews, Stephen A., and Tse-Chuan Yang. "Spatial Polygamy and Contextual Exposures (Spaces) Promoting Activity Space Approaches in Research on Place and Health." *American Behavioral Scientist*, 57, no. 8 (2013): 1057–1081.

Mayer, Jonathan D. "Geography, Ecology and Emerging Infectious Diseases." *Social Science and Medicine*, 50, no. 7-8 (2000): 937–952.

Meade, Melinda S. "Medical Geography." *The Wiley Blackwell Encyclopedia of Health, Illness, Behaviour, and Society* (Hoboken, NJ: John Wiley & Sons, Ltd., 2014): 1375–1381.

Mladenoff, David J. 'The Promise of Landscape Modeling: Successes, Failures, and Evolution," in *Issues and Perspectives in Landscape Ecology*, eds. John A. Wiens and Michael R. Moss (Cambridge: Cambridge University Press, 2005), 90–100.

Morita, Emi, S. Fukuda, Jun Nagano, N. Hamajima, H. Yamamoto, Y. Iwai, T. Nakashima, et al. "Psychological Effects of Forest Environments on Healthy Adults: Shinrin-Yoku (Forest-Air Bathing, Walking) as a Possible Method of Stress Reduction." *Public Health*, 121, no. 1 (2007): 54–63.

Moser, Edvard I., Emilio Kropff, and May-Britt Moser. "Place Cells, Grid Cells, and the Brain's Spatial Representation System." *Annual Review of Neuroscience*, 31 (2008): 69–89.

Moser, May-Britt, David C. Rowland, and Edvard I. Moser. "Place Cells, Grid Cells, and Memory." *Cold Spring Harbor Perspectives in Biology*, 7, no. 2 (2015): a021808.

Neely, Abigail H., and Alex M. Nading. "Global Health from the Outside: The Promise of Place-Based Research." *Health and Place* 45, (2017): 55–63.

O'Keefe, John, and D.H. Conway. "Hippocampal Place Units in the Freely Moving Rat: Why They Fire Where They Fire." *Experimental Brain Research*, 31, no. 4 (1978): 573–590.

Pickett, Steward T.A., and Mary L. Cadenasso. 'Landscape Ecology: Spatial Heterogeneity in Ecological Systems." *Science*, 269, no. 5222 (1995): 331–334.

Post, Eric, and Mads C. Forchhammer. "Climate Change Reduces Reproductive Success of an Arctic Herbivore through Trophic Mismatch." *Philosophical Transactions of the Royal Society B: Biological Sciences*, 363, no. 1501 (2008): 2367–2373.

Rickbeil, Gregory J.M., Txomin Hermosilla, Nicholas C. Coops, Joanne C. White, and Michael A. Wulder. "Barren-Ground Caribou (*Rangifer tarandus groenlandicus*) Behaviour After Recent Fire Events; Integrating Caribou Telemetry Data with Landsat Fire Detection Techniques." *Global Change Biology*, 23, no. 3 (2017): 1036–1047.

Robertson, Colin. 'Towards a Geocomputational Landscape Epidemiology: Surveillance, Modelling, and Interventions." *GeoJournal*, 82, no. 2 (2017): 397–414.

Robertson, Colin, Chiranjib Chaudhuri, Majid Hojati, and Steven A. Roberts. "An Integrated Environmental Analytics System (IDEAS) Based on a DGGS." *ISPRS Journal of Photogrammetry and Remote Sensing*, 162, (2020): 214–228.

Robertson, Colin, and Rob Feick. "Inference and Analysis Across Spatial Supports in the Big Data Era: Uncertain Point Observations and Geographic Contexts." *Transactions in GIS*, 22, no. 2 (2018): 455–476.

Robertson, Colin, Trisalyn A. Nelson, Ying C. MacNab, and Andrew B. Lawson. 'Review of Methods for Space-Time Disease Surveillance." *Spatial and Spatio-Temporal Epidemiology*, 1, no. 2–3 (2010): 105–116.

Robertson, Colin, Lauren Yee, Julia Metelka, and Craig Stephen. "Spatial Data Issues in Geographical Zoonoses Research." *The Canadian Geographer/Le Géographe canadien*, 60, no. 3 (2016): 300–319.

Rolls, Edmund T. "Spatial View Cells and the Representation of Place in the Primate Hippocampus." *Hippocampus*, 9, no. 4 (1999): 467–480.

Rothman, Kenneth J. "A Sobering Start for the Cluster Busters' Conference." *American Journal of Epidemiology*, 132, no. 1 (1990): S6–S13.

Russell, Roly, Anne D. Guerry, Patricia Balvanera, Rachelle K. Gould, Xavier Basurto, Kai M.A. Chan, Sarah Klain, et al. "Humans and Nature: How Knowing and Experiencing Nature Affect Well-Being." *Annual Review of Environment and Resources*, 38, (2013): 473–502.

Save, Etienne, Arnaud Cressant, Catherine Thinus-Blanc, and Bruno Poucet. "Spatial Firing of Hippocampal Place Cells in Blind Rats." *Journal of Neuroscience*, 18, no. 5 (1998): 1818–1826.

Thomashow, Mitchell. *Ecological Identity: Becoming a Reflective Environmentalist* (Cambridge: MIT Press, 1996).

Tilman, David, Peter Kareiva, E.E. Holmes, and M.A. Lewis. "Space: The Final Frontier for Ecological Theory." *Ecology*, 75, no. 1 (1994): 1.

Tuan, Yi-Fu. "Place: An Experiential Perspective." *Geographical Review*, 65, no. 2 (1975): 151–165.

Turner, Monica Goigel. "Landscape Ecology: The Effect of Pattern on Process." *Annual Review of Ecology and Systematics*, 20, no. 1 (1989): 171–197.

Vallée, Julie, Emmanuelle Cadot, Christelle Roustit, Isabelle Parizot, and Pierre Chauvin. "The Role of Daily Mobility in Mental Health Inequalities: The Interactive Influence of Activity Space and Neighbourhood of Residence on Depression." *Social Science & Medicine*, 73, no. 8 (2011): 1133–1144.

Vors, Liv Solveig, and Mark Stephen Boyce. "Global Declines of Caribou and Reindeer." *Global Change Biology*, 15, no. 11 (2009): 2626–2633.

Williams, Allison. "Therapeutic Landscapes in Holistic Medicine." *Social Science & Medicine*, 46, no. 9 (1998): 1193–1203.

WHO. "The Ottawa Charter for Health Promotion." World Health Organization, 1986. https://www.who.int/healthpromotion/conferences/previous/ottawa/en/ (accessed March 2020).

Zamin, Tara J., Steeve D. Côté, Jean-Pierre Tremblay, and Paul Grogan. "Experimental Warming Alters Migratory Caribou Forage Quality." *Ecological Applications*, 27, no. 7 (2017): 2061–2073.

9 Helping People Make Healthy Decisions for Themselves, Animals, and Nature

Craig Stephen

CONTENTS

Health promotion in a One Health world it is about empowering people, organizations, and governments to make decisions that promote reciprocal care of ourselves, animals, and ecosystems. Health promotion and harm reduction help people make choices and changes that increase their capacity to manage the complex of factors that influence health outcomes and create healthier environments. Health behaviour theories and theories of change can guide a systematic approach to promoting healthy change.

Our actions and choices determine if or how we translate the vision of reciprocal care into tangible change. Doing something about One Health problems, whether a zoonotic disease outbreak or food insecurity, will mean action, not only talk and research. There are often huge gaps between the available knowledge about healthy and sustainable living and actions towards it. Simply providing people with new knowledge is insufficient to promote action. Understanding the gaps between knowing and doing requires an understanding of what affects the choices people have, what enables or impedes their willingness or ability to act, and how well they can access and understand information available to them.

Change is hard. The outcomes of change are often less certain than the status quo. It can be threatening to confront uncertainty. People resist change if they believe they will lose something they value, or they won't be able to adapt to the change. People quickly build psychological defences against change when uncertainty exists or if the suggestion of change implies they are aren't acting in their own best interests or in the interests of others (Cohen and Sherman, 2014). Cultural entrenchment, dysfunctional management, and poor leadership can cause organizations to resist change (Hoag et al., 2002). Change requires patience, persistence, and a process. There are many models and theories of change. None are ideal and suited to all situations and context. Each is subject to limitations and questions. This chapter introduces some of the theories and frameworks that can guide strategies to help people, organizations, and communities make choices that lead to health-promoting actions for themselves, their communities, their animals, and Nature.

INDIVIDUAL CHANGE

Rarely is change a discrete or single event. It is a process. Not all people will be met at the same stage in the process of change (Bandura, 2004). Some people will have a high degree of motivation with a high sense of self-efficacy and positive expectations for a change. They can be induced or supported to change with minimal guidance. Other people will have self-doubts about their ability to change and the benefits of their efforts. They may start the process of change but can be quick to give up when they run into difficulties. They need interactive support and guidance to maintain their momentum to change. Still others will believe that the desired outcomes are beyond their control. They need a great deal of guidance in a structured program.

Theories can be useful in developing a systematic approach to helping people acquire the skills, capacities, and motivations to make decisions and take actions. Theories or models of change help us move beyond intuition to design and evaluate interventions based on understanding human behaviour (Box 9.1). Although there are differences among models and theories, they are complimentary with significant overlap. They all generally state that what people know and think affects how they act, but that knowledge is necessary, yet insufficient, to produce change. People's perceptions, motivations, skills, and environments are key influencers of change.

There are two broad aspects to change: thinking about it and doing it. Thinking about change is supported by helping people gain and understand information; becoming comfortable with the value, feasibility, and acceptability of change; and helping them believe they can change. Doing is supported by creating or finding opportunities to perform the new action, showing the value or benefits of those changes, and developing the social support to motivate maintaining the change.

BOX 9.1 CONSERVATION IS ABOUT INDIVIDUAL DECISIONS: A CASE STUDY OF FERAL CAT CONTROL

Major threats to biodiversity, such as habitat loss, overexploitation, climate change, and pollution, are the result of a cascade of earlier decisions that trace back to individual choices. The everyday decisions we make that influence these major threats are known as "target behaviours." Each target behaviour will be driven by a set of motivations, attitudes, knowledge, values, and barriers that are specific to each target audience. McLeod et al. (2019) applied social and behavioural sciences to design equitable and ethically acceptable interventions for free-roaming cat management. They used four principles:

1. Focus on human behaviour to understand how different stakeholders perceive and are affected by the problem, their priorities and interests, and potential conflicts with personal interests and what is collectively desirable.
2. Know your audience by collecting target-specific information on the barriers that impede or enable their engagement in the desired behaviour as well as those factors that drive action.
3. Link specific drivers and barriers to behaviour change techniques that are feasible and acceptable to the target audience.
4. Use a science-based evaluation to determine what works and why.

McLeod et al. (2019) used these principles to work with land managers, cat owners, and the public to develop and adopt best practices for managing free-roaming cats.

Transtheoretical Model

The Transtheoretical Model tells us that people will not change until they are ready to do so (Prochaska and Velicer, 1997), and that there are five stages of change: (i) precontemplation, (ii) contemplation, (iii) preparation, (iv) action, and (v) maintenance or termination. Interventions need to be customized to match the stage at which you find the people with whom you work (Table 9.1). This is reflected in the principle of "meeting people where they are" found in harm reduction thinking (see Chapter 6 for more on harm reduction). People may move back and forth between stages rather than follow an inevitable linear progression from inaction to a permanent change. Knowing where they are in the process will help you match your interventions to someone's readiness for change.

The Transtheoretical Model has been used to influence human behaviours linked to chronic diseases such as smoking, bad diets, and inactivity. Lessons

TABLE 9.1
The Stages of Change According to the Transtheoretical Model

Stage	Defining Features
Precontemplation	People do not intend to change in the foreseeable future. They may not see the need for change or be aware that the current situation can lead to negative consequences. The pros of the current situation are perceived to outweigh the cons
Contemplation	People recognize the current situation may be problematic. They start to consider both the pros and the cons of change but still have mixed or contradictory feelings about the need to change
Determination	People believe change is for the better and start making steps towards that change
Action	People have made the change and intend on maintaining the change for the foreseeable future
Maintenance	People are maintaining the actions needed to keep the change going. They work to prevent relapsing to the earlier state
Termination	People may not want to return to the old state, but they are sure the situation will not reoccur, so they stop the necessary actions

learned from overcoming such behaviours have informed strategies to discourage destructive environmental behaviour and inspire people to engage in pro-environmental actions (Nisbet and Gick, 2008). The Theory of Routine Mode Choice Decisions (Schneider, 2013), for example, was developed to promote sustainable transportation choices and seems influenced by the Transtheoretical Model. The five steps of the Theory of Routine Mode Choice are (i) making people aware of options and making them available, (ii) apprising people of the safety of the more sustainable options, (iii) making the more sustainable options convenient and cost-effective, (iv) making the sustainable option enjoyable, and (v) helping people sustain their choices. The Transtheoretical Model has been used in veterinary medicine to understand the decisions of animal owners or managers. For example, it was used to identify determinants of dairy farmers' biosecurity attitudes and behaviours in Great Britain (Richens et al., 2018). That study found a clear dichotomy between "precontemplaters" and "maintainers," with few farmers in the three other stages of change (Table 9.1). Many farmers saw the benefits of using biosecurity, so a lack of awareness seemed unlikely to be the reason behind lack of implementation. The investigators found farmer perceptions that diseases were inevitable and beyond their control to be an important deterrent to change. The use of a theory of change in this case helped the investigators understand how to guide precontemplative farmers towards change.

Health Belief Model

The Health Belief Model is widely used to guide human health promotion and disease prevention programs (Champion and Skinner, 2008). Table 9.2 presents some guiding question from this model to help in program design. The model

TABLE 9.2
Some Guiding Questions to Encourage Behaviour Change Derived from the Health Belief Model

From the Model	Guiding Questions
Perceptions of susceptibility, seriousness, benefits, and barriers	• Who/what is at risk? • Does the population at risk (or those in a position to make decisions on behalf of animals or ecosystems at risk) have access to accurate, trustworthy information presented in a manner suited to their circumstances and characteristics? • Have the benefits of the actions been explained in a manner relevant to those being asked to change? • Why do the perceived risks of change outweigh perceived benefits? • What is the nature of evidence upon which these perceptions are based? • Can the barriers be feasibly overcome?
Cues to action	• What is the best medium and method of providing cues to action to the target audience(s)? • Who is a trusted voice that can provide cues?
Self-efficacy	• Have people been shown how to perform the desired behaviour, or trained or assisted in implementing the change? • Is there a series of incremental steps that can encourage change?

argues that to change, people must first perceive that the problem they are trying to avoid is serious and that they are susceptible to the problem. They must believe that the proposed change can reduce risk and that the barriers to taking the action are outweighed by the benefits. These barriers can be logistic, financial, social, or others. This model recognizes that people need a cue to action before change will happen. These may be internal cues (e.g. pre-clinical symptoms of a diseases) or external cues (e.g. media messages or prompts from a health care professional). The last component of the model deals with self-efficacy: a persons' confidence in his or her capacity to act and make the change. Perceptions and beliefs influencing the various stages of this model are impacted by modifying variables such as age, gender, personality, socio-economic status, and knowledge. Perceived barriers and perceived susceptibility can be the most powerful predictors of the likelihood that a person will adopt a health behaviour.

The Health Belief Model has been used most extensively to understand why people don't adopt disease prevention strategies, undertake screening tests for early disease detection, or comply with treatment recommendations. The model has also been used as a framework to design animal health studies as well as to put the results of research into context. For example, investigators in Nigeria used the model to design questionnaires to identify social and cognitive factors predicting the practice used by meat handlers of eating parts of the lung from cattle visibly infected with bovine tuberculosis in order to convince customers to buy meat (Hambolu et al., 2013). The model has also been used to explore motivators or disincentives to adopt environmentally friendly practices such as recycling (Lindsay and Strathman, 1997) and well water conservation (Morowatisharifabad et al., 2012). Understanding perceptions, barriers, and self-efficacy is increasingly important when trying to promote actions to mitigate climate change (Gifford, 2011). There has been less uptake of this model in proactive conservation because it is concerned more with actions to avoid negative consequences than actions to promote healthy capacities and environments.

Integrated Behaviour Model

The Theory of Planned Behaviour and the Theory of Reasoned Action both assume the best predictor of a behaviour is an intention to act, which in turn is determined by attitudes towards the action, social norms, and perceived control over the action. An Integrated Behaviour Model draws on these two theories as well as well as others (Montano and Kasprzyk, 2008). This model, like others, recognizes that people must have the intention or motivation to change. They must feel positive about the change. A desire to comply with social pressures or norms that promote the change further motivates people to act. People must have the knowledge and skills to act and face few or no constraints to impede the action. The requested change must be pertinent or relevant to the person being asked to change. Finally, the change should be performed often enough that it becomes a habit and relies less on intentions in order to sustain the change.

Precaution Adoption Process Model

The Precaution Adoption Process Model lists seven distinct stages a person goes through from lack of awareness to adoption and/or maintenance of a behaviour (Weinstein and Sandman, 2002):

- Stage 1 – The person is unaware of the issue.
- Stage 2 – The person is aware of the issue but is unengaged by it.
- Stage 3 – The person faces a decision to act or not.
- Stage 4 – The decision to not act has been made.
- Stage 5 – The decision to act has been made.
- Stage 6 – This is the stage of action.
- Stage 7 – The action is maintained.

This model recognizes that people who are unaware of an issue, or are unengaged by it, face different barriers from those who have decided to act or not. People in stage 1 need basic information about the hazard and the recommended precaution. Moving from stage 1 to stage 2 will be influenced by a person's access to information, whether formally or informally. Moving from stage 2 to stage 3 will be similarly influenced but may be prompted by something that makes the threat and action seem personally relevant. For example, a farmer may be more likely to decide about biosecurity measures when a neighbouring farm is affected by a disease. Or a person may be more likely to decide to do something if a friend or family member experiences the negative outcome he or she is trying to avoid. Becoming aware that others are making up their minds may also motivate a decision. Moving from stage 3 to stage 4 or stage 5 will be heavily influenced by a person's beliefs about hazard likelihood and severity, personal susceptibility, effectiveness and difficulty of changing, social norms, and personal fears. People who have come to a definite position on an issue have different responses to new information and are more resistant to persuasion than people who have not formed opinions. The implications of some people saying that they have decided not to act are not the same as saying it is "unlikely" they will act or that they "might" act. People in stage 4 can be quite well informed but will tend to dispute or ignore information that challenges their decision not to act. Moving people from stage 5 to stage 6 means closing the gap between an intention to act and acting. Whereas detailed information on "how" to change may not be as important as "why" to change for people in earlier stages, information on how to change is essential for people at this stage. Understanding the time, skills, and effort needed to change; knowing how to find support for change; detailed "how-to" information; and cues for action are needed. Finally, the transition from stage 6 to stage 7 needs ongoing support and re-enforcement of the value of action to habituate the change.

SOCIAL AND ORGANIZATIONAL CHANGE

The World Health Organization's (WHO) Ottawa Charter for Health Promotion (1986) states that health is created and lived within the settings of everyday life. A healthy settings approach employs a "whole systems" view to integrate action across sectors, determinants, and risks. Changing whole socio-ecological systems can be a daunting task given their complex and dynamic nature. The literature on leading change in socio-ecological systems dictates that complex systems cannot be changed by the top-down, command and control forms of management but instead needs a collaborative, inclusive and participatory action and that the capacity of individuals to act independently and make their own free choices is critical for systems change (Westley et al., 2013). The frequent disconnect between what we know about the biology and epidemiology of a health issue and program success has led to a growing realization that political, social, or economic conditions are often the primary determinants of success of health interventions. Health promotion and harm reduction, therefore, need to influence change at the personal, interpersonal, institutional, community, systems, and policy levels to successfully implement and sustain change.

ORGANIZATIONAL CHANGE

Lewin's Theory of Change describes three stages of change in organizations: (i) people are ready to alter the status quo (unfreezing); (ii) a new state of affairs is created and customized to a setting through trial and error (changing); and (iii) new behaviours are implemented, creating a new organizational system (refreezing) (Kritsonis, 2005). Internal forces (such as an organization's structure, leadership, strategy, and personnel) and external factors (such a public policy or market forces) encourage or impede movement through these changes.

Diffusion Theory addresses how new ideas and practices spread (Dearing, 2009). Firstly, new ideas are generated. Secondly, advocacy and screening help evaluate if an idea matches core beliefs of an organization and compares its advantages and risks. Thirdly, the idea is incrementally implemented, tested, and adjusted to fit the organization. Fourthly, change is implemented with ongoing assessment of its usefulness. Finally, the change is institutionalized. Diffusion Theory has been used to study the adoption of a wide range of health behaviours such as condom use, smoking cessation, and use of new tests and medical technologies.

Kotter's eight steps to change (as described in Appelbaum et al., 2012) (Table 9.3) has been used to lead change in the private and public sectors. For example, it was used in Canada in 2018 to lead provincial, territorial, and federal government ministers to adopt a new Pan-Canadian Approach to Wildlife Health (Stephen, 2019). It has been used by hospital managers, clinicians, and nurses to change clinical practices in hospital settings. It has guided the implementation of conservation practices. Kotter's eight steps need to be adopted and adapted to meet where the organization is in its readiness for change and with the organization's culture. Like all theories of change, its use does not guarantee success, but it can help plan change management process in an explicit and systematic fashion.

TABLE 9.3
Steps Leading to Change

Step	Rationale
Create a sense of urgency	People will not change if they cannot see the need to do so
Create a guiding coalition	Assemble and support a group with energy and influence in the organization to lead the change
Develop a vision and strategy	Tell people why the change is needed and how it will be achieved. Change is more likely to happen if people know not just how but also why
Communicate the vision	Tell people, through multiple modes of communication and at every opportunity, about the why, what, and how of the changes
Empower broad-based action	Involve other people in the change effort. Help them see the advantages of change and how to achieve it rather than thinking about why they do not like the change and how to stop it
Make short-term wins	Seeing progress towards the larger vision and seeing that change is happening and having effects helps sustain involvement
Consolidate gains and produce more change	Building on successes of small changes invigorates people and helps them see themselves as agents of change
Solidify the changes in the organization's culture	Institutionalize the changes through advocacy and re-enforcement to prevent reversion to the old and comfortable ways of doing things

TABLE 9.4
Critical Questions to Answer When Promoting Organizational Change

Is the proposed change better than what it is meant to replace? (the criteria for judging better may differ from person-to-person and organization-to-organization)
Is the proposed change compatible with the organization's mandate, and does the organization have the capacity to change?
How easy will it be to implement the proposed change?
Can the change be tried and tested for being fit for the organization before it is fully implemented?
Will the change result in tangible results that are observable and easy to measure?

Source: Adapted from USDHHS, 2005.

The answers to some questions in Table 9.4 will affect whether change will be adopted and spread in an organization. Understanding who says what, in which ways, to whom, and with what effects as well as who are the formal and informal leaders in an organization can help tailor communications within an organization to disseminate the answers to these questions.

There are three essential elements needed to ensure an organization can act to promote health: (i) a mandate to act, (ii) a framework for action, and (iii) the capacity to act. It is worth remembering that it is people who change and not organizations. The concepts and theories described above for supporting individual change are also relevant to supporting community and organizational change.

Community Change

A community can be defined in geographical terms as a group connected by shared interests, as a community of collective identities, or as a group of interacting organisms that live in a shared habitat. Community structures and norms constrain individual health behaviours. The state, the market, the social environment, and the local environmental conditions establish the texture of daily life. Not all communities are the same, so there is no one-size-fits-all approach to community interventions. It takes time and effort to understand a community, especially if you are not a member of it. But that effort will help you build trust with the community as well as understand the community's strengths, challenges, and prevailing attitudes and opinions. Time, effort, and resources can be misdirected without understanding the underlying reasons behind the problem and the opportunities and willingness to change.

Community change happens when local people work together to transform the conditions and outcomes that matter to them. A key to community change is finding out what is important to people in the community and helping them reach their goals rather than organizing people to do something you think should be done. This principle underlies the harm reduction approach described in Chapter 6. Those wishing to enable community change need to understand the community

TABLE 9.5
Five Stages of Community Change Described by the U.S. Centers for Disease Control and Prevention (USCDC, 2018)

Stage	Features
Commitment	A coalition of community members is assembled and partnerships with other agencies are established to give participants ownership of the process and to create a pool of fiscal and human resources to support change strategies
Assessment	Data and input on what the community needs are gathered in a manner that gives the community a voice in the process. This stage helps to organize the community around the issue and can significantly influence program design
Planning	The community coalition works with key partners to collectively develop a plan to implement the change
Implementation	Stakeholders and partners in collaboration with the community team implement and maintain the change by securing commitment and ownership of the actions needed to lead to the desired change
Evaluation	Evaluation runs throughout the entire change process. It serves to determine if you are implementing the right strategies and if the desired impacts are being realized

context, be adept at inspiring and maintaining collaborative and participatory planning, and have the leadership skills to attract and sustain participation.

The Centers for Disease Control and Prevention in the United States described five phases in the process of community change (Table 9.5) (USCDC, 2018).

A good community action plan maps a clear course of action, including roles and responsibilities and specific outcome and output targets. There can be an unlimited number of possible deterrents and motivators for behaviour change within a community. The more that are included as targets for an action plan, the more complex and complicated it will be to achieve the plan's goals. Working with community partners can help identify those actions that are feasible within what we know and the resources at hand that are acceptable to the community and are technically feasible. While it is reasonable to include recommendations to make fundamental changes in knowledge, governance, or regulations, few of such changes come to fruition in a timely manner. Threats of harms can prevail and action on shared goals can be delayed if actions wait until new knowledge, regulations, or technology can be secured. Strategic collaborations are needed in community action plans to promote actions to reduce the more immediate harmful consequences of an activity through pragmatic, realistic, and low-threshold programs feasible under current conditions.

CHANGING PUBLIC POLICY

Policy is everywhere. A policy is simply a plan of what to do in a situation that has been agreed to by a group of people. There can be a household policy on screen time, a farm policy on milking hygiene, or a government policy on land use. Public policy is what a government does or does not do about a problem that is in the public interest. Policies can come as laws, regulations, incentives, programs, or services. Something must be a collective problem rather than a purely individual problem for it to make it onto the government policy agenda. Health promoters need to be aware of the social and political context of change so that the requested policy change can be framed in a manner that resonates with the departments, agencies, and politicians who can influence the policy agenda as well as with the communities being served (Beland and Katapally, 2018). Framing your desired change in a way that addresses the strategic priorities of government and community partners will have more success than simply asking them to change policy to address your priorities.

The Stages Model (Anon, 2003) describes a policy change process. It starts with agenda setting wherein the problem is seen to be of public interest. From a policymaker's perspective, a new policy is appropriate when (i) the community asks for it, (ii) an issue has reached crisis proportions, and/or (iii) there is a long-standing major issue with little progress (CCHD, 2020). To get on the policy agenda, the situation needs to be recognized as being problematic, proposed solutions should be available, and someone needs to engage with government to influence the situation. The next step is to draft the policy. At this stage, political power dynamics plays an influential role. Government actors see how the problems and solutions match their mandate while advocates for change try to get their priorities high profile. After the decision is made to implement the policy, success can be influenced by the type and complexity of the problem, the size and speed of the expected change, the human and financial resources available to implement the policy, and the government structures and regulations in place to enforce or encourage policy implementation. The Stages Model helps health promoters identify different steps in the public policymaking so that they can appropriately adapt information sharing, persuasion, and action strategies.

Because health is influenced by factors beyond health care, the creation of health promoting public policy depends on the collaboration with and among multiple government departments to achieve health goals. Such a collaboration is not easy to establish or continue. There are eight key elements that can lead to positive partnership: (i) a shared mission aligned to the partners' individual or institutional goals; (ii) a broad range of participation from diverse partners and a balance of human and financial resources; (iii) leadership that inspires trust, confidence, and inclusiveness; (iv) effective multi-way communication; (v) clear formal and informal roles; (vi) trust; (vii) attention to the political, economic, cultural, social, and organizational impacts; and (viii) evaluation for continuous improvement (Corbin, 2017).

HEALTH AND ENVIRONMENTAL LITERACY

Improving people's access to information and their capacity to use it effectively is the key to empowerment (Nutbeam, 2000). Health literacy is concerned with helping people obtain, process, and understand health information and services

TABLE 9.6
Some Influences on Different Aspects of Health or Environmental Literacy

Literacy Aspects	Influences
Ability to seek, find, and obtain health information	• Knowing how to access information • Impediments and opportunities to access information • Timing and medium for information delivery • Trustworthiness of information sources
Ability to comprehend the accessed information	• Perceived utility of the information • Relevance of outcomes and information to the individual • The use of plain language or jargon
Ability to interpret, filter, judge, and evaluate accessed information	• Complexity of the information • How cause-effect relations are perceived or understood • Ability to weigh different types of information
Ability to communicate and use the information to make a decision to act	• Comprehension of the information and its relevance to change • Knowing how to overcome impediments to action

Source: Adapted from Nutbeam, 2000.

needed to make appropriate decisions. Environmental literacy is similarly concerned with a person's understanding, skills, and motivation to make decisions that consider his or her relationships with natural systems, communities, and future generations. Effective health and environmental literacy involve more than ensuring people can access and read information. They also involve developing skills and attitudes that motivate people to seek and use knowledge and improve their self-efficacy and confidence. Many factors influence health literacy (Table 9.6). As with all the theories and models introduced in the chapter, interventions to promote health or environmental literature need to be tailored to the personal and situational factors in the populations of interest.

SUMMARY

Helping people make decisions that are good for themselves, their communities, their animals, and their ecosystems needs a systematic approach to developing their health and environmental literacy and creating social environments that will make change acceptable, feasible, and effective. Leading positive change requires us to understand what motivates and empowers people to move from the status quo to an unknown future. This chapter has only scratched the surface of the plethora of theories and models that have been developed to understand the process of change. Chapter 20 provides a case example of how these theories have been used to combat plastic pollution to benefit wildlife and communities in the Caribbean. Other case studies presented in this book implicitly or explicitly illustrate that to make change, we have to be more effective at helping people make good choices and doing so requires more than only giving them new information in the hopes they will do what we think is right.

REFERENCES

Anon. "Public Policy Models and Their Usefulness in Public Health: The Stages Model." *Quebec Institut National de Santé Publique*. 2003:11pp.

Appelbaum, Steven H., Sally Habashy, Jean-Luc Malo, and Hisham Shafiq. "Back to the Future: Revisiting Kotter's 1996 Change Model." *Journal of Management Development*, 31, no. 8 (2012): 764–782.

Bandura, Albert. "Health Promotion by Social Cognitive Means." *Health Education and Behaviour*, 31, no. 2 (2004): 143–164.

Beland, Daniel, and Tarun R. Katapally. "Shaping Policy Change in Population Health: Policy Entrepreneurs, Ideas, and Institutions." *International Journal of Health Policy and Management*, 7, no. 5 (2018): 369–373.

CCHD. "Section 3. Social Planning and Policy Change." Center for Community Health and Development at the University of Kansas, University of Kansas, 2020. https://ctb.ku.edu/en/table-of-contents/assessment/promotion-strategies/social-planning-policy-change/main (accessed April 2020).

Champion, Victoria L., and Celette Sugg Skinner. "The Health Belief Model," in *Health Behaviour and Health Education: Theory, Research, and Practice*, 4th Edition, eds. Karen Glanz, Barbara K. Rimer, and Kasisomayajula Viswanath (San Francisco: Jossey-Bass, 2008), 45–65.

Cohen, Geoffrey L., and David K. Sherman. "The Psychology of Change: Self-Affirmation and Social Psychological Intervention." *Annual Review of Psychology*, 65 (2014): 333–371.

Corbin, J. Hope. "Health Promotion, Partnership and Intersectoral Action." *Health Promotion International*, 32, no. 6 (2017): 923–929.

Dearing, James W. "Applying Diffusion of Innovation Theory to Intervention Development." *Research on Social Work Practice*, 19, no. 5 (2009): 503–518.

Gifford, Robert. "The Dragons of Inaction: Psychological Barriers that Limit Climate Change Mitigation and Adaptation." *American Psychologist*, 66, no. 4 (2011): 290.

Hoag, Bruce G., Hans V. Ritschard, and Cary L. Cooper. "Obstacles to Effective Organizational Change: The Underlying Reasons." *Leadership & Organization Development Journal*, 23, no. 1 (2002): 6–15.

Hambolu, Dupe, Jenny Freeman, and Henock B. Taddese. "Predictors of Bovine TB Risk Behaviour Amongst Meat Handlers in Nigeria: A Cross-Sectional Study Guided by the Health Belief Model." *PLoS One*, 8, no. 2 (2013): e.56091.

Kritsonis, Alicia. "Comparison of Change Theories." *International Journal of Scholarly Academic Intellectual Diversity*, 8, no. 1 (2005): 1–7.

Lindsay, James J., and Alan Strathman. "Predictors of Recycling Behavior: An Application of a Modified Health Belief Model 1." *Journal of Applied Social Psychology*, 27, no. 20 (1997): 1799–1823.

McLeod, Lynette J., Donald W. Hine, and Aaron B. Driver. "Change the Humans First: Principles for Improving the Management of Free-Roaming Cats." *Animals*, 9, no. 8 (2019): 555. doi:10.3390/ani9080555.

Montano, Daniel E., and Danuta Kasprzyk. "Theory of Reasoned Action, Theory of Planned Behavior, and the Integrated Behavioral Model." *Health Behavior: Theory, Research and Practice*, 4th Edition, eds. Karen Glanz, Barbara K. Rimer, and Kasisomayajula Viswanath (San Francisco: Jossey-Bass, 2008), 67–96.

Morowatisharifabad, Mohammad Ali, Mahdieh Momayyezi, and Mohammad Taghi Ghaneian. "Health Belief Model and Reasoned Action Theory in Predicting Water Saving Behaviors in Yazd, Iran." *Health Promotion Perspectives*, 2, no. 2 (2012): 136.

Nisbet, Elizabeth K.L., and Mary L. Gick. "Can Health Psychology Help the Planet? Applying Theory and Models of Health Behaviour to Environmental Actions." *Canadian Psychology/Psychologie canadienne*, 49, no. 4 (2008): 296–303.

Nutbeam, Don. "Health Literacy as a Public Health Goal: A Challenge for Contemporary Health Education and Communication Strategies into the 21st Century." *Health Promotion International*, 15, no. 3 (2000): 259–267.

Prochaska, James O., and Wayne F. Velicer. "The Transtheoretical Model of Health Behavior Change." *American Journal of Health Promotion*, 12, no. 1 (1997): 38–48.

Richens, I.F., Jonathan Houdmont, Wendela Wapenaar, Orla Shortall, Jasmeet Kaler, Hd O'Connor, and Marnie L. Brennan. "Application of Multiple Behaviour Change Models to Identify Determinants of Farmers' Biosecurity Attitudes and Behaviours." *Preventive Veterinary Medicine*, 155, (2018): 61–74.

Schneider, Robert J. "Theory of Routine Mode Choice Decisions: An Operational Framework to Increase Sustainable Transportation." *Transport Policy*, 25 (2013): 128–137.

Stephen, Craig. "The Pan-Canadian Approach to Wildlife Health." *The Canadian Veterinary Journal*, 60, no. 2 (2019): 145.

USDHHS. *Theory at a Glance: A Guide for Health Promotion Practice*, 2nd Edition (U.S. Department of Health and Human Services, National Institute of Health, 2005), 52.

USCDC. "Community Change Process and the CHANGE Tool." The United States Centers for Disease Control and Prevention, 2018. https://www.cdc.gov/nccdphp/dnpao/state-local-programs/change-tool/community-change-process.html (accessed April 2020).

Weinstein, Neil D., and Peter M. Sandman. "The Precaution Adoption Process Model and Its Application," in *Emerging Theories in Health Promotion Practice and Research*, eds. Ralph J. DiClemente, Richard A. Crosby, and Michelle C. Kegler (Jossey-Bass: San Francisco, 2002), 16–39.

Westley, Frances R., Ola Tjornbo, Lisen Schultz, Per Olsson, Carl Folke, Beatrice Crona, and Örjan Bodin. "A Theory of Transformative Agency in Linked Social-Ecological Systems." *Ecology and Society*, 18, no. 3 (2013): 27.

10 Expanding the Concept of Healthy Public Policy for Animals, Health, and Society

Craig Stephen

CONTENTS

INTRODUCTION

Healthy public policy is one of five foundational strategies for health promotion (see Chapter 4). It is different than health policy. Health policy includes decisions, plans, and actions undertaken to achieve specific health care goals. Healthy public policy recognizes that health is determined by more than health services. It pays attention to a wider range of economic, social, environmental, and political policies that influence environments that enable health to happen.

A general introduction to policy is needed before discussing healthy public policy. Policies guide actions. They can be rules, regulations, laws, principles, or guidelines. Governments have policies, as do companies and families. Policies describe what is to be done, who is to do it, how to do it, and for (or to) whom it is to be done. Policies can be made in response to a problem or to an opportunity. They set roles and responsibilities. Policies can be written (like a government regulation) or unwritten (like a family policy of no computer use after 9 pm). Public policy deals with governments' actions towards some issue. It helps establish the ideas and values that guide decisions. Policy objectives and principles are the basis for ensuing strategies, actions, and decisions.

There are three tiers to public policy: (i) laws created by a governing body; (ii) rules, guidelines, principles, or methods created by agencies with regulatory authority; and (iii) rules or practices established within an agency or organization. Public policies are influenced by political ideologies, social values, competing interests, and evidence. Good public policy leads to measurable, efficient, and positive solutions to public problems in a just way. A good public policy has few unintended side effects, and it is acceptable to the public it serves. Many One Health professionals undervalue, ignore, or misunderstand policy because their education empathized biomedical, ecological, or other social dimensions of health. But policies shape what individuals, groups, and societies do and are therefore essential One Health tools.

AN OVERVIEW OF HEALTHY PUBLIC POLICY AND HEALTH IN ALL POLICIES

Public policies have improved population health throughout history. Laws and regulations governing clean water and sanitation, food safety regulations, and labour laws, for example, made substantial impacts on public health. Policies on livestock housing and husbandry have had major impacts on animal disease control and animal welfare. Land use policies and natural resource use policies affect many wildlife determinants of health. The concept of healthy public policy was prominent in the World Health Organization's (WHO) Ottawa Charter for Health Promotion (WHO, 1986). The Charter noted that to achieve its goals, health promotion had to be concerned with opportunities and obstacles to creating healthy environment caused by policies in non-health sectors. The WHO 2010 Adelaide Statement on Health in All Policies (WHO, 2010) built on the Ottawa Charter. The aim of Health in All Policies is to improve health outcomes through collaboration between health practitioners and those non-traditional partners who influence the determinants of health. The Adelaide Statement noted that government objectives are best achieved when all sectors include health and well-being as a key component of policy development. An evolution of the Adelaide Statement to adapt it to a One Health context could be: "the fundamental aim of Health in All Policies is to improve health outcomes through collaboration between health practitioners and those non-traditional partners who have influence over vulnerability and adaptive capacity of people and animals inhabiting the same ecosystem."

Healthy public policy is by nature intersectoral because the health sector needs to collaborate to affect policy in other sectors. Healthy public policy requires governmental processes to coordinate policymaking across sectors through strategic plans that set out common goals, integrated responses, and shared accountability. This, in turn, requires institutionalized processes supporting and enabling cross-sector problem solving and redressing power imbalances. Careful attention to the roles and responsibilities of the various players influencing health determinants and outcomes is required. The Adelaide Statement noted that for agencies to work collaboratively with each other, civil society and the private sector, attention needed to be placed on integrated leadership, mandates, incentives, budgetary commitments, and sustainable mechanisms. These

TABLE 10.1
Conditions for Making the Health in All Policy Approach Work as Described in the Adelaide Statement on Health in All Policies (WHO, 2010)

There is a clear mandate making it imperative that the health sector systematically engages across government, with other sectors, with civil society and private interests to address the health and well-being dimensions of their activities
Cross-sector initiatives must build partnerships and trust through openness and full consultative and practical approaches
The interactions across sectors are systematically identified and considered
There are mechanisms to mediate across interests
There are explicit accountability, transparency, and participatory processes in place
Experimentation and innovation are encouraged to find new ways to integrate social, economic, and environmental goals

attributes and others (Table 10.1) mirror the conditions for successful One Health (Stephen and Stemshorn, 2016; Stephen, 2020).

Policy tools direct, manage, and shape behaviours through laws, regulations, strategic plans, standard-operating plans, and frameworks. The approaches described below summarize the U.S. Department of Disease Prevention and Health Promotion's cross-sectoral policy approach to promote fruit and vegetable access and intake (ODPHP, 2020). Removing taxes from fruit and vegetables and imposing them onto foods of minimal nutritional value incentivizes consumers to buy more fruit or vegetable. Local government licensing regulations on food purveyors can influence stocking practices to make sure fruits and vegetables are consistently available. Requiring food sellers to post calories of food helps people make more informed food choices. Local food safety policies that allow fruits or vegetables from home gardens to be donated to food banks increase food accessibility. Land use policies and agriculture subsidies influence the nature of local farming practices. Policies that reduce permit costs for farmers' markets can affect accessibility to local products. Monitoring and enforcement may be needed to guarantee compliance with the policy tools. This example illustrates how achieving policy goals (access to fruit and vegetables in this case) needs a mix of policy tools, often delivered by more than one level of government and/or more than one policy partner.

The healthy public policy and Health in All Policy approaches seem well suited to One Health and EcoHealth, but, to date, they have largely been used for human health promotion. Policy approaches for fish and wildlife tend to be restricted to environmental or natural resource departments, while livestock, poultry, and aquaculture health policies fall into agriculture or fisheries legislation. Animal health still largely falls under the purview of one sector. It is often the case that, even within one sector, the regulations and programmes intended to protect a species' health are separate from other programmes within the same agency that influences that species' determinants of health. Because One Health, EcoHealth, or kindred approaches require attention to a wider suite of mechanisms to promote and protect health, it is usually beyond the scope of one organization to fulfil all requirements for a comprehensive scope of activity, especially when both public and private interests intersect. An "all hands-on deck" perspective built on intersectoral partnerships is an essential attribute of effective healthy public policy leaders in the One Health world.

That health is influenced by a wide suite of policies outside of health policies is not a uniquely human situation. Policies explicitly dealing with salmon health in Canada, for example, are almost exclusively concerned with a subset of infectious and parasitic diseases, whereas the programs that influence the determinants of salmon health, like migration fidelity, clear water, fishing pressures and climate change, reside in other policies (Wittrock et al., 2019). However, principles of a healthy public policy approach for animal health or One Health are hard to find. This in part is due to the historic legislative and research focus on infectious and parasitic disease management and the usual role for private rather than public actors in designing and implementing domestic animal health management programmes. Despite this, principles for good One Health policies can be derived by integrating lessons from standard textbooks and publications on herd health and human population health (including Kahn et al., 2005; Wittrock et al., 2019; Bhattacharya and Bhatt, 2017; Ibrahim et al., 2001; Radostits et al., 1994; FAO, 2007; FAO, 1991) (Table 10.2). These principles serve as a foundation for developing and adapting healthy public policy and Health in All Policies for One Health.

HEALTHY POLICY FORMULATION AND IMPLEMENTATION

There is growing interest and debate about evidence-based policymaking. Evidence-based policy ideally translates into saying people use evidence to make policy. Two words in that last sentence cause some problems: use and evidence. Let us deal with evidence first. Few would argue that it would be wrong to base policy on irrefutable data that had been rigorously and systematically assessed for relevance, reliability, and effectiveness before paying the costs to implement the policy. But there are challenges to implementing an evidence-based approach to public policy. The first challenge comes with what is considered as evidence. In some circles, evidence is restricted to knowledge produced in accord with the standards of a relevant academic discipline. In other cases, evidence can be knowledge, skills, and practices developed by and sustained between generations within a community. Others consider evidence as what they observe and document using their own senses. An early lesson in the debates of evidence-based policy for me came during an environment impact assessment wherein a community group was incensed about the statement that "there was no evidence that this practice killed wildlife." This statement was based on a critical review of the scientific literature to determine if there were data to fulfil well-established postulates for causation that the practice resulted in the death of free-ranging wildlife. The postulates

TABLE 10.2
Five Principles and Their Implications for Developing Healthy Public Policy in One Health

Principle	Implication
Policies are outcome based	There must be a clear definition of the preferred health outcomes across species, how they are measured, and thresholds of acceptability in order to assess programme effectiveness
	There must be clarity on the population(s) being considered and the ecological or social expectations that determine the appropriate outcomes
Policies are evidence based	There must be a standard acceptable to partners on the nature of biological, ecological, and social evidence to consider when making decisions and how to accommodate varying levels of reliability, validity, and representativeness of that evidence
	Health management priorities need to be identified through health needs assessments that consider the health of the interacting species influenced or influencing the issue of concern and input from affected communities and stakeholders to align organizational strategic planning and the management
Policies emphasize all levels of prevention consistent with the population health model involving the determinants of health	Policies need to be attentive to how to prevent long-term harm or premature losses after events, harmful events resulting in health impacts, risk factors leading to harmful events, and other populations from becoming at risk, as well as consider how the needs for prevention vary across species
	Policies require a variety of mechanisms to achieve outcomes, including directives, plans, and courses of action that may be required by law or proffered for implementation of norms and practices
Policies must be adaptable to heterogeneity resulting from different ecological, biological, or social conditions	There must be a balance between standardization and customization of interventions and how policy actions will vary with species and context
Policies must be attentive to unanticipated or secondary consequences on subsections of the population(s) of concern and other ecological, biological, or social attributes impacted by the policies	Policies and their associated actions should not impede the health status or access to health determinants for the entire population nor those populations unintentionally or intentionally impacted by health management decisions
	The evaluation of health policy includes both normative and empirical inquiries that may be applied to all its component considerations

could not be fulfilled, partly due to data speaking against the association, partly due to conflicting data, and partly due to data gaps. The community group, on the other hand, rejected this conclusion because "Billy saw it happen." The lesson that different people accept different types and standards of evidence upon which to make policy decisions was made crystal clear in this case.

The second concern with the statement "people use evidence to make policy" comes from the word use. Do decision-makers use all available evidence in an objective and balanced way, or do they only select evidence that supports their view or distort evidence with arbitrary safety factors to suit the needs? How do political agendas, economic motivators, or cultural values influence people's willingness to use the various forms of evidence available to them, without prejudice or favour? Allegations of technical and political bias in the production and use of evidence in policymaking needs to be balanced with the good intentions of those who seek to take ideology and politics out of the policy process by regularly and systematically bringing research and knowledge into the policymaking process. The key take away message is that evidence is increasingly playing a central and influential role in policymaking, but policymakers use other types of information besides research evidence. What is regarded as evidence and how it is used will depend on the nature of the policy and the political situation.

Strehlenert et al. (2015) presented a four-stage process for evidence-informed public policy formulation and implementation. Their model mirrors the typical policy process. It starts with Agenda Setting. In this stage, the "problem" finds its way onto the formal policy agenda. A lot needs to happen before this stage is initiated. A problematic situation needs to be identified by a person or group. An understanding of why the situation is problematic and some options for what should be done need to be in place. There next needs to be some activities by a proponent for a policy change to get government engaged, helping them to see they need to be engaged, either alone or with other groups. Some problems will inevitably not make it onto the government agenda due to competing interests, lack of time, competition with other problems, or lack of an apparent government role.

Once a problem is on the government's agenda, the next stage is Policy Formulation when possible actions are proposed, debated, and drafted. Policy goals and outcomes along with indicators by which they will be achieved and measured are developed at this stage. Evidence, dialogue, and advocacy not only help get a problem on the policy agenda but also help set realistic and reliable goals and indicators during the policy formulation stage. After a policy is adapted, a plan must be made for the next stage, Implementation. Stakeholders' views, expert and professional opinions, values, traditions, pressure groups, and the pragmatic realities of the situation policymakers face will influence how or if a policy is implemented. Target audiences must be made aware of the new policy and outreach and support provided to ensure its uptake. Resources and partnerships must be secured and adapted to enable implementation under realistic and sometime heterogeneous settings and circumstances.

The next stage of policy formulation is Maintenance. This is the stage where the policy tools and actions become part of normal business. Incentives, disincentives, education, outreach, and other approaches are needed to maintain these behaviours. Chapters 9, 12, and 20 talk more about methods and approaches to promoting and maintaining behaviour changes. The final stage in policymaking is Evaluation. This is a critical but sometimes forgotten stage of the policymaking process. Evaluation is needed to see if the policy's effects are aligned with the policy objectives and if any unanticipated or unintended effects arose. It should examine if the policy is still relevant and useful. The results of the policy evaluation should cycle back to the Agenda Setting stage when the evaluation finds a need to amend or discontinue a policy, starting a new policy cycle.

SEEKING WIN-WIN-WIN SOLUTIONS FOR HEALTHY PUBLIC POLICY

There are many examples of a policy success in one of the three domains of One Health that resulted in negative or unanticipated effects in the other (Table 10.3 provides some examples). The push for affluence as a major goal of human development has been one of the main drivers of the declining quality, quantity, and sustainable of nature (Dietz et al., 2007). Destruction of mangrove swamps in the global rush to develop jobs, revenue, and food from shrimp farming (Martinez-Alier, 2001) and deformities and deaths in wildlife due to agricultural chemicals (Köhler and Triebskorn, 2013) are two examples that quickly come to mind when contemplating the many ways where good intentions in one sector lead to tremendous harms in another. When policymakers stay sharply focused on the goals and needs of their own issues, without consideration of the implications of a policy change for other parts of their own sector or broader impacts outside of their sector, unanticipated and unintended consequences can follow. Managers have three options when faced with conflicting goals: (i) manage one goal and accept the collateral damage, (ii) abandon management of their goal and accept its impacts, and (iii) seek a strategy that allows both goals to be attained (Buckley and Han, 2014).

Fortunately, there is a growing case load of win-win and even win-win-win outcomes between people, animals, and ecosystems. The United Nations 2030 Agenda for Sustainable Development produced 17 interrelated and interconnected goals (Figure 10.1). These goals recognize the dependencies between resilient biodiversity, healthy ecosystems, and human well-being. Although the ideal of collaborative, concurrent, and equitable delivery on all 17 goals is yet to be fully realized and implemented, the creation of these goals communicated the need for a reciprocal care approach to policymakers and heightened awareness on the necessity of coordinate action around the world. The UN's Convention on Biological Diversity goals of conserving biological diversity, the sustainable use of its components, and the fair and equitable sharing of its benefits strives to manage health of people, animals, and ecosystems as an integrated

TABLE 10.3
Win-Lose Examples of Public Policy Outcomes Benefiting One Domain of One Health While Harming Another

The Issue	Winners	Losers
Waste management and pathogen translocation (Chipman et al., 2008)	One state imports solid waste from surrounding American states to alleviate waste disposal limitations in the region and to generate local revenue	Seven of the states exporting waste to Ohio had endemic racoon rabies and raccoons were translocated long distances in waste disposal vehicles, risking significant impacts on the local rabies situation in Ohio
New conservation opportunities and impacts on local communities (Bocarejo and Ojeda, 2016)	Creation of a national park in Columbia provided conservation benefit and new ecotourism revenue	Peasants living in the park were categorized as illegal occupants and invaders, leading to eviction from the park and loss of crops
Recommendations in economically developed countries to eat more fatty fish and secondary public health and conservation impacts (Brunner et al., 2009)	The low-saturated fatty acids, selenium, and long-chain omega-3 fatty acids in some fatty fish species provide heart health benefits to people	Fatty fish are also a source of methylmercury and polychlorinated biphenyls, which are a health risk. Increased demands for fish protein put pressures on fish stocks and limited access to fish for poorer people
Early warning in response to an emerging infection and impacts on animal hosts (Wetli, 2020)	New genetic methods allow rapid identification of potential sources of new zoonotic diseases, allowing prompt isolation measures to reduce public exposure	The early association of pangolins as possible sources of the COVID-19 virus caused conservation concerns that a fearful public would kill this species, further pushing these threatened animals towards extinction
Global fisheries and emerging zoonotic infections (Khan and Sesay, 2015)	West Africa fish stocks generate income and employment for African and European companies	Food insecurity increased as fish were sent to external rather than local markets. This drew people to wildlife as sources of protein, which increases opportunities for exposure to pathogens like Ebola virus

whole. Chapter 19 illustrates this approach through a case study wherein local efforts of a small group of dedicated individuals protected the habitat of uncharismatic, yet endangered species, which in turn promoted better climate change planning for water security, new recreational opportunities, and sustained habitat protection that benefited a wide suite of wild species and people. The explicit reference to sustainability in food guides produced by Germany, Brazil, Sweden, the United Kingdom, and Qatar help consumers make choices

FIGURE 10.1 The 17 UN sustainable development goals. (From https://www.un.org/sustainabledevelopment/sustainable-development-goals/. The content of this publication has not been approved by the United Nations and does not reflect the views of the United Nations or its officials or Member States.)

that can decrease their ecological footprint, climate impacts, and effects on biodiversity when they make their food choices (Lee et al., 2017). Making health and well-being the motivation for responding to the climate emergency has the potential to build social consensus for climate action that would benefit people, animals, ecosystems, and economies (Comeau, 2019). Urban forests and green spaces can help offset wildlife habitat loss while at the same time provide cooling, storm water, and pollution control services to a city's human inhabitants (Livesley et al., 2016).

Despite the increasing research and social preference for win-win-win solutions, they are hard to develop and harder to implement. It is rarely someone's job to worry about what happens outside of the purview of their policy domain. Policy planning tends to deal with individual problems assigned to organizations with narrowly defined responsibilities. Implementing Health in All Policies is like implementing many, if not all, intersectoral actions (Molnar et al., 2016). Elements of success include increasing awareness and confidence that the policy actions are acceptable and feasible, developing a shared language to facilitate communication between different sectors, recognizing and communicating the costs and benefits to all three of the One Health domains, using multiple types of evidence to give confidence in the need for and effectiveness of the win-win-win approach, and using systematic and integrated assessments to give credibility to policies being collaboratively developed by diverse policy sectors. The development of a language shared between sectors and embedding multiple outcomes into projects can facilitate conversations to find synergies and previously unrecognized opportunities (Kokkinen et al., 2019). Five guiding questions can help those leading efforts to promote

TABLE 10.4
Five Guiding Questions to Help Find Win-Win-Win Solutions

What are the main goals of the different interests and parties who are debating or can be affected by the different policy options?
What are the alternatives for action for the groups dealing with the problem(s)?
What are the relationships between the various alternatives and goals?
Are there alternatives that when implemented collaboratively and cooperatively can result in better outcomes than if the alternatives were implemented in isolation? (i.e. the win-win-win alternative)
Is the proposed win-win-win alternative feasible and acceptable given the existing circumstances?

win-win-win solutions facilitate and direct these conversations (Table 10.4 adapted from Nagel, 2000). When conversations fail to find solution, conflict management comes into play.

UNINTENDED POLICY CONSEQUENCES

How can one develop a policy for something previously thought to be impossible or unimaginable? In some cases, we might have certainty about a threat but uncertainty about its impact. In other cases, we lack knowledge of the possibility of the unimagined or unintended threat. Multiply these uncertainties across numerous species sharing the same landscape, and the likelihood of not thinking about an unintended consequences seems inevitable.

The first step in avoiding harmful unintended or unanticipated consequences is to be alert to their possibility. Increased understanding of the origins, impacts, and consequences of a policy decisions on non-policy targets requires an understanding of the influences of policy actions on the determinants of health of the non-targets actors before policies are implemented. The second step is to be open to the signals and clues from outside of your sector.

There are three common scenarios wherein the relationships and attitudes needed for you to be aware of and open to signals outside of your sector breakdown: (i) bureaucratic conflicts and inadequate protocols make your organization unreceptive to warning signals outside of its usual scope of practice, leading to the breakdown of cross-sectoral communication; (ii) politics and overcrowded agendas discourage collaborative actions on determinants of vulnerability that extent beyond immediate interests; and (iii) insensitivity to new information, perceived power dynamics, cognitive overload, and wishful thinking lead to failure to recognize and act on very early warning signals (Stephen, 2019). Overconfidence in big science and artificial intelligence to provide warning early enough and convincing enough to result in action creates additional vulnerabilities when trying to prepare for a surprising and rapidly changing world. Human intelligence is still needed to link surveillance of threats and outcomes with reconnaissance of population relationships and vulnerabilities to recognize situations prone to harm.

An unanticipated event occurs when there is a gap between one's expectations about what is plausible and what occurs. Experience shapes expectations, and incompletely informed expectations produce surprises. New forms of cross-sectoral co-learning to better anticipate threats and consequences require people embedded within a health agency to facilitate transdisciplinary intelligence gathering by promoting collaboration and knowledge exchange across fields and between knowledge generators and knowledge users. This will allow for better anticipation of consequences across and among populations and species and the implementation of rapid responses.

No person or organization can anticipate and prepare for all unintended and unimaginable outcomes of a policy decision. The best way to cope with an unanticipated event is to improve coping mechanisms, either in preparation for surprise or for a response to the surprise once it occurs (Streets and Glantz, 2000). Mounting evidence recognizes the role for healthy ecosystems and resilient communities for risk reduction, recovery, and resilience to unanticipated threats. The ecological services provided by nature are critical for community resilience. Future One Health leaders will require broad cross-cultural competencies to be build collaborative policies between multiple actors at multiple administrative levels to allow different aspects of a problem to be seen and managed. By exploring these differences, impacts and actions that go beyond traditional health perspectives may be found. The likelihood of innovative policy responses will increase as leaders diversify their professional network and in doing so increase their exposure to non-redundant information, skills, and support (Tortoriello et al., 2015).

REALIZING RECIPROCAL CARE HEALTHY PUBLIC POLICY

Policies shape the social environments that allow and enable people to make win-win-win choices. But policies only shape the intention for action. They only make a difference if they are effectively implemented. Policies must be turned into actions that influence how we act and the decisions we make. If the ideals of One Health, EcoHealth, or related fields are to be realized, there needs to be a shift from making polices in isolation to a policy agenda that explicitly looks across sectors to find opportunities and efficiencies and avoid obstacles to action that build health for animals, people, and ecosystems. Rare is the case, and many are the challenges to having one sector win without another sector losing something. But by being attentive to the implications and opportunities for policies to promote actions that create benefits beyond an individual sector, less harms and more wins will be realized. The long-term goal of creating public policy for the reciprocal care of human, animal, and environmental health is dependent on sustained policy advocacy for a new view of healthy public policy.

REFERENCES

Bhattacharya, Dru, and Jay Bhatt. "Seven Foundational Principles of Population Health Policy." *Population Health Management*, 20, no. 5 (2017): 383–388.

Brunner, Eric J., Peter J.S. Jones, Sharon Friel, and Mel Bartley. "Fish, Human Health and Marine Ecosystem Health: Policies in Collision." *International Journal of Epidemiology*, 38, no. 1 (2009): 93–100.

Bocarejo, Diana, and Diana Ojeda. "Violence and Conservation: Beyond Unintended Consequences and Unfortunate Coincidences." *Geoforum*, 69 (2016): 176–183.

Buckley, Yvonne M., and Yi Han. "Managing the Side Effects of Invasion Control." *Science*, 344, no. 6187 (2014): 975–976.

Chipman, R., D. Slate, Christine Rupprecht, and M. Mendoza. "Downside Risk of Wildlife Translocation," in *Towards the Elimination of Rabies in Eurasia*, eds. B. Dodet, A.R. Fooks, T. Muller, N. Tordo, and the Scientific and Technical Department of the OIE (Developmental Biology 131, 2008), 223–232.

Comeau, Louise. "Protecting Public Health Should Drive Climate Policy." Policy Options Politique, 2019. https://policyoptions.irpp.org/magazines/august-2019/protecting-public-health-should-drive-climate-policy/ (accessed May 2020).

Dietz, Thomas, Eugene A. Rosa, and Richard York. "Driving the Human Ecological Footprint." *Frontiers in Ecology and the Environment*, 5, no. 1 (2007): 13–18.

FAO. "Guidelines for Strengthening Animal Health Services in Developing Countries." Food and Agriculture Organization, 1991. http://www.fao.org/3/u2200e/u2200e00.htm#Contents (accessed May 2020).

FAO. "Aquaculture Development. Health Management for Responsible Live Movements of Live Aquatic Animals." *FAO Technical Guidelines for Responsible Fisheries Series*, 5, Suppl. 2 (2007): 31.

Ibrahim, Michael A., Lucy A. Savitz, Timothy S. Carey, and Edward H. Wagner. "Population-Based Health Principles in Medical and Public Health Practice." *Journal of Public Health Management and Practice*, 7, no. 3 (2001): 75–81.

Kahn, Cynthia M., Scott Line, and S. E. Aiello. *The Merck Veterinary Manual* (Whitehouse Station, NJ: Merck & Co. Inc., 2005).

Khan, Ahmed S., and Sanie SS Sesay. "Seafood Insecurity, Bush Meat Consumption, and Public Health Emergency in West Africa: Did We Miss the Early Warning Signs of an Ebola Epidemic?" *Maritime Studies*, 14, no. 1 (2015): 1–12.

Köhler, Heinz-R., and Rita Triebskorn. "Wildlife Ecotoxicology of Pesticides: Can We Track Effects to the Population Level and Beyond?" *Science*, 341, no. 6147 (2013): 759–765.

Kokkinen, Lauri, Alix Freiler, Carles Muntaner, and Ketan Shankardass. "How and Why Do Win–Win Strategies Work in Engaging Policy-Makers to Implement Health in All Policies? A Multiple-Case Study of Six State-and National-Level Governments." *Health Research Policy and Systems*, 17, no. 1 (2019): 1–11.

Lee, Angela, Heather McLeod-Kilmurray, and Nathalie Chalifour. "Food Choices Govern Food Production, Which Has Environmental Impacts in a Time of Global Climate Change." Policy Options Politique, 2017. https://policyoptions.irpp.org/magazines/january-2017/canadas-food-guide-update-needs-to-address-sustainability/ (accessed May 2020).

Livesley, S. J., E. G. McPherson, and C. Calfapietra. "The Urban Forest and Ecosystem Services: Impacts on Urban Water, Heat, and Pollution Cycles at the Tree, Street, and City Scale." *Journal of Environmental Quality*, 45, no. 1 (2016): 119–124.

Martinez-Alier, Joan. "Ecological Conflicts and Valuation: Mangroves Versus Shrimps in the Late 1990s." *Environment and Planning C: Government and Policy*, 19, no. 5 (2001): 713–728.

Molnar, Agnes, Emilie Renahy, Patricia O'Campo, Carles Muntaner, Alix Freiler, and Ketan Shankardass. "Using Win-Win Strategies to Implement Health in All Policies: A Cross-Case Analysis." *PLoS One*, 11, no. 2 (2016): e0147003.

Nagel, Stuart. "Win-Win Policy for Developing and Industrial Nations: Six Examples." *International Journal of Urban Sciences*, 4, no. 2 (2000): 139–151.

ODPHP. "Law and Health Policy: Policy Levers Graphic." Office of Disease Prevention and Health Promotion, U.S. Department of Health and Human Services, 2020. https://www.healthypeople.gov/2020/law-and-health-policy-policy-levers-graphic.

Radostits, Otto M., K.E. Leslie, and J. Fetrow. *Herd Health: Food Animal Production Medicine,* 2nd Edition (Philadelphia: W.B. Saunders, 1994), 648.

Stephen, Craig. "Rethinking Pandemic Preparedness in the Anthropocene." *Healthcare Management Forum* (Los Angeles: SAGE Publications, 2019), 0840470419867347.

Stephen, Craig. "The Practice of One Health: Lessons Learned," in *One Health: The Theory and Practice of Integrated Health Approaches*, 2nd Edition, eds. Jakob Zinsstag, Esther Schelling, Lisa Crump, Maxine Whittaker, Marcel Tanner, and Craig Stephen (Oxfordshire: CABI, 2020).

Stephen, Craig, and Barry Stemshorn. "Leadership, Governance and Partnerships Are Essential One Health Competencies." *One Health*, 2 (2016): 161–163.

Streets, David G., and Michael H. Glantz. "Exploring the Concept of Climate Surprise." *Global Environmental Change*, 10, no. 2 (2000): 97–107.

Strehlenert, Helena, Linda Richter-Sundberg, Monica Elisabeth Nyström, and Henna Hasson. "Evidence-Informed Policy Formulation and Implementation: A Comparative Case Study of Two National Policies for Improving Health and Social Care in Sweden." *Implementation Science*, 10, no. 1 (2015): 169–179.

Tortoriello, Marco, Bill McEvily, and David Krackhardt. "Being a Catalyst of Innovation: The Role of Knowledge Diversity and Network Closure." *Organization Science*, 26, no. 2 (2015): 423–438.

Wetli, Patty. "If Pangolins Are Linked to Coronavirus, Conservationists Fear for Animal's Safety." Window to the World (WTTW) News, 2020. https://news.wttw.com/2020/02/12/if-pangolins-are-linked-coronavirus-conservationists-fear-animal-s-safety.

WHO. "The Ottawa Charter for Health Promotion." World Health Organization, 1986. https://www.who.int/healthpromotion/conferences/previous/ottawa/en/

WHO. "Adelaide Statement on Health in all Policies Moving Towards a Shared Governance for Health and Well-Being." World Health Organization, 2010. https://www.who.int/social_determinants/hiap_statement_who_sa_final.pdf

Wittrock, Julie, Michele Anholt, Michael Lee, and Craig Stephen. "Is Fisheries and Oceans Canada Policy Receptive to a New Pacific Salmon Health Perspective?" *FACETS*, 4, no. 1 (2019): 615–625.

11 Practical Approaches to Leadership and One Health

David Butler-Jones

CONTENTS

Many examples of leadership over the centuries have drawn from the military, political, and business realms where a strong leader has some ability to compel others to act in a certain way, to work on a particular product or project, or even to go to war. This approach has often carried over into other areas as we seek strong leadership with the answers we perceive we need. Even in democracies, there is a tendency for political leaders to want to control the actions and messages of others, both within and outside their political party. The power to withhold power and influence from others enforces conformity. The risk adverse may seek to avoid responsibility and then blame others, rather than finding needed solutions.

An increasing challenge is that most of our "wicked problems" for which no one seems to have a specific answer are incredibly complex and often occur at the margins or interface of different sectors, professions, and organizations. One Health concepts and approaches can help us to not lose focus on possible solutions even though they may not be predictable nor within our individual mandates or capabilities. One Health leadership brings collective knowledge and expertise to shared problems in animal and human health, the environment, and the economy. It requires skills in influence, not control; cooperation and collaboration, not compulsion; and building coalitions across disciplines and sectors, not bureaucratic or academic empires. It is a successful leadership approach less familiar to most of us working to plan for and respond to the next pandemic, mitigate and adapt to climate change, or address other wicked problems. It is the combined skills, talents, and resources of the many partners that allow us to have an outcome such as Canada's success in the 2009 H1N1 influenza pandemic, rather than repeating the errors of severe acute respiratory syndrome (SARS 2003). The what and how of leadership matters.

THE CASE FOR LEADERSHIP

Leaders tend to rise to a situation. Rarely is one leadership style good in all situations. There are many academic and popular press books, articles, and news items on what makes someone a good leader. Winston Churchill was the right leader to bring Britain through World War II, but it did not take long in peacetime for him to be ousted. Abraham Lincoln was the right person for the demanding job of United States president trying to keep the Union together. Sadly, he was assassinated, and we will never know if that same support would have persisted in the period of reconstruction. Leadership styles following World War I demanded reparations and harsh treatment of the Axis powers, which then influenced the conditions that brought about World War II. Post–World War II leadership created different conditions such that former enemies are now allies.

This chapter is not about the kind of leadership brought about by force, nor what we may think of typically in a military, bureaucratic, or corporate management sense, even though its approaches can enhance effectiveness in those settings. It is about the challenge of leadership in complex situations where others cannot be compelled to work together or share resources, yet where collaboration and sharing are necessary to achieve common ends.

In modern democracies, in the age of the Internet, it is virtually impossible to control information or the message completely as censors have tried in the past. Different political jurisdictions and even departments within those governments can easily find ways to avoid or subvert involvement if they wish or see it in their interest to do so. They must be convinced that not only is it the right thing to do, but that they too can benefit, or at least not lose power, influence, or resources; and the information or story must be both factual and compelling when exposed to public scrutiny. Maturity and experience can help to discern the noise from legitimate criticism, and a measure of humility to accept the importance of changing course or ideas if necessary.

One of the things I tell managers is that it is more important to get it right than to be right; and to surround yourself with smart people willing to tell you what you need to hear, not what you want to hear. Then, you can seek the best ideas and understandings both inside and outside the organization to deal with whatever decisions need be made. The risk is otherwise missing or suppressing solutions, insecurity and poor decisions, or worse, not making intentional decisions for fear of error, which in itself becomes a decision as events carry on.

WHAT DO WE MEAN BY LEADERSHIP?

One can find many a wide variety of lists of the necessary qualities for effective leadership.

BOX 11.1 NECESSARY QUALITIES FOR EFFECTIVE LEADERSHIP

Flexible	Adaptive	Responsive
Creative	Eclectic	Clear
Honest	Respectful	Sensible
Humble	Inclusive	Transparent
Concise	Accountable	

Most would include some or all of what is in Box 11.1. Most readers would recognize these as admirable qualities. The challenge is how to consistently apply them in an understandable way that others can buy into and perhaps emulate.

Two historic leadership examples that have relevance to One Health come out of experiences addressing two national crises in the United States: (i) Abraham Lincoln's style of leadership which exhibited pragmatic approaches to dealing with complicated issues and (ii) the early 21st century concept of Meta-Leadership following the 9/11 attack on the World Trade Center in New York.

Some Lincoln Lessons

Lincoln was the president of the United States just prior to and during the American Civil War. He demonstrated much of what it takes to lead in a complex system. He was certain when a decision needed to be made, to do so. He surrounded himself and kept in his cabinet those who thought differently than himself, including some of his political rivals, in essence "smart people willing to tell you what you need to hear, not just what you want to hear." He would delegate to others and expect them to do their best but not without having clear conversations on what was needed and expected. He recognized his ultimate responsibility and would accept it even when the decision that was in error was not his. His focus was on finding a solution, and he would not throw colleagues under the proverbial bus. Lincoln managed to find principled compromise where necessary and used science and evidence as essential to decision-making. He was known as Honest Abe, maintaining his integrity in all things and not lying to avoid responsibility. He also was strategic in finding the right time to push an agenda, working towards a yes, rather than demanding an answer when it almost certainly would be a no. It allowed him to ultimately eliminate slavery, even though many would have hoped he did so earlier, and while that may have been possible, it appeared at the time to be very unlikely to be successful and risked further negative entrenchment.

Meta-Leadership: A Response to Crises

Following the attack on the World Trade Center, it became apparent that some of the national security structures in the United States operated too often in silos, as key information was not shared across departments that may have pointed to proactive opportunities to intervene or respond quickly to the terrorist attack. As a result of the 9/11 experience, the Harvard School of Public Health and the Kennedy School began developing Meta-Leadership as a way of training and working in a way more conducive to work across sectors and departments. While the focus was largely on national security and emergency preparedness and response, their principles and approach have much in common with what is needed in other, important, complicated challenges, or wicked problems. They focused on three dimensions: the person, the situation, and connectivity across both vertical dimensions of hierarchy and horizontally to peers and external partners. It is an approach to breaking down barriers to collaborative responses.

SEVEN LESSONS FROM FAILED LEADERSHIP

There is much to be learned from failed leadership.

1. *Loose thinking is not the same as freedom of thought.*
   One Health leadership is about a well thought out reasoned position and not necessarily just the first thoughts that come to mind. It is about both validating and challenging ideas with others.
2. *We can spend too much time looking for certainty in decision-making.*
   There is an old expression "Do not let the best be the enemy of the good." By postponing a decision because we are uncomfortable making it, we risk the decision being made for us or events overtaking us. It is one thing to delay for a specific purpose, such as getting another perspective or evidence; however, at a certain point, the risks of not deciding will far outweigh the risks of making the best decision under the circumstances. For example, debates during the 2009 H1 avian influenza pandemic over the ideal guidelines for schools continued up until it was almost time for students to return to school, which risked putting them at greater risk due to a lack of prevention and care guidance.
3. *Some insist on wanting to win every battle. However, the risk is that you may end up losing the ultimate goal.*
   It can be strategically valuable to compromise or give in on some less critical areas to gain what is essential. If one is always seen as inflexible, others may be unwilling to collaborate at critical times. What is most critical is to not lose sight of what matters most to partners, so that they too can see benefit in the collaboration.
4. *Some leaders are unwilling to admit mistakes, and as soon as they receive a promotion, they are already planning towards their next move.*

It is important to stay in a position long enough to see "your own bombs go off," to learn from them and to fix them. Moving on too quickly leaves the mistakes for those who follow, who then gain the learnings and may resent your poor decisions, as you progress to making more mistakes, but higher levels of impact. I believe that as soon as it is about me, rather than the organization or purpose, I am no longer able to lead as effectively. One should not ignore one's own needs, but when making a critical choice, it is important to remember there is something more important than oneself. Otherwise it is time for someone else to lead.

5. *The rush to follow the decision sequence of "It's an important problem, we must do something, this is something, let's do this" can lead to erroneous decisions.*
   For example, during the SARS outbreak, it was decided to set up temperature scanners at some airports. This hasty decision failed to take all aspects of the decision into account, such as the following: (i) affected individuals are not infectious until extremely ill, and unlikely to be mobile; (ii) there are many common causes of fever and taking something to suppress fever could bypass the screening; and (iii) individuals who develop a fever later but do not understand the natural history of the disease may assume that as they were screened; it must not be SARS and therefore may not take appropriate action.

   Much better would have been to train border services on who to refer to the quarantine officers, and to provide information to passengers as to what to do should they develop the appropriate symptoms later, as was done during the H1 influenza pandemic in Canada. Such an approach would have covered a range of possibilities not just current fever with many false positives, few if any true positives and many false negatives.
6. *A particularly frustrating pattern of behaviour in organizations is, "we cannot solve a real problem, because there might be a small side effect ... which then will be ours. However, if we don't solve it, it wasn't our problem in the first place."*
   Nothing is completely without risk and not taking even a calculated risk to improve the situation means not succeeding. For example, some governments' failure to act on their public health responsibility of harm reduction in response to the opioid crisis resulted in needless premature deaths.
7. *There are those who in effect say, "We won't do anything, but we'll be damned if we let anyone else."*
   They would rarely say it that way but in effect they choose not to take on or at least partially address an issue; but they then expend more energy fighting others from working on an issue they perceive as their mandate than it would have taken to actually do it.

NINE PRACTICAL ONE HEALTH LEADERSHIP SKILLS

1. *Leading with*
   In dealing with complex situations where collaboration is essential, and cooperation cannot be guaranteed, it is important for leaders to assume

a non-authoritarian style of leadership that understands their and other's responsibilities, roles, and perspectives. As it is in effect a team with many coaches and quarterbacks (depending on the analogy you would like to use), open and transparent information sharing appropriate to the situation is essential. For example, in Canada's response to the 2009 H1 influenza pandemic, provincial and territorial ministers, deputy ministers, and senior public health officials would be advised of what the federal public health agency knew shortly after it was known. No one was told what to say in terms of public messaging, but the key issues were discussed and information shared such that a consensus of best advice and information could be made public. This illustrates the value of intelligent people with accurate information and a common purpose when striving for a reasonable consensus.

2. *Getting it right*
   It is less important "to be right" than it is "to get it right." It is impossible for any single individual to have all the knowledge and experience to consistently make the best decisions. Even organizations may not have within them all the necessary skill sets. The focus must not be on one's ability to be right but rather to seek out the needed expertise or other perspectives whether from within or outside the organization, to get it right.
3. *Framing issues*
   It is helpful to be able to understandably frame issues, including the underlying principles on which decisions are based. For example, given the challenges in dealing with complex issues such as climate change, ecosystem preservation, or the determinants of health, it is easy for people to say they cannot do anything meaningful to make things better. What practically can an individual or organization do? One way I found helpful is using the acronym PACEM for different possible types of action.
   - **P**artner: Finding others of similar interest to work together on those things we can do better together than individually.
   - **A**dvocate: It is not necessarily protesting in the streets, or writing letters, but more so bringing evidence to bear on the debates and discussions.
   - **C**heerlead: This may sound trite, but at times it has been felt that for various reasons we are criticizing ourselves more than organizing a common response to address the major issues. For example, in public health, the shifting terminology from health education to health promotion to population health became more a debate about how one concept is superior to the others, rather than how each has something to contribute to our overall understanding.
   - **E**nabling: What are the things we can do within our own organization or programmes to make them more compatible with the principles we espouse, such as providing service accessibly in the affected community rather than a hospital, or ensuring our own practices have reduced negative environmental impact.

- **M**itigation: This is addressing those things that reduce the impacts of the risk. For example, in northern Saskatchewan, Canada, the problems of poor water quality, sanitation, and overcrowding in housing meant periodic outbreaks of Hepatitis A. While working towards addressing these underlying risks, the use of Hepatitis A vaccine for children and youth in affected communities mitigated the outbreak by lowering the rate of Hepatitis A in children and youth to less than the general population.

4. *Sometimes it is critical to see for yourself*
Getting to know the communities, populations, or systems you are responsible for or involved with can reveal important things otherwise missed. Just reading others' reports or papers may not be enough. For example, a community outbreak of Giardia was most likely from what the community believed was safe protected spring water source that the community had used for decades. By visiting the site, it was clear that the spring water was actually surface waters coming through the side of the hill. In another case, a community stated that its water treatment plant was functioning normally. A site visit found that the filters were not working, and they mistakenly thought that raising the chlorine level would eliminate any risk, not knowing that *Cryptosporidium* spp. spores would be resistant to the chlorine. It is critical to talk to those affected to better understand what they view as the issues and what matters. For example, in Amazonia, the government had established a programme to encourage sustainable planting, and not burning, of the rainforest. Participating communities would receive additional funds. One community requested a paved path from the village to the river, which the government administrators thought odd and frivolous. However, when the governor asked the question of why, the community response was that they bathe in the river, but by the time they get back to the village they are dirty from the muddy paths.
5. *Building plans and relationships before a crisis*
Eisenhower would say: "Peace-time plans are of no particular value, but peace time planning is indispensable." The reality is that emergencies by their nature are full of surprises. Thus, it is important to plan for the knowable possibilities and build the relationships with needed partners in advance, so that time is not wasted in trying to build trust amid a crisis. Energy can then be focused on solving the unanticipated. Understanding the governance and organization of your emergency response team ahead of time can clarify roles and responsibilities and ensure a more coordinated and effective response. For example, it is wise to have one team focused on operations and another to focus on strategic aspects of the response. Establishing the role of collaborating agencies and how communications will occur within and across response teams is an essential pre-emergency activity. This can help avoid miscommunication and help each partner to better understand the other's situation.

6. *Think vertically and horizontally at the same time*
   Many of our most important issues or problems occur at the margins, or intersections that require multisectoral, multidisciplinary, and multijurisdictional approaches. Often, however, our assessments tend to be linear in trying to understand that x causes y, or we give up as the issues are beyond our individual control. In complex systems, such as nature and human interactions, the solutions are also more complex and rarely linear, let alone having a single cause. Variation and connectivity are hallmarks of biology, whether in genetics, immunity, exposure, dose, social structure, or other influencing factors and differences. When trying to understand causes in the hope to intervene, limiting our thinking to linear solutions risks serious untoward effects or unintended consequences. More details on complex systems and unintended consequences can be found in Chapter 13. The dynamic nature of these relations requires solutions to change over time. For example, waste and pollution management through dilution was feasible when human populations were small, but not feasible as populations grew and additional harms reduced the capacity of the receiving environment to cope with our wastes.
7. *Working through decisions*
   There are two principles that I believe are essential in seeking the best, safest, and most appropriate options whether in clinical, programme, or policy spheres. The first is "first do no harm." The principle is that the essential obligation is to not make things worse. And while there is always the potential for unwanted side effects or adverse outcomes, the primary consideration is a focus on improving health or reducing disability or their correlates in other sectors. The second principle is "do what is the least intrusive, most effective with the fewest side effects." It is an additional ethical and practical view that interventions always have risks, but these can be minimized by utilizing those that are less invasive or violating. We should always aim for the best possible outcome and in order to do so, seek options shown or likely to be most effective. Respecting the concept of "first do no harm" requires us to seek options with the least side effects. These principles are particularly important in facing problems that are less than catastrophic or fatal, and are as important in designing an environmental, social, or economic policy as they are in deciding on a clinical course.
8. *Working with coalitions*
   There are some key principles that help to make effective coalitions.
   i. Respect
      It is difficult to influence who or what we do not respect. We need not like them or want to socialize with them, but we must respect who they are, the role they play, and that they have a perspective that we need to understand, even if we do not agree. I have yet to see where a harangue has successfully changed a person's long-held perspective. Rather it may entrench their views, or at the very least cause them to disengage, awaiting a chance to escape the encounter.

ii. Make it practical
Great ideals without substance or do-ability only tend to frustrate. For policymakers or others, there is always a long line of advocates, academics, or others happy to point out the problems, and the need for someone to do something. The challenge is finding practical, acceptable, and achievable solutions that help to move us in that direction. It is particularly well received if they also help meet objectives that the decision-maker values.

iii. Rule of three
Committees, coalitions, or negotiations often bog down around things that different governments, disciplines, or individuals will never agree on. Yet given they are at the table, there are likely far more issues that they might agree on. As such, I try to break down issues into three categories. Those we can agree with we should just do. Those solutions that we may not have perfect agreement on but could live with, we also do. Then, those few that we will never agree on, while not ignoring them completely, we do not waste all our time trying to convince others of what is unlikely to change for jurisdictional, situational, or any number of reasons. The nice irony is that when we successfully tackle the first two categories together, it is easier at times to find some accommodation or agreement on the third.

iv. "Having something to offer"
Many good ideas flounder as they are passed from department to department or organization to organization because no one has the authority, mandate, or capacity to take it all on. However, if we have something we can offer, or if the other organizations could contribute another part, sometimes that logjam can be broken. For example, wanting to have health and social impacts included in environmental assessments was seen by colleagues in environment as one more complication for which they did not have the expertise or capacity. However, by offering to collaborate and take on responsibility for that aspect of the work, the conversation changed.

v. Coalitions and collaborations form and reform as needed
Organizations or individuals need not agree on everything to work together on the things upon which they do agree.

vi. Other expertise or perspective
There is always a tendency to consult with "the usual suspects," so we develop blind spots to both problems and potential. Many organizations, including governments, will work away internally to figure out how to address an issue, then when it is finally public get defensive against all the other perspectives or alternate suggestions that present. It is possible to explore options without tipping one's hand or betraying government secrets in order to ensure that things are considered more fully by the time decisions need to be made. There will be items that must be done in secret, but that excuse may be overextended into

areas that would do better with transparency. For example, the many advisors I had assisting with the Annual Reports on the State of Public Health when I was Canada's Chief Public Health Officer (which always were in the context of the complex relationships of Determinants of Health and One Health perspectives) were initially required to not only sign confidentiality agreements but also to return any numbered drafts upon which they commented. I was eventually able to get that changed as the purpose of these reports to parliament were to provide the best understanding of the issues and possible solutions, which required hearing from multiple perspectives and expertise, and were not a matter of national security. One of the nice things about working in Public Health and on One Health or Health Determinants issues is that the more people know and understand what makes a difference to human and animal health, the environment, the economy, and what we can do about it, the better.

9. *Crisis communications*
   There is much that has been studied and written on risk management and crisis communications. While there may be many nuanced differences in situations and the appropriate response, the five core messages below must always be kept in mind and articulated clearly. Too often I hear officials and others responsible for communicating messages miss the point by effectively dismissing the fears or concerns of the public by only eating a hamburger to declare it safe from BSE, attending a restaurant during SARS in Toronto, drinking a glass of water to demonstrate safety, or simply saying something is safe, or the risk is low. The challenge for the hearer is what does that mean? How low? You may eat that one burger, but I have been eating that meat or drinking that water daily for weeks or months.

There are five core messages that need to be part of crisis communication:

- What we know
- What we don't know
- What we are doing to find out
- What we are doing to address the issue
- What you can do to protect yourself and others

These five core messages in and of themselves do not make up crisis communication, but they are critical to helping people understand the real nature of risk. No matter what else is said, and whenever possible, it is essential to offer people something that they can do to protect themselves and others. It was striking as part of the response to the 2009 H1 influenza pandemic that people took personal action. I am told you could identify Canadians in international airports because they were the ones that would cough or sneeze into a tissue or their arm rather than contaminating their hands. This is something I had advocated early, and it was reinforced in diverse settings from schools to television comedy skits. In the most remote parts of the country I travelled, there were sanitizing gels in shops and public places. People increased the washing of hands, avoided others if ill, had antivirals readily available should they become ill in pregnancy, and almost half the population was immunized in the autumn, sufficient to stop the pandemic in Canada before Christmas. Thus, for the first time in history, the course of an influenza pandemic was fundamentally and positively altered. At the same time, food establishments and hospitals increased their focus on hygiene, hand washing, and cleanliness, dramatically diminishing the usual hospital-acquired infections and food-borne outbreaks.

SUMMARY

Dealing with the wicked problems we face in One Health requires a broad range of disciplines, sciences, evidence, expertise, sectors, organizations, and disciplines. It demands a different kind of leadership than the typical approaches and structures when a leader has some measure of control over those in their company, military, government, or organization. It is less about power or command and more about influence, persuasion, collaboration, and modelling of effective strategies.

There is not as much written about this type of leadership, perhaps as it is less certain, and more working in the grey area of motivation. And yet much can be effectively accomplished, with experience and practice being reinforcing. It can be very satisfying to see positive change effected whether one's personal leadership in it is recognized or not, for it is a team sport. I am regularly reminded of the quote by Samuel Johnson paraphrased as, "It is amazing what can be accomplished when no one has to assume the credit."

12 Bridging the Knowing to Doing Gap to Support One Health Action

Craig Stephen

CONTENT

If you only rely on your knowledge, your options are already limited.

Tremendous volumes of important new knowledge that could improve the quality of our and animals' lives and reduce our impacts on the planet are regularly being produced. But little will come of it without the capacity to effectively get this new knowledge to the people who can act to make the necessary changes. The gaps between knowing and doing are well known and long-standing. This has been called the knowing-to-doing gap or the implementation gap. These gaps are created when there is a difference between the evidence for the most effective actions and their actual implementation in practice. Despite the considerable amount of money spent on research to understand how and why a problem occurs, comparatively little effort is dedicated to ensuring that research findings can be and are implemented in policy or practice.

There are many examples of ineffective or inefficient policies and practices that are widely used, thus wasting resources and placing additional burdens on overstretched services. Why this happens is not always obvious. In some cases, differences in organizational or individual practices come from differences in the nature, quality, and depth of what they know. In other cases, it may be affected by the ability to translate the knowledge into action. Never in human history have we been able to produce and share so much information, making the knowing versus not knowing gap less pressing than the gap between knowing but not doing.

One Health professionals are challenged by the scope and scale of the problems presented to them and by the number and variety of actors with whom they need to interact to facilitate change. Closing the knowing-to-doing gap requires processes that create and support conditions and relationships to effectively enable access, exchange, use, and evaluation of knowledge to support decisions and actions. Chapter 9 explored factors that affect how individuals make decisions to act. This chapter is concerned with ways to put knowledge into action for the purpose of changing policies and practices.

MOBILIZING, TRANSLATING, BROKERING, AND EXCHANGING – WHAT ARE WE DOING WITH KNOWLEDGE?

"If we keep on doing what we have been doing, we are going to keep on getting what we have been getting." (Wandersman et al., 2008)

New evidence will not change outcomes unless people apply it in practice and in policy. Unfortunately, the transfer of evidence into practice is often unpredictable, slow, or haphazard. The gap between knowledge and action has been highlighted in a range of One Health relevant fields such as conservation, ecosystem management, and public health. The literature on how to close the knowing-to-doing gap is growing and can at times be confusing. Many disciplines are producing their own bodies of literature on knowledge management and mobilization. Table 12.1 defines some key terms as they are used in this chapter.

Simply telling people about evidence and urging them to change what they do is ineffective (Levin, 2008). Practices are socially determined by many elements such as norms, cultures, and habits. Knowledge takes shape and has effect in a wide variety of ways but is always mediated through social and political processes (Levin, 2008). One Health knowledge-to-action strategies need to meet the various requirements of different, heterogeneous knowledge users.

Interdisciplinary knowledge-to-action work is difficult because different fields of study have different ways of approaching problems and because knowledge-sharing networks often exist within rather than across disciplines because people tend to form ties to those who are like themselves in terms of socially significant behaviour and attitudes (Perry et al., 2018).

WHAT IS THIS THING CALLED KNOWLEDGE THAT WE WANT TO SHARE?

Knowledge, pragmatically speaking, results from familiarity, awareness, and understanding of information gained through experience or study. It is created by making comparisons, identifying consequences, and making connections. But, as with many of the core concepts in this book, defining knowledge is not always pragmatic

TABLE 12.1
Knowledge-to-Action Terminology

Term (Source)	Description
Knowledge dissemination (NCCMT 2020)	The process of distributing information, adapting the content, and developing the means for delivery that are appropriate for the intended audience
Knowledge synthesis (CIHR 2016)	Reproducible and transparent methods to contextualize and integrate research findings of individual studies within the larger body of knowledge on the topic
Knowledge exchange (CMHA 2020)	The process of engaging researchers, policy developers, and decision-makers in the development and application of research knowledge to effect timely, relevant, and evidence-based decisions. It involves dynamic information sharing and exchange among all stakeholders, with those traditionally considered the users of information as active participants rather than just passive recipients of knowledge
Knowledge broker (Ward et al., 2009)	People at the interface between the worlds of knowledge creation and knowledge use who help find, assess, and interpret evidence. They help information and knowledge flow back and forth between creators and users. They facilitate interactions and identify emerging questions
Knowledge user (CIHR 2016)	Those who are likely to be able to use the knowledge generated to make informed decisions about policies, programmes, and/or practices
Knowledge translation (OCCYMH, 2020)	The process of translating knowledge from one format to another so that the receiver can understand it. A dynamic and iterative process that includes the synthesis, dissemination, exchange, and ethically sound application of knowledge to improve outcomes, provide more effective services, and strengthen systems
Knowledge mobilization (SSHRC 2020)	A range of activities relating to the production and use of knowledge, including knowledge synthesis, dissemination, transfer, and exchange. It involves the reciprocal and complementary creation, flow, and uptake of knowledge between knowledge creators, brokers, and users in such a way that may benefit users and create positive impacts
Evidence-informed practice (OCCYMH, 2020)	Combines the best available research with the experience and judgement of practitioners to deliver measurable benefits
Evidence-based policy (Salafsky et al., 2019)	Use available body of facts and other information indicating whether a belief or proposition is true or valid to identify programmes and policies that produce desirable outcomes

or straightforward. Epistemologists study how we know things. They have been doing so since the times of Plato and beyond. They recognize that there is not just one type of knowledge (Turri, 2012). For example, there is perceptual knowledge we gain by using our senses. There is introspective knowledge we gain by using our reflective abilities. There is rational knowledge we deduce by exercising our powers of reasoning. There is traditional knowledge that has been developed, sustained, and passed on from generation to generation within a community. None of these, or other types of knowledge, are better or worse than another. No single type of knowledge alone provides a complete picture, but taken together, they offer a more balanced understanding of what can be done under specific circumstances.

Knowledge derived from rigorous peer-reviewed information is widely acknowledged as a foundation for evidence-informed decision-making. However, people's willingness to act is affected by scientific, tacit, traditional, and personal forms of knowledge (see Chapter 9). The scope of information used to make a decision must be tailored to the context of the decision-maker. It may include the perspectives and values of people being affected by the decisions, the broader impacts on society, an assessment of the social costs and benefits of in/action, as well as the research findings. The traditional paradigm of professional expertise being the sole source of enough information to inspire actions is no longer valid.

There is a recurring temptation to impose standards on what it means to know something, such as in criteria for truth in philosophy, postulates for causation in epidemiology, and standards of evidence in law. It becomes challenging to apply those standards in interdisciplinary settings where people of different intellectual heritages hold different standards for knowledge and truth. Much time can be spent having debates on whose standards are right. Agreed-upon One Health methods to judge, weigh, and integrate different kinds of knowledge are not yet available or are in the early stages of development. There are, unfortunately, too many pressing One Health problems in the world to delay action until these disagreements are resolved. One way to deal with this conundrum is to impose a shared set of standards across the One Health team. This might work when the differences in standards are not large (for example, between a microbiologist and an epidemiologist) but can be problematic when the differences are greater (for example, between a politician and a molecular biologist). One such example from the author's own practice involved a multi-stakeholder meeting to recommend federal policies to reduce the ecological risk from transferred pathogens between wild and farmed animals while protecting the economic benefits of farming to farmers and their communities. There was no explicit standard for weighing or assessing evidence agreed to at the outset of the meetings. As the process unfolded, the federal regulator increasingly turned to the opinions of its scientific staff, who used a different causal model and criteria for "good" evidence than did community members and opponents to the farming activities. While the groups agreed on the risks that required management, the lack of a shared vision on how to judge and integrate different forms of knowledge resulted in an acrimonious process ending without agreement on how to interpret the emerging and abundant scientific information and community experience available for their consideration. As such, no change in policy happened. Because beliefs and justifications used to accept information and transpose it into knowledge can be tightly connected to our sense of self, imposing a standard for knowing can be divisive if the fundamental differences in what it means to know something are large. Deft facilitation and negotiation skills are needed to navigate this gap.

Another strategy is to find ways to accommodate multiple ways of knowing into knowledge mobilization planning. For example, the concept of two-eyed seeing refers to learning to "see from one eye" the strengths of Indigenous knowledge and ways of knowing, and from the other "eye" the strengths of Western knowledge and ways of knowing. The overlapping perspectives of the "two-eyed" approach provide a wider, deeper, and more generative field of view (Bartlett et al., 2012). Similarly, evidence-based medical practitioners need both a sophisticated appreciation of patients' personal knowledge and scientific knowledge when developing effective and acceptable clinical interventions. Effective One Health practitioners require broad cross-cultural competencies to be able to build collaborations between multiple actors at multiple administrative levels to allow different aspects of a problem to be seen and, by exploring these differences, find solutions that go beyond traditional perspectives. Cross-sectoral co-learning to develop shared ways of knowing requires people skilled in bridging different types of knowledge to facilitate transdisciplinary conversations. As we build such bridges, our knowledge networks expand and diversify, which, in turn, increase our exposure to a range of non-redundant information, thus increasing the likelihood of innovative thinking (Tortoriello et al., 2015).

Various sources of knowledge need to be part of the knowing-to-doing process as equal players to optimize the breadth and quality of the information that can influence policy and practice. Take for example a project in sustainable food systems in post-war Sri Lanka (Stephen et al., 2013). Fish are a vital food source in Sri Lanka, comprising over 50% of animal protein of many people's diets. Knowledge sources and expertise on fish production were poorly connected after the civil war, limiting the diffusion of knowledge on techniques, structures, and strategies for sustainable production to farmers in war-torn regions. Multi-Criteria Decision Analysis (MCDA) (see Box 12.1) was used to enhance

BOX 12.1 AN OVERVIEW OF MULTI-CRITERIA DECISION ANALYSIS (MCDA)

What is it: MCDA is a systematic approach for ranking action options against a range of decision criteria. The various social and scientific criteria can be weighted to reflect their relative importance to participants in the decision-making process. MCDA can combine expert judgement, stakeholder preferences, and multiple dimensions in decision-making.

When to use it: MCDA is a structured framework for combining expert judgement and stakeholder preferences. It is used in situations where a combination of a wide variety of factors must be considered in ranking decision options, even where quantification is challenging or limited. The approach supports stakeholder's engagement in identifying, scoring, and weighing decision criteria.

What do you get: Depending on the specific MCDA approach used, the method produces a single most preferred option, ranked options, a short list of options for further appraisal, or a characterization of acceptable or unacceptable possibilities.

knowledge connectivity by bringing together a suite of experts ranging from Sri Lankan and international scientists, farmers, community members, aquaculture sector salespeople, and government agents. The approach involved participatory decision-making processes to identify preferred, yet locally feasible, production systems to support culture-based fisheries. The process provided an opportunity for community members and farmers to share their beliefs and knowledge with each other and with government agents to develop widely accepted action plans. Understanding how various ways of knowing synergize or conflict with each other across the wide suite of players in One Health problems is essential for mobilizing knowledge into action.

GUIDING QUESTIONS AND FRAMEWORKS FOR CLOSING THE GAP

A useful question to ask when developing strategies to mobilize knowledge into action is: "What is the knowledge resource base needed to understand the enablers and impediments to action?" A knowledge resource base can be constructed by first mapping the people who can inform and influence action. This mapping exercise should bring together people who hold information on the causes and effects of the problem of concern, the nature and influence of the social context on decision-making and action, and experience of those who will affect or will be affected by action. It should also include customary or traditional knowledge that will affect people's acceptance of information and recommendations to act. This wider field of knowledge can underpin dialogue that will lead to a more balanced and comprehensive understanding of the problem and what must occur to promote action. This approach is used in Chapter 22 to better understand monkey-human conflicts in St. Kitts and Nevis. Bridging the knowing-to-doing gap requires reciprocal and iterative flows of information from both knowledge producers and knowledge users prior to research initiation and beyond its completion. The goal is not to only distribute knowledge but also to share it in such a way that it is easily accessible, useful, and used.

Theories and frameworks can provide modifiable guides to systematically understand problems and suggest entry points to developing and evaluating action strategies. There are many theories that try to answer the question: "How does successful change happen?" Some view humans as goal-oriented actors who will follow a rational plan, while others view decisions as irrational and influenced by emotions and feelings. Some theories overlap, others conflict. The following questions, abstracted from various theories of change, can help a One Health team develop strategies for moving knowledge into action.

Do You Have a Theory of Change?

A theory of change is a comprehensive description of how and why a desired change is expected to happen in a particular context. It aims to build a bridge between what we know, what we want to achieve, and the activities it will take to get there. A theory of change helps identify the approach that should be taken to effectively address the causes of problems that hinder progress. A change theory requires the involvement of knowledge creators, planners, beneficiaries, and stakeholders at the start to develop consensus on the shared goals by explicitly documenting different views and assumptions and by helping people see how sharing their knowledge contributes to long-term positive impacts.

A theory of change can help to systematically think through the nature and interactions of the underlying causes of a problem and identify actionable steps that would logically make incremental progress towards shared goals. A well-developed theory of change helps to see the connections between short-term action and long-term goals, making explicit what we know, what we assume, and what we can feasibly do. For example, in 2017, the Canadian Wildlife Health Cooperative initiated a multi-stakeholder, multi-governmental effort to transform how wildlife health is protected by facilitating the development of the Pan-Canadian Approach to Wildlife Health (Stephen, 2019). The first step was to create a theory of change to set a "big picture" strategy that linked the desired change to the needs of wildlife health knowledge users. Reviews of international and national legislative obligations, conversations with decision-makers to identify the problems they faced and the changes they desired, and an assessment of what was feasible with existing resources and partnerships informed this first step. This was followed by scholarly work to develop the evidence base for the desired change. The first step outlined the need for change and considered the ecological, social, economic, political, and institutional processes that enabled or impeded change. From this first step came a clarity of purpose that led to a shared vision and mission. The second step was to create a logical framework to depict how this strategy could be implemented into programmes (Figure 12.1). The logical framework connected programme activities to outputs which lead to outcomes and the goal. This theory of change helped garner support for the concept by translating the concept into change. By 2018, all federal, provincial, and territorial ministers in charge of biodiversity and the environment endorsed the Pan-Canadian Approach.

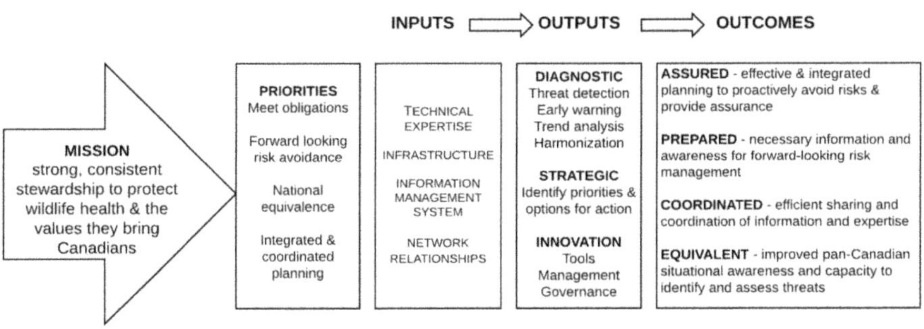

FIGURE 12.1 The logic model for change for a Pan-Canadian Approach to Wildlife Health used in 2018.

What Helps or Prevents People from Adopting New Ideas?

The barriers to more effective movement of knowledge into action are multiple and well documented (Levin, 2008). They include lack of access to enough high-quality evidence, lack of interest among potential knowledge users, low trust in the evidence, lack of capacity or ability to find and interpret evidence, lack of support for knowledge mobilization, strong forces that resist change, and pressures of various kinds pushing against the available evidence. The large and growing volume of research evidence, lack of time to read and thoughtfully review the evidence, structural barriers (e.g. financial disincentives), organizational barriers (e.g. lack of facilities or equipment), and peer group barriers (e.g. social norms that are not in line with desired action) can further widen the implementation gap.

The characteristics of the potential adopters, organizational or systems characteristics, can be enablers of barriers to change (Grol and Wensing, 2004). Some individuals are more apt to adopt new ideas or innovation than others. Some are willing to take risks and adopt new ideas quickly, while others lag and are more conservative or traditional in their practices. Chapter 9 discusses some key theories and factors that describe how individuals and organizations change. Regardless of their willingness to change, people go through five steps before adopting a new idea or innovation: (i) they need to become aware of the new idea, (ii) they need to become motivated and able to find out more, (iii) they need to see how the change applies to their own needs and circumstances, (iv) they decide to try (or reject) the new idea, and (v) they need to confirm that their decision helped meet their goals to continue its application (Kaminski, 2011). The potential adopter's perceptions of a change strongly affect the rate of adoption. Differences in individuals' ways of thinking and knowing, their motivations, and their beliefs in their ability to change will influence willingness and ability to adopt new ideas. An innovation is more likely to be adopted and spread if its advantages can be demonstrated to those who adopt it, it is consistent with social norms, and it can be feasibly applied.

Change can be enabled or dissuaded by the capacity, services, and resources of an organization. Implementation is more likely to succeed when there is consensus on the types and quality of evidence needed, the leadership and culture of an organization is more receptive and conducive to the integration of new information into practice, and there are facilitators who can help individuals and teams understand what they need to change and how they need to change it in order to apply evidence to practice or policy (Rycroft-Malone et al., 2002).

How Does Knowledge Spread?

The spread of information is influenced by social networks. Social networks are influenced by the strengths and diversity of ties between different network members. Social learning networks consist of the connections between people who change an individual's knowledge or motivation, resulting in behaviour change and alteration of practices (Wu et al., 2016). A person in a network will be more motivated to seek information from within his or her network if he or she (i) knows

BOX 12.2 SOCIAL LEARNING NETWORKS AND SHRIMP FARMING SUSTAINABILITY IN SRI LANKA: A CASE EXAMPLE BASED ON Wu et al. (2016)

The context: How to increase the adoption of best management practices for disease control in smallholder shrimp farms to increase farm profitability, food production, and sustainability in Sri Lanka.

The findings:

- Social learning networks differed based on geographic location and ethnicity.
- About one-fifth of the farmers were isolated from the knowledge on best management practices being disseminated through the shrimp farming network.
- The people whom most farmers accessed for knowledge lacked training in best management practices for shrimp farming.
- Farmers with larger social learning networks were wealthier but had farms that were less ecologically sustainable.

The recommendations:

- Strategies to increase farmers' uptake of best management practices to reduce the impact of disease included (i) encouraging the flow of accurate knowledge through existing farmer-to-peer networks, (ii) strengthening farmer-to-expert networks, and (iii) engaging farmers who are isolated from existing networks.
- A "one size fits all" intervention would likely not succeed due to differential effect of wealth and ethnicity within the network.

what others in the network might know, (ii) values what others know, (iii) can gain timely access to that knowledge, and (iv) perceives that seeking information would not be too costly (Borgatti and Cross, 2003). Understanding the structure, relationships, and characteristics of people within a social learning network can help identify key influencers and conduits to diffuse information. This requires a dedication to understanding your audience's needs and ensuring you have the strategies and tools in place to engage, inform, and motivate them. Box 12.2 highlights the applications of social network thinking to post-war food sustainability planning in Sri Lanka, this time for shrimp farming.

How Do People Learn?

Helping people learn about a need for and options to change is cardinal goal of knowledge mobilization activities. There is a growing dissatisfaction in contemporary societies for people to uncritically accept the explanation of an authority figure as the sole basis for action. Most adult learners resist having information arbitrarily imposed on them. They are open to learning from others rather than solely from traditional authoritative teachers. They value role models and peer-to-peer learning, have well-established cognitive frameworks, and like to know why what they are learning matters (Kenner and Weinerman, 2011). It is, therefore, important to frame knowledge-sharing strategies in a way that allows learners to see the purpose of learning. This will require an awareness of the different learning styles of a target audience and framing learning strategies in immediately useful ways.

One can help in the learning process in four keys ways. Firstly, enable people to access the necessary information. This can include centralizing and making open the access to the primary information available and building capacity in critical assessment of evidence. Secondly, one can help knowledge users make sense of the available information by serving as a knowledge translator. Thirdly, one can bridge the knowledge users and creators, building relationships that help tailor inquiry and knowledge creation to the context of the targeted users. Fourthly, one can help shape the wider context influencing effective and efficient knowledge uptake and use.

What Does It Take to Be an Agent of Change?

People will either be a change agent or a change target in knowing-to-doing strategies. A change agent has the skill and capacity to stimulate, facilitate, and coordinate the change. There are some key features that have been associated with effective agents of change (Lunenburg, 2010). Most of these factors are related to how well the change agent is connected to and understands those he or she is trying to help change. Understanding and empathizing with the change targets helps communicate the need for and the value of change. Strong linkages to or similarity with the knowledge users helps increase acceptance of the knowledge mobilization messages. Respected peers can effectively be used to promote the use of research knowledge to influence change. The change agents and change targets need to be able to hear from, respond to, and influence each other. A clear plan co-developed by the change agents and change targets helps increase acceptance of the recommended actions. All these characteristics of a change agent need to be complemented with people or organizations with the openness and energy to change, both of which can be augmented by helping them see the rewards of change. Readiness for change requires a resolve to implement a change and a belief in the capability to do so.

ARE THERE FRAMEWORKS THAT CAN HELP IN KNOWLEDGE MOBILIZATION PLANNING?

The Knowledge-to-Action Framework

The Knowledge-to-Action Framework was developed to help create and sustain evidence-based actions (Graham et al., 2006). It is an explicit process to determine what knowledge needs to be translated, how it is translated, by whom, when, and why. It was generated through a synthesis of planned action theories about the process of change, largely in the health field. The Knowledge-to-Action Framework recognizes that every situation depends on the players involved and the context in which they operate. It is composed of two components: knowledge creation and the action cycle. Each component involves several phases which influence each other (Graham et al., 2006). A critical first step is the cultivation of the trust and relationships between knowledge creators and knowledge users to establish a common understanding of needs and process.

The three stages of Knowledge Creation start with the inquiry phase, which is characterized by a diversity of studies of variable quality that are distributed throughout a variety of sources and locations. Next comes knowledge synthesis, which uses reproducible methods, such as a systematic literature review or meta-analysis, to identify, assess, and synthesize information relevant to specific questions. The third phase of knowledge creation is the production of tools, such as guidelines, that are clear, concise, and user-friendly to facilitate knowledge uptake and implementation. Knowledge is best co-created by researchers and those who need to use the knowledge, but there are tasks in the knowledge creation processes that may be better suited to some subgroups than others. As ones goes through the three stages of knowledge creation, the information being assembled becomes more and more tailored to local needs and circumstances.

The Action Cycle portion of the Knowledge-to-Action Framework describes an eight-part process that leads to knowledge implementation (Table 12.2). Each step can influence the other and can be influenced by the knowledge creation component.

PRECEDE-PROCEED as a Planning Model

The PRECEDE-PROCEED model is another framework to help plan knowledge mobilization efforts. The model was created to assess needs for designing, implementing, and evaluating health promotion and public health programmes

TABLE 12.2
Eight Questions Arising from the Action Cycle of the Knowledge-to-Action Framework Described by Graham et al. (2006)

What is the knowledge-to-action problem that needs to be addressed?
What is the nature and quality of the knowledge needed to address the problem?
What is the best way to adapt available knowledge to local needs, knowledge users, and circumstances?
What is currently blocking the movement of knowledge to action?
What is the best way to get the message to those who need to know in order to implement change?
Is the knowledge that has been shared being used?
Have practices or policies changed after the knowledge was shared?
Is the knowledge implementation effectively and efficiently being sustained?

(Crosby and Noar, 2011). It can provide some guidance for developing knowledge-to-action activities.

PRECEDE stands for **P**redisposing, **R**einforcing, and **E**nabling **C**onstructs in **E**ducational **D**iagnosis and **E**valuation. This phase is used to plan programmes. It involves (i) determining the problems and needs of a given population and identify desired results; (ii) identifying the determinants of the identified problems and set priorities and goals; (iii) analyzing determinants that predispose, reinforce, and enable actions; (iv) identifying administrative and policy factors that influence implementation; and (v) matching appropriate interventions that encourage the desired and expected changes.

PROCEED stands for **P**olicy, **R**egulatory, and **O**rganizational **C**onstructs in **E**ducational and **E**nvironmental **D**evelopment. This phase is used to assess the effects of the knowledge-to-action activities in four steps: (i) designing an implementation assessment plan before implementing the programme, (ii) determining if the programme is following the plan and adaptively managing the plan or activities accordingly, (iii) evaluating if there was a change in action, and (iv) determining if the actions were associated with changes in the target goal. The process guides the planner to think logically about the desired end point and work backwards to achieve that goal.

SUMMARY

Knowledge mobilization is the push and pull of data, information, and knowledge in multiple directions, between individuals and groups, for mutual benefit. It is not easy to effectively communicate the need to and means to change to a wide variety of audiences typical of One Health problems, with messages that are clear, simple, and relevant. Regardless of the end users of knowledge, be they policymakers, resource managers, health practitioners, or the public, knowledge mobilization requires a relentless dedication to understanding the users' needs and strategies to engage, inform, and motivate them under the circumstances they find themselves.

Good knowledge mobilization takes time. Rarely will one meeting, or one paper, be enough to change what people do. Investing in the time to build the relationships and understand the context for change before planning underlies all activities to effectively inspire change. Successful knowledge-to-action plans are built on ongoing collaborations that build capacity and readiness for change. A clear plan that outlines roles, responsibilities, and authority to support action and secures the partnership and resources needed to sustain change are essential. Often the "personal touch," like face-to-face meetings, can facilitate a better understanding of the context for change and help tailor knowledge products to specific audiences and local strategic priorities. Knowledge brokers and trusted champions of change should be identified early and brought into the knowledge-to-action process. A final but essential part of closing the implantation gap is evaluation. There is no one method to evaluate knowledge-to-action efforts, but systematically collecting and analyzing information to see if a programme or a project is doing what it set out to do helps identify changes that need to be made along the way to you achieving your goals.

REFERENCES

Bartlett, Cheryl, Murdena Marshall, and Albert Marshall. "Two-Eyed Seeing and Other Lessons Learned within a Co-Learning Journey of Bringing Together Indigenous and Mainstream Knowledges and Ways of Knowing." *Journal of Environmental Studies and Sciences*, 2, no. 4 (2012): 331–340.

Borgatti, Stephen P., and Rob Cross. "A Relational View of Information Seeking and Learning in Social Networks." *Management Science*, 49, no. 4 (2003): 432–445.

CIHR. Canadian Institutes for Health Research. Knowledge Translations Definition. 2016. https://cihr-irsc.gc.ca/e/29418.html#2

CMHA. Canadian Mental Health Association. Knowledge Exchange: A Framework for Action by the Canadian Mental Health Association. 2020. https://cmha.ca/documents/knowledge-exchange-a-framework-for-action-by-the-canadian-mental-health-association

Crosby, Richard, and Seth M. Noar. "What Is a Planning Model? An Introduction to PRECEDE-PROCEED." *Journal of Public Health Dentistry*, 71 (2011): S7–S15.

Graham, Ian D., Jo Logan, Margaret B. Harrison, Sharon E. Straus, Jacqueline Tetroe, Wenda Caswell, and Nicole Robinson. "Lost in Knowledge Translation: Time for a Map?" *Journal of Continuing Education in the Health Professions,* 26, no. 1 (2006): 13–24.

Grol, Richard, and Michel Wensing. "What Drives Change? Barriers to and Incentives for Achieving Evidence-Based Practice." *Medical Journal of Australia*, 180 (2004): S57–S60.

Kaminski, June. "Diffusion of Innovation Theory." *Canadian Journal of Nursing Informatics*, 6, no. 2 (2011): 1–6.

Kenner, Cari, and Jason Weinerman. "Adult Learning Theory: Applications to Non-Traditional College Students." *Journal of College Reading and Learning*, 41, no. 2 (2011): 87–96.

Levin, B. "Thinking About Knowledge Mobilization. A Discussion Paper Prepared at the Request of the Canadian Council on Learning and the Social Sciences and Humanities Research Council." Prepared for the Canadian Council on Learning (2008), 32.

Lunenburg, Fred C. "Managing Change: The Role of the Change Agent." *International Journal of Management, Business, and Administration,* 13, no. 1 (2010): 1–6.

NCCMT. "Knowledge Translation: Definition." National Collaborating Centre for Methods and Tools, 2020 https://cihr-irsc.gc.ca/e/29418.html#2

OCCYMH. "Glossary." Knowledge Mobilization Toolkit, Ontario Centre for Excellence for Child and Youth Mental Health, 2020. http://www.kmbtoolkit.ca/glossary-glossaire#undefined

Perry, Brea L., Bernice A. Pescosolido, and Stephen P. Borgatti. *Egocentric Network Analysis: Foundations, Methods, and Models*. Structural Analysis in the Social Sciences, 44 (Cambridge: Cambridge University Press, 2018), 350.

Rycroft-Malone, Jo, Alison Kitson, Gill Harvey, Brendam McCormack, K. Seers, A. Titchen, and C. Estabrooks. "Ingredients for Change: Revisiting a Conceptual Framework." *British Medical Journal Quality and Safety,* 11, no. 2 (2002): 174–180.

Salafsky, Nick, Judith Boshoven, Zuzana Burivalova, Natalie S. Dubois, Andres Gomez, Arlyne Johnson, Aileen Lee et al. "Defining and Using Evidence in Conservation Practice." *Conservation Science and Practice*, 1, no. 5 (2019): e27.

SSHRC. "Definitions of Terms." Social Sciences and Humanities Research Council, 2020. https://www.sshrc-crsh.gc.ca/funding-financement/programs-programmes/definitions-eng.aspx#km-mc

Stephen, Craig. "The Pan-Canadian Approach to Wildlife Health." *The Canadian Veterinary Journal,* 60, no. 2 (2019): 145.

Stephen, Craig, Tim DeJager, Sam Daniel, Sunil Jayakody, and Jennifer Dawson-Coates. "Promoting Rural Income from Sustainable Aquaculture Through Social Learning in Sri Lanka." Final Report to the International Development Research Centre (2013): 57.

Tortoriello, Marco, Bill McEvily, and David Krackhardt. "Being a Catalyst of Innovation: The Role of Knowledge Diversity and Network Closure." *Organization Science*, 26, no. 2 (2015): 423–438.

Turri, John. "Is Knowledge Justified True Belief?" *Synthese*, 184, no. 3 (2012): 247–259.

Wandersman, Abraham, Jennifer Duffy, Paul Flaspohler, Rita Noonan, Keri Lubell, Lindsey Stillman, Morris Blachman et al. "Bridging the Gap Between Prevention Research and Practice: The Interactive Systems Framework for Dissemination and Implementation." *American Journal of Community Psychology*, 41, no. 3–4 (2008): 171–181.

Ward, Vicky, Allan House, and Susan Hamer. "Knowledge Brokering: The Missing Link in the Evidence to Action Chain?" *Evidence & Policy: A Journal of Research, Debate and Practice*, 5, no. 3 (2009): 267–279.

Wu, Jessica., Theresa Burns, P. K. Kawadugama, Tim DeJager, Trishia A. Westers, Sylvia Checkley, Carl Ribble et al. "Linkages Between Social-Learning Networks and Farm Sustainability for Smallholder Shrimp Farmers in Sri Lanka." *Asian Fisheries Society*, 29 (2016): 96–111.

13 Complex Systems Thinking in Health

David Stephen, Craig Stephen, Luis Pedro Carmo, and John Berezowski

CONTENTS

The concepts of systems, complexity, and chaos are appearing with increasing frequency in the health literature. Much health research and practice have, however, historically been successfully guided without evoking complexity but rather by applying a linear, reductionistic paradigm. That paradigm views reality as the sum of components that can be separated and studied as isolated entities. The reductionistic approach has been extremely successful in combating many diseases. Diseases caused by single elements, like a vitamin deficiency or a bacterial infection, can be remedied by targeting that element alone, saving countless lives, relieving much animal suffering, and improving the well-being of many. The challenges of chronic diseases, however, highlighted the limitations of the one-to-one linear model of disease causation. Lifestyle, genetic, environmental, and social factors interact in complicated ways over varying time scales that make prediction of clinical outcomes of many chronic diseases challenging and limits the value of attacking only one component cause as a management strategy. Epidemiologists began to ask if "the challenge of studying causality [can] be adequately addressed if emphasis continues to be placed on using tools and methods that are geared towards looking at … systems from a linear paradigm?" (Philippe and Mansi, 1998). Things get further complicated as our attention turns from disease to health, where health is characterized as the cumulative effect of capacities and resources derived from interacting individuals and social and environmental determinants necessary to adapt to, respond to, or control life's challenges and changes (see Chapter 2 on defining health). The potential for additive and multiplicative interactions between determinants, synergistic and antagonistic relationships, and varying social perceptions influencing how we weigh and value these interactions make it hard to accept that health is a topic well suited to the linear, reductionist paradigm. People working in ecosystem-based management, EcoHealth, and the sustainability and health agendas are now thinking in terms of systems in which the parts and whole are inevitably connected (Waltner-Toews and Kay, 2005). To make matters even more complicated, One Health now asks us to consider interactions between different types of health for interacting species and over multiple generations that each change over time, as do the hazards and harms they experience. Health, as it is experienced in real life, is complex and messy.

A complex systems approach has been evoked as a response to calls for an alternative to the linear, reductionistic paradigm. Throughout this book, there are many instances where authors have referred to health and One Health problems as being complicated, complex, or wicked. It is hoped and hypothesized that a complex systems approach would broaden the spectrum of methods to design, implement, and evaluate One Health interventions. Yet concepts like complexity and chaos are foreign to many health scientists and "there is some looseness in how they have been translated from their origins in mathematics and physics, which is leading to confusion and error in their application" (Rickles et al., 2007). In this chapter, we introduce complex systems and how complexity thinking can be relevant to One Health. Our goal is to lay a foundation from which interested readers can dive into the growing literature on complexity and health.

WHAT IS A COMPLEX SYSTEM?

A complex system consists of many parts interacting in a nonlinear fashion. Here, nonlinear means that a change in one part does not have a fixed effect on the whole system, rather it depends on the current state of that part and other parts in the system (Strogatz, 2015). Unlike simple, linear systems, a complex system cannot be understood by extrapolating the behaviour of the individual parts. The parts, and the organized structures they create, change dynamically, interdependently, and often unpredictably over time. In this section, we will describe some phenomena that occur in complex systems, and the implications they have on how we must adapt our usual linear, reductionist approaches to encompass complex behaviour. A more extensive glossary of concepts related to complex systems, and their relation to health sciences, is given by Rickles et al. (2007).

A defining characteristic of complex systems is the notion of a feedback loop. This occurs when the interactions between parts results in a collective behaviour that feeds back into the behaviour of the individual parts, thus dampening or amplifying changes made to the system. In the former case, the system can display rigidity or adaptability in the face of a changing environment. In the latter case, small changes such as internal stochastic effects or external perturbations can be amplified to have a drastic impact on the system, such as causing a sudden transition into another regime of stability.

Another hallmark of complex systems is the existence of chaos. Since chaos is often incorrectly used synonymously with complexity, we need to clarify its meaning. Chaos refers to a well-defined mathematical notion in which two initial conditions that start arbitrarily close to each other diverge exponentially fast in time. This is colloquially known as the butterfly effect. In a chaotic system, the extreme sensitivity to initial conditions can make system change appear irregular and even random, even though the system is evolving deterministically according to some simple rules. Chaos appears in many mathematical models, even simple ones with a single variable. Chaos theory reminds us that, even in a completely deterministic system, it can be impossible to accurately predict its future state, and that apparently random behaviour need not be due to stochastic external effects.

Complex systems operate on different scales over time and space, and each scale can exhibit entirely new properties that cannot be extrapolated from the behaviour on smaller scales (Anderson, 1972). This is due to the principle of emergence, which arises from the intricate network of interactions between the many parts of a system. Disease systems, for example, can be simplified as a multilevel nested hierarchy (Ceddia et al., 2013). The "ground-floor" level is where the disease occurs. Below that is the level of individual decision and relationships affecting the day-to-day small-scale actions and relationships that influence disease dynamics. Above is the level of collections of individuals and institutions that enable or dissuade actions and decisions. Above that is the biophysical level where ecosystem processes influence how the other constituent parts of the system interact. The level at which one encounters (or studies) such a disease system will affect the perspective one has of the system. This means that when viewed from different perspectives or disciplines at different times and scales, the "same" complex system can be described differently.

A comprehensive understanding of a health outcome requires us to look at multiple variables interacting across all levels and across different spatio-temporal scales. This seems an overwhelming task, one to which people are increasingly evoking complexity theory as help. There is, however, no single "complexity theory" per se. The term complexity theory refers to several fields of study which all aim to (i) understand which obstacles prevent us from predicting the evolution of a system, such as those described above, and (ii) find ways around these obstacles (e.g. using mathematical or statistical methods). A complexity-based approach involves questions different than asking: "Does pathogen A cause disease B?" or "What risk factors are associated with the transmission of infection?" (Pearce and Merletti, 2006). Instead, it is better used to ask, "Are there circumstances where certain subpopulations are more vulnerable to a disease?" or "Are there situations where surveillance resources would be more likely to detect an emerging issue?" or "Which upstream intervention should we target knowing that there are many intervening variables between the intervention and the health outcome that could modify its impact?"

The interdependence of humans, animals, and their shared environments, the co-evolution of their interactions, the emergent properties of their behaviours, feedback loops within the system which enable or constrain further behaviours, the networked nature of relationships, and the different scales of socio-ecological systems, all point out that One Health operates in the sphere of complex systems. There are implications to thinking about complexity in One Health (based on Preise et al., 2018). Firstly, it shifts our attention from components of the system to the system as a whole. In a systems approach to a problem, the emphasis shifts from the parts that make up the system to their interrelationships (Pohl and Hadron, 2008). This requires us to pay attention to organizational processes, connections, and emergent behavioural patterns of the system. This shift asks us to spend less time looking for causal pathways and more time characterizing relationships and interactions that influence patterns of system behaviour to facilitate our understanding of how systems transform and how emerging characteristics arise. Secondly, adopting this perspective requires the use of different tools and methods adept at capturing and characterizing relationships such as network analysis, participatory systems analysis, and transdisciplinary methods. These methods need to be adaptable, they should cultivate social learning, and they should favour synthesis over isolated analysis. The tools and methods need to be able to capture spatial and temporal dynamics, be attentive to surprises, and identify critical thresholds and tipping points. System dynamics models (Keeling and Eames, 2005), agent-based modelling, and time-series analysis are candidate methods. Finally, we need to be aware that the application of our tools and methods could affect the system and that some boundaries for our investigations will need to be imposed or constructed, which in turn will influence what we measure.

WHY THINK ABOUT COMPLEX SYSTEMS IN ONE HEALTH?

Our current methods and tools are not well equipped to integrate information that span environment, animal, and social and health concerns, making it hard to "see" an integrative way, or to benefit from the full range of knowledge and experience relevant for creating healthier circumstances. Many of the models, perspectives, and methods used to study human-animal-environmental health systems were created to answer questions about the health of individuals or groups of individuals. Berezowski et al. discuss this challenge from a surveillance perspective in Chapter 7. In the literature, it is more common to find metaphors and analogies adapted from the natural or social sciences that justify the use of complexity and chaos thinking in health sciences than it is to find evaluated cases of the value of this thinking in One Health.

The hope for complex systems thinking in health is that by considering the changing context, its key actors, and their interactions over time, we can more effectively understand and improve health (Rusoja et al., 2018). Better understanding of how multiple human and animal components interact in a non linear fashion to produce highly context-dependent outcomes could, for example, help identify situations predisposed to emerging diseases or find synergies to produce more efficient use of health resources for shared human, animal, and environmental health benefits. Understanding how the multiple determinants of a problem within a human-animal-environment system relate with each other and with the history of that system could help identify new ways to deploy health intelligence resources to find vulnerable situations in advance of harm (see Chapter 7 for details). Yet promises such as these have largely been untested and unfulfilled.

Complex systems thinking is compatible with several One Health aspirations. The focus on understanding boundaries of systems, exploring interrelationships, and seeing unique perspectives of a system from different people's viewpoint could help define the scope of a One Health problem that a group wishes to tackle, identify common sets of indicators of success, and open group members' eyes to opportunities to influence systems' outcomes in ways they may not have from their own vantage point. Recognizing that a complex system cannot be fully understood from only one perspective can reveal the need for multiple collaborators to learn from each other to develop acceptable and effective One Health actions.

While the authors firmly believe that thinking about One Health from a systems perspective is an excellent teaching metaphor and a framework to plan One Health research and actions, we cannot provide the reader with published evidence of its utility, feasibility, and acceptability. What follows are two scenarios that illustrate the promise for complex systems thinking for One Health.

Scenario 1 – Emerging Diseases and Epidemics

The fact that the number of times the conditions conducive to epidemics exist without an epidemic occurring far outnumber the few occasions when one occurs (Stephen et al., 2004) supports the conclusion that epidemics are emergent phenomena of complex adaptive systems. The long and sometimes convoluted causal chains between upstream and downstream animal, human, and environmental determinants of disease emergences and our poor success rate at forecasting an emerging disease with sufficient precision to inspire local action call out for a new way to think about how we prepare for emerging diseases.

Ideally, one would like to solve a mathematical model that describes a system to the extent that we are able to exactly predict the state of a system in the future given its current state (i.e. predict precisely the combination of system attributes that lead to an emerging disease). The idea of studying or modelling emergence as a complex system phenomenon can seem an elusive if not impossible task. However, we need not always capture all underlying behaviours of a complex systems to understand the relevant characteristics. Complexity approaches may not provide the level of precision to provoke local biomedical interventions, but they have been used to "diagnose" when a system will change from one state to another. Instead of asking "can we predict which pathogen will emerge on which day, in which locations," complex systems perspectives may help us ask "what are the circumstances that tip a system from being unconducive to conducive to an emerging disease, epidemic or pandemic?"

Epidemics as Critical Transitions

Epidemics occur in complex systems involving the interactions between reservoirs and hosts, exposure pathways and transmission rates, and environmental and social factors that are inherently interrelated and unpredictable (Wilcox and Colwell, 2005). Yet we can use a single number (R) to describe a situation when a disease can spread in a population. R is the effective reproduction number, which is the average number of secondary infections resulting from a single infected case. R is affected not only by the nature of the pathogen but also by interacting host and environmental factors. The point $R = 1$ can be viewed as a transition point between the disease states: one where a population can sustain an epidemic ($R > 1$) and one where the epidemic is likely to die out ($R < 1$). In the study of dynamic systems, transitions between two regimes such as this are called critical transitions. Therefore, disease (re)emergence, epidemics, and elimination can be conceived as critical transitions (Drake et al., 2019).

Critical transitions are a fundamental aspect of non linear systems, described by the mathematical notion of bifurcation, in which more than one steady state exists for a system. If system parameters are tuned in a certain way, it is said to be at a tipping point, wherein small perturbations to a system's state or its parameters can trigger a cycle of positive feedback, causing the system to dramatically shift from one steady state to another. Critical transitions play an important role in all complex systems, including ecosystems, economics, and human biology (Scheffer, 2009).

Transition Early Warning Systems

How can we apply the immense body of understanding of critical transitions to the scenario of disease emergence and epidemics? It would be of great benefit to epidemic and pandemic prevention programmes to be able to characterize and diagnose systems that are vulnerable to critical transitions. A systems approach to early warning would aim to detect the characteristic behaviour of systems near critical transitions using spatiotemporally ordered data (Boettiger et al., 2013; Scheffer et al., 2012). One such behaviour is known as critical slowing down, which describes the fact that near a tipping point, a system will recover more slowly from perturbations. This can be observed via the increase of autocorrelation and variation in time-series data of the relevant variable, such as the number of infected in the case of disease. A key advantage of early warning systems is their genericity: since they are based on general features of non linear dynamical systems, they do not require precise knowledge of the feedback loops that may trigger the transition. Even without a full understanding of the nonlinearities present in a system, there are diagnostics which can be used as early warning signals for such critical transitions (Scheffer et al., 2012). However, different early warning systems may be better suited to different types of transitions, so their proper use can still be informed by such knowledge (Brett et al., 2017; Boettiger et al., 2013; Dakos et al., 2015).

Transition early warning systems have been successfully applied in many fields, such as in ecology (Scheffer et al., 2015), where the critical transition may involve species extinction or population regime shifts (including large-scale experimental verification in lake ecosystems; see Carpenter et al.,2011 for example); human health (Rikkert et al., 2016), where transitions coincide with sudden events (Maturana et al., 2020), such as seizures or heart attacks and long-term changes such as the onset of depression (van de Leemput et al., 2014); and climate science, where, for example, ancient abrupt climate changes were shown to be preceded by critical slowing down (Dakos et al., 2008). While the theory and application of transition early warning systems is successful in these fields, their presence and use in the context of emerging disease and epidemics are in their infancy. One important reason for this is the fundamental difference between the nature of disease emergence, in which the number of infected individuals grows continuously albeit explosively, versus more commonly studied critical transitions where the transition between two states is discontinuous, or "catastrophic" (Drake et al., 2019). This difference means that different early warning systems are required; their theoretical development is just now occurring (Drake et al., 2019; Boettiger et al., 2013; Dakos et al., 2015). Very recently, these ideas were applied to real-world data by Harris et al., 2020, who studied the re-emergence of malaria in Kenya, triggered by the slow development of parasite resistance to treatment, and showed that it could be detected by early warning systems several years prior to the transition. Crucial to the successful application of transition early warning systems are consistent high-quality epidemiological data, which are needed to reduce statistical uncertainty to a point where early warning systems are reliable enough to motivate proactive behaviour (NRC 2001).

Adapting a transition early warning system approach would allow us to complement the typical "surveillance and response" approach to disease emergence, where emerging diseases are detected early and rapid intervention is used to prevent their spread, with the approach of "prediction and prevention," where a system is monitored and characterized to determine susceptibility to outbreaks, and preventative action is taken before emergence (NRC 2001). While yet to be realized in practice, recent work is showing the theoretical possibility of analytically identifying critical transitions associated with disease elimination and emergence (e.g. O'Regan and Drake, 2013; Brett et al., 2017). Various works have moved these findings closer to implementation by showing that some approaches to early warning that use signals of critical transition were robust to imperfect epidemiological data (Brett et al., 2018), increasing model complexity and dimensionality (Brett et al., 2020), and the inclusion of social factors (Phillips et al., 2020).

Scenario 2 – Engineering Resilience and Fostering Change

One Health practitioners are often asked to identify actions to modify human, animal, and environmental relationships to foster resilience or promote a change that reduces vulnerability even when those relationships are dynamic, complex, and not fully accounted for. Complex systems thinking can provide new tools to rise to these tasks.

Promoting Resilience

Two main characteristics that determine a network's resilience are diversity and connectivity (Brett et al., 2017; O'Regan and Drake, 2013). A network which is diverse and modular (connected primarily into smaller sub-networks) is resilient to critical transitions, as external perturbations can be compensated for by the greater variety of negative feedback loops provided by the diverse composition, and are less likely to propagate throughout the entire system due to the modularity. Such a network can gradually adapt in the face of external changes. A network which is highly connected and homogeneous, on the other hand, might initially have a rigidity that resists external perturbations. However, at a critical stress level, the components of the network may undergo a sudden collective change that propagates throughout the whole network, resulting in a critical transition.

Increasing diversity and modularity can help protect a complex system against critical transitions. This observation has a long history in ecology and has also been explored in economics (Scheffer et al., 2012; Haldane and May, 2011). Modularity is an important factor to prevent the spread of disease. The temporal dynamics of modules (such as communities) have been associated with epidemic processes (Nadini et al., 2018). Sah et al. (2018), for example, concluded that "high fragmentation and high subgroup cohesion, which are both associated with high modularity in social networks, induce structural delay and trapping of infections that spread through [animal social] networks, reducing disease burden." The effect of biodiversity is less clear. Morand and Lajaunie (2018) concluded that "empirical studies and often-correlative analyses show that biodiversity is a source of pathogens, but increases in epidemics and risks of emergence are associated with decreased biodiversity." Luis et al. (2018) found that biodiversity can dilute, amplify, or have no effect on zoonotic disease transmission and risk. Local factors, such as changes in habitat connectivity and edges, or access to health protection resources, can modify these relationships. General theories of the relationship between biodiversity and the risk of epidemic or pandemic diseases are awaiting.

Diversity has, however, been noted as a hallmark of resilient systems. For example, increased biodiversity on farms has been associated with increased resilience to extreme weather events associated with climate change (Altieri et al., 2015). Another study found that marine protected areas were better able to meet their conservation goals where there was a diversity of interconnected incentives that arose from diverse institutional governance systems (Jones et al., 2013). The functional redundancies that come with species biodiversity have been investigated as a contributor to coral reef ecosystem resilience (Micheli et al., 2014).

Network theory is being increasingly used to investigate how social diversity, connectivity, and modularity influence disease spread and control in human and animal networks (Pastor-Satorras et al., 2015). In health, it has been used to study disease transmission, information transmission, the influence of personal and social networks on health behaviour, and interorganizational structure of health systems. Wu et al. (2016), for example, examined how social learning networks influenced shrimp farmer disease control behaviour in Sri Lanka. Wittrock et al. (2019) used network analysis to visualize how experts conceived the interrelationships of multiple determinants of fish and wildlife health. Examination of social networks and contacts has been used to study the spread of disease in people, wildlife, and livestock (e.g. Nöremark et al., 2011; Hamede et al., 2009; Latkin et al., 2013).

Fostering Change

What are we trying to do when we launch a One Health intervention? From a simple systems point of view, we are asking person A to apply intervention B in situation C. For example, there is an infectious disease outbreak that is vaccine preventable, so we ask farmer A to use vaccine B in his susceptible animals on farm C. From a complexity perspective, what we are trying to do is initiate a time-limited series of events that interact with the social context of the system to change the trajectory of a socio-ecological system to a state that is not conducive to a specific rate of occurrence of an infectious disease. Implementing change is not straightforward in an unpredictable system.

Critical to scaling up or sustaining an intervention is understanding how the social and environmental characteristics and circumstances surrounding the implementation interact, influence, modify, facilitate, or constrain the intervention and its implementation (May et al., 2016). The Context and Implementation of Complex Interventions framework is an example of a tool that tries to reflect on systems attributes of the intervention with the space and context they take place in, in order to better understand whether and how interventions work (Pfadenhauer et al., 2017). Complexity thinking helped Braithwaite et al. (2018) identify that while change can be stimulated in many ways, a triggering mechanism is needed, such as legislation or widespread stakeholder agreement; that feedback loops are crucial to continue change momentum; that extended sweeps of time are involved, typically much longer than believed at the outset; and that taking a systems-informed, complexity approach, having regard for existing networks and socio-technical characteristics, is beneficial.

As another example, Leykum et al. (2007) showed how complex adaptive systems thinking helped plan more effective organizational interventions for type II diabetic management. Specifically, they found the ability of patients to modify practices based on forces internal and external to the clinical setting (co-evolution), and paying attention to interconnections affecting client communications had the strongest relationship with an intervention effect.

Implementing change in complex systems may not always have the intended effect. Sterman (2006) used the term "policy resistance" to refer to the "tendency for interventions to be defeated by the system's response to the intervention itself." For an example, consider hospital waiting times, which have been historically resistant to efforts to reduce them. One mechanism for this was given by Smethurst and Williams (2002), who showed that reduced waiting times may lead to an increase in referrals, such that the waiting times are unaffected in the end. In certain cases, the system's response to an intervention can cause unintended negative effects. Efforts to eliminate all forest fires, for example, can cause the build-up of undergrowth and old or dead trees, thereby greatly increasing the risk of large-scale fires (Malamud et al., 1998).

We can try to diagnose this kind of resistance to change by searching for patterns in observable data that are characteristic of complex systems. In the case of hospital waiting times, if one observes the quarter-to-quarter variations in waiting times, one finds that the frequency of occurrence $N(x)$ of a variation of size x scales as a power law, $N(x) \sim x^a$ (Papadopoulos et al., 2001; Smethurst and Williams, 2001). This means that as is the case for earthquakes and forest fires (Malamud et al., 1998), the distribution is "fat-tailed" in that variations of significant magnitude are more likely to occur than would be the case in a standard bell curve having a characteristic scale, like human height. This lack of a characteristic scale, or "scale invariance," is often (but not solely) associated with systems in the vicinity of a critical transition, at "criticality" (Gisiger, 2001). This is a powerful, albeit controversial (Frigg, 2003), concept that may underlie many complex systems because systems at criticality show an optimal balance between robustness and adaptability (Munoz, 2018). Because of this, implementing changes in such systems can have a much larger, or smaller, effect than intended, and caution must be exercised. On the other hand, it suggests that large-scale change can be possible with relatively small intervention efforts (Fullilove et al., 1997). The practical consequence for the example of hospital waiting times is that occasional long times may be an inevitability, and points of evaluation and change should rather focus on, for example, quality of care (Smethurst and Williams, 2001).

Another example where scale invariance appears is epidemics in small, isolated populations of susceptible people, where outbreaks occur with dramatic variation of size and are separated by long periods of disease absence. Here, both the size and duration of epidemics can be fit to power laws as above. This behaviour has been found in isolated island populations (Rhodes et al., 1997), outbreaks of cholera (Roy et al., 2014) and dengue fever (Saba et al., 2014), and measles cases in populations with declining vaccine usage (Jansen et al., 2003). Conventional epidemiological models are unable to capture this scaling behaviour. In fact, these types of epidemics are better described by forest fires models, which are paradigmatic examples of systems at criticality (Rhodes et al., 1997).

The apparent robustness of these scaling laws in such epidemic scenarios suggests that short-term interventions like treatment and vaccination programmes may not be effective in controlling and preventing the epidemics due to their self-regulating nature. Rather, one should focus on eliminating the conditions that enable the persistence of the disease. The existence of the power-law scaling gives some limited predictive power: it is possible to infer the frequency of large outbreaks from that of small outbreaks (Rhodes et al., 1998) as is routinely done for earthquakes (Aki, 1981). It is also possible to estimate the distance to the critical transition by fitting the scaling exponent a in the power law (Jansen et al., 2003; Roy et al., 2014).

FUTURE PROSPECTS

At the time of writing this chapter, the world was facing the COVID-19 pandemic. In response to the immense scale and impact of the pandemic, researchers from all fields were being drawn to the issue and were applying their own techniques for dealing with complex systems in efforts to understand and suppress the disease. Neuroscientists, for example, were applying models that have been used to map the brain, one of the most complex systems we know of, to the pandemic (Friston et al., 2020). Artificial intelligence was also being applied for the purposes of tracking and prediction, diagnosis and prognosis, and the development of treatments and vaccines (Naudé, 2020). Interesting is the impact of social media, both in the rapid spread of (mis)information (the so-called infodemic) and its ability to help track the spread of the disease (Eysenbach, 2009; Cinelli et al., 2020). Due to the scale of the pandemic, enormous amounts of data are available, and attempts to make sense of this data were ongoing on various fronts (Latif et al., 2020). Indeed, entire journals were being created to collect the unprecedented amounts of new research.

How the lessons learned from the massive investment to understand the COVID-19 pandemic will influence or advance the use of complexity theory in One Health cannot be forecast at the time of writing this chapter, but we hope that we have shown a place for complexity in conceiving, communicating, planning, and implementing One Health research and action. This chapter has not exhaustively reviewed and presented all potential avenues for complex systems thinking in One Health. Instead, we hope that the ideas and examples presented herein motivate readers to learn more about how to match the way we study One Health systems to the ways they exist in nature as complex, messy sets of dynamic interactions.

REFERENCES

Aki, Keiiti. "A Probabilistic Synthesis of Precursory Phenomena." *Earthquake Prediction: An International Review*, 4 (1981): 566–574.

Altieri, Miguel A., Clara I. Nicholls, Alejandro Henao, and Marcos A. Lana. "Agroecology and the Design of Climate Change-Resilient Farming Systems." *Agronomy for Sustainable Development* 35, no. 3 (2015): 869–890.

Anderson, P.W. "More Is Different." *Science*, 177, no. 4047 (1972): 393–396.

Boettiger, Carl, Noam Ross, and Alan Hastings. "Early Warning Signals: The Charted and Uncharted Territories." *Theoretical Ecology*, 6, no. 3 (2013): 255–264.

Braithwaite, Jeffrey, Kate Churruca, Janet C. Long, Louise A. Ellis, and Jessica Herkes. "When Complexity Science Meets Implementation Science: A Theoretical and Empirical Analysis of Systems Change." *BMC Medicine* 16, no. 1 (2018): 63.

Brett, Tobias S., John M. Drake, and Pejman Rohani. "Anticipating the Emergence of Infectious Diseases." *Journal of the Royal Society Interface*, 14, no. 132 (2017): 20170115.

Brett, Tobias S., Eamon B. O'Dea, Éric Marty, Paige B. Miller, Andrew W. Park, John M. Drake, and Pejman Rohani. "Anticipating Epidemic Transitions with Imperfect Data." *PLoS Computational Biology*, 14, no. 6 (2018): e1006204.

Brett, Tobias, Marco Ajelli, Quan-Hui Liu, Mary G. Krauland, John J. Grefenstette, Willem G. van Panhuis, Alessandro Vespignani et al. "Detecting Critical Slowing Down in High-Dimensional Epidemiological Systems." *PLoS Computational Biology* 16, no. 3 (2020): e1007679.

Carpenter, Stephen R., Jonathan J. Cole, Michael L. Pace, Ryan Batt, W. A. Brock, Timmothy Cline, Jim Coloso et al. "Early Warnings of Regime Shifts: A Whole-Ecosystem Experiment." *Science*, 332, no. 6033 (2011): 1079–1082.

Ceddia, Michele Graziano, Nicholas O. Bardsley, Robin Goodwin, Garth Holloway, Giuseppe Nocella, and Antonio Stasi. "A Complex System Perspective on the Emergence and Spread of Infectious Diseases: Integrating Economic and Ecological Aspects." *Ecological Economics*, 90 (2013): 124–131.

Cinelli, Matteo, Walter Quattrociocchi, Alessandro Galeazzi, Carlo Michele Valensise, Emanuele Brugnoli, Ana Lucia Schmidt, Paola Zola et al. "The COVID-19 Social Media Infodemic." *arXiv preprint arXiv:2003.05004* (2020).

Dakos, Vasilis, Stephen R. Carpenter, Egbert H. van Nes, and Marten Scheffer. "Resilience Indicators: Prospects and Limitations for Early Warnings of Regime Shifts." *Philosophical Transactions of the Royal Society B: Biological Sciences*, 370, no. 1659 (2015): 20130263.

Dakos, Vasilis, Marten Scheffer, Egbert H. van Nes, Victor Brovkin, Vladimir Petoukhov, and Hermann Held. "Slowing Down as an Early Warning Signal for Abrupt Climate Change." *Proceedings of the National Academy of Sciences*, 105, no. 38 (2008): 14308–14312.

Drake, John M., Tobias S. Brett, Shiyang Chen, Bogdan I. Epureanu, Matthew J. Ferrari, Éric Marty, Paige B. Miller et al. "The Statistics of Epidemic Transitions." *PLoS Computational Biology*, 15, no. 5 (2019).

Eysenbach, Gunther. "Infodemiology and Infoveillance: Framework for an Emerging Set of Public Health Informatics Methods to Analyze Search, Communication and Publication Behavior on the Internet." *Journal of Medical Internet Research*, 11, no. 1 (2009): e11.

Frigg, Roman. "Self-organised Criticality: What It Is and What It Isn't." *Studies in History and Philosophy of Science Part A*, 34, no. 3 (2003): 613–632.

Friston, Karl J., Thomas Parr, Peter Zeidman, Adeel Razi, Guillaume Flandin, Jean Daunizeau, Oliver J. Hulme et al. "Dynamic Causal Modelling of COVID-19." *arXiv preprint arXiv:2004.04463* (2020).

Fullilove, Robert E., Jennifer Choe Edgoose, and Mindy Thompson Fullilove. "Chaos, Criticality, and Public Health." *Journal of the National Medical Association*, 89, no. 5 (1997): 311.

Gisiger, T. "Scale Invariance in Biology: Coincidence or Footprint of a Universal Mechanism?." *Biological Reviews,* 76, no. 2 (2001): 161–209.

Haldane, Andrew G., and Robert M. May. "Systemic Risk in Banking Ecosystems." *Nature*, 469, no. 7330 (2011): 351–355.

Hamede, Rodrigo K., Jim Bashford, Hamish McCallum, and Menna Jones. "Contact Networks in a Wild Tasmanian Devil (Sarcophilus Harrisii) Population: Using Social Network Analysis to Reveal Seasonal Variability in Social Behaviour and Its Implications for Transmission of Devil Facial Tumour Disease." *Ecology Letters*, 12, no. 11 (2009): 1147–1157.

Harris, Mallory J., Simon I. Hay, and John M. Drake. "Early Warning Signals of Malaria Resurgence in Kericho, Kenya." *Biology Letters*, 16, no. 3 (2020): 20190713.

Jansen, Vincent A.A., Nico Stollenwerk, Henrik Jeldtoft Jensen, M.E. Ramsay, W.J. Edmunds, and C.J. Rhodes. "Measles Outbreaks in a Population with Declining Vaccine Uptake." *Science*, 301, no. 5634 (2003): 804–804.

Jones, P.J.S., W. Qiu, and E.M. De Santo. "Governing Marine Protected Areas: Social-Ecological Resilience through Institutional Diversity." *Marine Policy*, 41 (2013): 5–13.

Latif, Siddique, Muhammad Usman, Sanaullah Manzoor, Waleed Iqbal, Junaid Qadir, Gareth Tyson, Ignacio Castro et al. "Leveraging Data Science to Combat COVID-19: A Comprehensive Review." (2020). doi 10.36227/techrxiv.12212516.v1. https://www.techrxiv.org/articles/Leveraging_Data_Science_To_Combat_COVID-19_A_Comprehensive_Review/12212516

Keeling, Matt J., and Ken TD Eames. "Networks and Epidemic Models." *Journal of the Royal Society Interface* 2, no. 4 (2005): 295–307.

Latkin, Carl A., Melissa A. Davey-Rothwell, Amy R. Knowlton, Kamila A. Alexander, Chyvette T. Williams, and Basmattee Boodram. "Social Network Approaches to Recruitment, HIV Prevention, Medical Care, and Medication Adherence." *Journal of Acquired Immune Deficiency Syndromes,*63, no. 0 1 (2013): S54.

Leykum, Luci K., Jacqueline Pugh, Valerie Lawrence, Michael Parchman, Polly H. Noël, John Cornell, and Reuben R. McDaniel. "Organizational Interventions Employing Principles of Complexity Science Have Improved Outcomes for Patients with Type II Diabetes." *Implementation Science* 2, no. 1 (2007): 28.

Luis, Angela D., Amy J. Kuenzi, and James N. Mills. "Species Diversity Concurrently Dilutes and Amplifies Transmission in a Zoonotic Host-Pathogen System through Competing Mechanisms." *Proceedings of the National Academy of Sciences of the United States of America*, 115, no. 31 (2018): 7979–7984.

Malamud, Bruce D., Gleb Morein, and Donald L. Turcotte. "Forest Fires: An Example of Self-Organized Critical Behavior." *Science*, 281, no. 5384 (1998): 1840–1842.

Maturana, Matias I., Christian Meisel, Katrina Dell, Philippa J. Karoly, Wendyl D'Souza, David B. Grayden, Anthony N. Burkitt et al. "Critical Slowing Down as a Biomarker for Seizure Susceptibility." *Nature Communications*, 11, no. 1 (2020): 1–12.

May, Carl R., Mark Johnson, and Tracy Finch. "Implementation, Context and Complexity." *Implementation Science*, 11, no. 1 (2016): 141.

Micheli, Fiorenza, Peter J. Mumby, Daniel R. Brumbaugh, Kenny Broad, Craig P. Dahlgren, Alastair R. Harborne, Katherine E. Holmes et al. "High Vulnerability of Ecosystem Function and Services to Diversity Loss in Caribbean Coral Reefs." *Biological Conservation* 171 (2014): 186–194.

Morand, Serge, and Claire Lajaunie. "Loss of Biological Diversity and Emergence of Infectious Diseases." *Biodiversity and Health* (2018): 29.

Munoz, Miguel A. "Colloquium: Criticality and Dynamical Scaling in Living Systems." *Reviews of Modern Physics*, 90, no. 3 (2018): 031001.

Nadini, Matthieu, Kaiyuan Sun, Enrico Ubaldi, Michele Starnini, Alessandro Rizzo, and Nicola Perra. "Epidemic Spreading in Modular Time-Varying Networks." *Scientific Reports*, 8, no. 1 (2018): 1–11.

Naudé, Wim (2020). Artificial Intelligence vs COVID-19: Limitations, Constraints and Pitfalls. *AI & Society*, 1–5. https://doi.org/10.1007/s00146-020-00978-0

Nöremark, Maria, Nina Håkansson, Susanna Sternberg Lewerin, Ann Lindberg, and Annie Jonsson. "Network Analysis of Cattle and Pig Movements in Sweden: Measures Relevant for Disease Control and Risk Based Surveillance." *Preventive Veterinary Medicine* 99, no. 2–4 (2011): 78–90.

NRC (National Research Council). *Under the Weather: Climate, Ecosystems, and Infectious Disease* (Washington, DC: National Academies Press, 2001).

Olde Rikkert, Marcel G.M., Vasilis Dakos, Timothy G. Buchman, Rob de Boer, Leon Glass, Angélique O.J. Cramer, Simon Levin et al. "Slowing Down of Recovery as Generic Risk Marker for Acute Severity Transitions in Chronic Diseases." *Critical Care Medicine*, 44, no. 3 (2016): 601–606.

O'Regan, Suzanne M., and John M. Drake. "Theory of Early Warning Signals of Disease Emergence and Leading Indicators of Elimination." *Theoretical Ecology*, 6, no. 3 (2013): 333–357.

Papadopoulos, Marios C., Michalis Hadjitheodossiou, Costas Chrysostomou, Carl Hardwidge, and B. Anthony Bell. "Is the National Health Service at the Edge of Chaos?" *Journal of the Royal Society of Medicine*, 94, no. 12 (2001): 613–616.

Pastor-Satorras, Romualdo, Claudio Castellano, Piet Van Mieghem, and Alessandro Vespignani. "Epidemic Processes in Complex Networks." *Reviews of Modern Physics*, 87, no. 3 (2015): 925.

Pearce, Neil, and Franco Merletti. "Complexity, Simplicity, and Epidemiology." *International Journal of Epidemiology*, 35 (2006): 515–519.

Pfadenhauer, Lisa M., Ansgar Gerhardus, Kati Mozygemba, Kristin Bakke Lysdahl, Andrew Booth, Bjørn Hofmann, Philip Wahlster et al. "Making Sense of Complexity in Context and Implementation: The Context and Implementation of Complex Interventions (CICI) Framework." *Implementation Science*, 12, no. 1 (2017): 21.

Phillips, Brendon, Madhur Anand, and Chris T. Bauch. "Spatial Early Warning Signals of Social and Epidemiological Tipping Points in a Coupled Behaviour-Disease Network." *Scientific Reports*, 10, no. 1 (2020): 1–12.

Philippe, P., and Omaima Mansi. "Nonlinearity in the Epidemiology of Complex Health and Disease Processes." *Theoretical Medicine and Bioethics*, 19, no. 6 (1998): 591–607.

Pohl, Christian, and Gertrude Hirsch Hadorn. "Methodological Challenges of Transdisciplinary Research." *Natures Sciences Sociétés*, 16, no. 2 (2008): 111–121.

Preise, Rika, Reinette Biggs, Alta De Vos, and Carl Folke. "Social-ecological Systems as Complex Adaptive Systems: Organizing Principles for Advancing Research Methods and Approaches." *Ecology and Society*, 23, no. 4 (2018): 46.

Rhodes, Chris J., Henrik J. Jensen, and Roy M. Anderson. "On the Critical Behaviour of Simple Epidemics." *Proceedings of the Royal Society of London. Series B: Biological Sciences*, 264, no. 1388 (1997): 1639–1646.

Rhodes, Christopher J., A.R. Butler, and Roy M. Anderson. "Epidemiology of Communicable Disease in Small Populations." *Journal of Molecular Medicine*, 76, no. 2 (1998): 111–116.

Rickles, Dean, Penelope Hawe, and Alan Shiell. "A Simple Guide to Chaos and Complexity." *Journal of Epidemiology & Community Health*, 61, no. 11 (2007): 933–937.

Roy, Manojit, Richard D. Zinck, Menno J. Bouma, and Mercedes Pascual. "Epidemic Cholera Spreads Like Wildfire." *Scientific Reports*, 4, no. 1 (2014): 1–7.

Rusoja, Evan, Deson Haynie, Jessica Sievers, Navonil Mustafee, Fred Nelson, Martin Reynolds, Eric Sarriot et al. "Thinking about Complexity in Health: A Systematic Review of the Key Systems Thinking and Complexity Ideas in Health." *Journal of Evaluation in Clinical Practice*, 24, no. 3 (2018): 600–606.

Saba, H., J.G.V. Miranda, and M.A. Moret. "Self-Organized Critical Phenomenon as a Q-Exponential Decay: Avalanche Epidemiology of Dengue." *Physica A: Statistical Mechanics and Its Applications*, 413 (2014): 205–211.

Sah, Pratha, Stephan T. Leu, Paul C. Cross, Peter J. Hudson, and Shweta Bansal. "Unraveling the Disease Consequences and Mechanisms of Modular Structure in Animal Social Networks." *Proceedings of the National Academy of Sciences of the United States of America*, 114, no. 16 (2017): 4165–4170.

Scheffer, Marten. *Critical Transitions in Nature and Society*, Vol. 16 (Princeton University Press, 2009).

Scheffer, Marten, Stephen R. Carpenter, Timothy M. Lenton, Jordi Bascompte, William Brock, Vasilis Dakos, Johan Van de Koppel et al. "Anticipating Critical Transitions." *Science*, 338, no. 6105 (2012): 344–348.

Scheffer, Marten, Stephen R. Carpenter, Vasilis Dakos, and Egbert H. van Nes. "Generic Indicators of Ecological Resilience: Inferring the Chance of a Critical Transition." *Annual Review of Ecology, Evolution, and Systematics*, 46 (2015): 145–167.

Smethurst, Dominic P., and Hywel C. Williams. "Are Hospital Waiting Lists Self-regulating?" *Nature*, 410, no. 6829 (2001): 652–653.

Smethurst, Dominic P., and Hywel C. Williams. "Self-regulation in Hospital Waiting Lists." *Journal of the Royal Society of Medicine*, 95, no. 6 (2002): 287–289.

Sterman, John D. "Learning from Evidence in a Complex World." *American Journal of Public Health*, 96, no. 3 (2006): 505–514.

Stephen C., Harvey Artsob, William Bowie, Michael Drebot, Erin Fraser, Ted Leighton et al. "Perspectives on Emerging Zoonotic Disease Research and Capacity Building in Canada." *Canadian Journal of Infectious Diseases and Medical Microbiology*, 15, no. 6 (2004): 339–344.

Strogatz, Stephen. *Nonlinear Dynamics and Chaos: With Applications to Physics, Biology, Chemistry, and Engineering (Studies in Nonlinearity)*, 2nd Edition (Boulder: Westview Press, 2015), 512.

van de Leemput, Ingrid A., Marieke Wichers, Angélique O.J. Cramer, Denny Borsboom, Francis Tuerlinckx, Peter Kuppens, Egbert H. van Nes et al. "Critical Slowing Down as Early Warning for the Onset and Termination of Depression." *Proceedings of the National Academy of Sciences of the United States of America*, 111, no. 1 (2014): 87–92.

Waltner-Toews, David, and James Kay. "The Evolution of an Ecosystem Approach: The Diamond Schematic and an Adaptive Methodology for Ecosystem Sustainability and Health." *Ecology and Society*, 10, no. 1 (2005).

Wilcox, Bruce A., and Rita R. Colwell. "Emerging and Reemerging Infectious Diseases: Biocomplexity as an Interdisciplinary Paradigm." *EcoHealth*, 2, no. 4 (2005): 244.

Wittrock, Julie, Colleen Duncan, and Craig Stephen. "A Determinants of Health Conceptual Model for Fish and Wildlife Health." *Journal of Wildlife Diseases*, 55, no. 2 (2019), 285–297.

Wu, J., Theresa Burns, P.K. Kawadugama, Tim DeJager, Trishia Westers, Sylvia Scheckley, Carl Ribble et al. "Linkages Between Social-Learning Networks and Farm Sustainability for Smallholder Shrimp Farmers in Sri Lanka." *Asian Fisheries Society*, 29 (2016): 96–111.

Part 3

Actions and Applications

14 Everything Is Connected: Integrating First Nations Perspectives and Connection to Land into Population Health Reporting

Lindsay Beck, Daniéle Behn-Smith, Maya Gislason, Dawn Hoogeveen, Harmony Johnson, Krista Stelkia, Evan Adams, Perry Kendall, and Bonnie Henry

CONTENTS

> So the first question; how does connection to land, water and territory influence health and wellness, mental, physical, spiritual and emotional, that question doesn't make any sense to me personally. That's who I am. I am the land. I am the water. I am the territory. Part of, parts of me and the whole of me. There's no disconnect between me and whatever territory, no matter where I am on Earth. (Marie Oldfield, Knowledge Keeper, Vancouver Coastal, Kitasoo/Xai'xais First Nation)

INTRODUCTION

For First Nations in Canada, land reflects a connection to ancestors, a resource for living, a link to culture and teachings, and a gift for future generations. First

Nations is a term used to describe aboriginal people in Canada who are ethnically neither Métis nor Inuit. In this chapter, we provide an overview of the groundbreaking British Columbia (BC) First Nations health governance structure and describe how it enables a collaborative, two-eyed seeing approach (Iwama et al., 2009) to population health reporting that aims to privilege and elevate the importance of ecological health and connection to land. (The term "land" as used throughout this chapter includes land, water, and the animals and plants as well as other beings that live on this Earth.) Indigenous knowledge systems and ways of being in the world, which emphasize interdependence and reciprocal stewardship with all of our relations in the natural world, can offer solutions for advancing health promotion, equity, and sustainable development now and for future generations (Ratima et al., 2019). This case study explores early lessons from our efforts to integrate First Nations perspectives, notably connection to land, within a population health-reporting framework.

CONNECTIONS BETWEEN LAND AND HEALTH FOR FIRST NATIONS

In January 2020, the BC First Nations Health Authority (FNHA) and the Provincial Health Officer (PHO) released a First Nations Population Health and Wellness Agenda (PHWA) that foregrounds a framework of interconnected indicators that collectively illuminate elements that nourish health and well-being. This includes the relationship between land and health. We quote the report here to ground our case and demonstrate the journey the FNHA has been on to champion First Nations philosophies and worldviews and decolonize Western health care systems in the province of BC:

> Land, water, and territory permeate all aspects of First Nations wellness, as they are sources of healing, and of mental, physical, spiritual, and emotional health and wellness. Land and water nourish culture, form the basis of First Nations languages, systems of governance, and identities. Lands hold stories that guide people how to live a good, healthy life. First Nations people draw sustenance, healing, and medicines from the land. Animals that co-exist on the land and in the water provide teachings, food, clothing, and regalia. Access to healthy lands is required to exercise inherent rights as First Nations – they are stewards of the land and have sacred responsibilities to the land. Given that First Nations health and wellness is borne from a connection to the land, it is seen as a key determinant of health or "root of wellness." Without jurisdiction, access, and continuity of relationship with the land, First Nations peoples cannot fully exercise self-determination. (First Nations Health Authority and the Office of the Provincial Health Officer, 2020).

INTEGRATING "CONNECTION TO LAND" INTO BC FIRST NATIONS POPULATION HEALTH REPORTING

Created by and for BC First Nations, the province-wide FNHA is the first of its kind in Canada. FNHA pursues a vision of "healthy, self-determining and vibrant BC First Nations children, families and communities" through effective health system partnership and integration, as well as management and funding of First Nations health programmes. An important marker of FNHA success is the advancement of First Nations governance and self-determination and the hard-wiring of First Nations perspectives and values into the broader health system, such as the importance of connection to land and the integrity of First Nations territories (First Nations Health Authority, 2019).

To support efforts to change the health system through First Nations governance and self-determination, the FNHA created a First Nations leadership position responsible for monitoring and reporting on First Nations people's health in BC and tracking progress against health and wellness indicators. As a result, the FNHA implemented the only Chief Medical Officer (CMO) in Canada that works from the governance authority of First Nations (First Nations Data Governance Initiative, 2019). In many coastal BC First Nations cultures, the traditional role of the Watchmon is to watch over, protect, and guide their people through difficulties and challenges. Because of its importance, the Watchmon is depicted in carvings and atop poles in many communities. Inspired by these teachings, FNHA's CMO holds an important role as a Watchmon who monitors and acts as a guardian of the health and wellness of First Nations people province-wide. The CMO does so by drawing upon the richness of First Nations knowledge and teachings, alongside the best of mainstream population and public health approaches. The key roles of the CMO as Watchmon are as follows:

- **See and hear**, by gathering information and data from various sources to capture the story of health and wellness of First Nations and Indigenous people in BC
- **Report**, by sharing the story of health and wellness of First Nations and Indigenous people in BC to support improvements in health and wellness
- **Guide**, by providing two-eyed seeing leadership to facilitate a strategic response to what was seen, heard, and reported

Prior to the establishment of the FNHA, BC's PHO held the sole responsibility for reporting on the health of Indigenous peoples living in BC. In 2014, following the establishment of FNHA's Office of the CMO, the two organizations took on a shared responsibility for First Nations population health reporting. Through their commitment to the First Nations perspective of health and wellness (Figure 14.1), strength-based, wellness-focused, and two-eyed seeing approaches in their reporting processes, First Nations population health reporting was now able to create space for, and attend to, First Nations worldviews. For the first time in BC's provincial health reporting history, First Nations were driving their own data and telling their own story in a way that reflected their governance, philosophies, strengths, and resilience.

The FNHA's CMO and the BC PHO developed a First Nations Population Health and Wellness Agenda, which established a suite of health and wellness indicators reflective of First Nations understandings that "everything is connected"

FIGURE 14.1 The First Nations Population Health and Wellness agenda. (Reprinted with permission from the FNHA.)

and which supports a paradigm shift from a sickness-based to a wellness-based philosophy. By bringing together Indigenous knowledge and ways of knowing, and Western knowledge and ways of knowing, the PHWA provides an "eagle eye view" of the health and wellness of First Nations people living across BC. The development of the PHWA was an act of self-determination – as First Nations were controlling their health narrative in a way that reflects their strengths and resilience, and what is important to them.

A key aspect of the PHWA is a focus on the foundational roots of First Nations wellness, in particular the importance of connection to land in achieving FNHA's vision. However, the partners were challenged to hold space and find ways to honour this root of wellness within their population health reporting process, especially in the face of highly entrenched Western views of the environment and land. In general, our extensive review found that existing ecologic and health status reports typically:

1. Characterize the environment as a source of harm or hazard (e.g. pollutants or boil water advisories);
2. Rely on Western values of disturbance, commodification, and productivity (e.g. natural resources (Raibmon, 2018), hectares of productive old growth forest); or
3. Use measurements that are highly disconnected from the ecosystems that support them (e.g. numbers of caribou or salmon in a given area).

These three factors capture a fraction of the picture but none reflect First Nations perspectives which tend to view the environment as all-encompassing and a source of health and healing with connection to identity and ancestors. As described by the Potawatomi botanist Robin Wall Kimmerer (2013),

> In the settler mind, land was property, real estate, capital, or natural resources. But to our people, it was everything: identity, the connection to our ancestors, the home of our nonhuman kinfolk, our pharmacy, our library, the source of all that sustained us. It was a gift, not a commodity.

Integrating First Nations perspectives into a public health agenda means foregrounding water, land, and human and non-human life as interconnected kin.

Studies linking ecological change to human health outcomes are rare in the literature and insufficient to determine an ecological monitoring approach that is valid, understandable, and repeatable (Stephen and Wittrock, 2017). Our search of the literature found no Western evidence-based way to select the appropriate scale, frequency, or ecological variables to construct an ecological indicator; there is much complexity, debate, and inconsistency in how ecological indicators are selected, assessed, and measured (Stephen and Wittrock, 2017). The predominant biophysical focus of ecological approaches fails to consider socio-cultural elements such as the importance of spirituality and the impact of colonialism (Czyzewski, 2011) on ecological health and connection to land.

Following this initial exploration of "Western eye" ecological health indicators, the partners turned to their "First Nations eye" to re-ground. They recognized the need to situate themselves as learners and seek teachings from First Nations Elders, knowledge keepers, and youth on the connections between land and health. The land itself was seen as a profound teacher. While there is a small growing body of literature exploring connection to land, water, and territory for First Nations peoples in Canada, there is still very little empirical literature exploring this within a BC context (Yazzie and Baldy, 2018). To examine this integral connection and to support the development of the PHWA, the We Walk Together project was formed.

SEEKING KNOWLEDGE TO HONOUR CONNECTION TO LAND IN HEALTH REPORTING

The We Walk Together study was initiated in partnership between the FNHA, Office of the BC PHO, and academic partners. Its purpose was to explore the connection between land, water, and territory as an Indigenous determinant of health for BC First Nations. Gatherings were held or planned across diverse areas of the province. The gatherings were land-based and relied heavily on the sharing circle and land-based methodologies to enable First Nations Elders, Knowledge Keepers, and youth to share teachings and co-create their conceptions of the connections between land, health, and wellness.

Preliminary findings revealed a diversity of perspectives and worldviews among BC First Nations peoples, which are varied and specific to peoples and place. However, several overarching commonalities emerged, including a common belief of interrelatedness of all existence and a relationship between the spiritual and physical worlds. The lessons learned from the gatherings demonstrated that connection to the land is an integral component of health and well-being for individuals, families, and communities. This integral connection was described by participants as involving complex, personal relationships with land and all living and non-living things. The intersections between health and connection to land, along with cultural identity, spirituality, ancestral knowledge, and Indigenous ways of living were profound. The preliminary findings tell that connection to land, water, and territory for First Nations in BC is complex and interconnected and can only be captured using Indigenous ways of knowing and being frameworks.

FORWARD DIRECTIONS

It became clearer that the task of describing the complexity of land and human health connections at multiple scales and through various systems does not fit neatly into the logic of wellness indicators. Several key conceptual, methodological, and epistemological issues emerged in relation to the notions of reductionist indicators and wholistic First Nations understandings of land and environment. Careful consideration of these issues led to reflecting on the relevance of indicators, which are artefacts of settler colonial systems and practices. It left us with the following questions:

- How does our understanding and design of indicators need to transform to produce one that aligns with worldviews that value sustainability and the health of all our relations?
- How is it possible to work from First Nations understandings of "everything is connected," "all my relations," and "togetherness" and translate it into a Western scientific framework that works by dichotomizing concepts,

decontextualizing, and studying aspects of larger issues using in-depth Western expertise developed through siloed disciplinary approaches?
- Despite the challenges, is there a role for an ecological indicator that signals and alerts the Watchmon to trends that can be used to catalyze systems change?

Attempting to integrate a connection to land indicator within the PHWA, despite the tensions and challenges, has gone far beyond the previously imagined benefits of sharing data. It has reinforced the continued need to elevate First Nations philosophies to their rightful place in understanding the world (Ratima at al., 2019). This means meaningful representation of First Nations understandings in governance and decision-making processes at multiple levels, including local planning as well as system-wide cross-sectoral work that address the determinants of health. The emphasis on "interdependence and reciprocal stewardship with all of our relations in the natural world" (Ratima et al., 2019) offers solutions not only for advancing health promotion but also for stewardship, the development of public health, health service agendas and interventions in politics, economics, and social life. We hope to have demonstrated how the FNHA is contributing to a critical intervention in a historically Western public health lens. "Hardwiring" of integrated land, health, and wellness perspectives will nourish the roots of wellness not only for First Nations but also for all British Columbians.

AUTHORS' SELF-IDENTIFICATION

- Lindsay Beck, MSc, fifth-generation Canadian settler, American, Norwegian, and English ancestry. Born and raised in the territories of the Yellowknives Dene and North Slave Métis in Yellowknife, NWT.
- Daniéle Behn-Smith, MD, Eh Cho Dene (Big Animal People) of the Fort Nelson First Nations in British Columbia, with Métis roots in the Red River Valley.
- Maya Gislason, PhD, seventh-generation Canadian settler, Icelandic and Scottish ancestry. Born and lives in British Columbia on unceded Coast Salish territory.
- Dawn Hoogeveen, PhD, third-generation Canadian settler of Dutch and British ancestry, raised on Anishinaabe territory on Williams Treaty lands near Peterborough, Ontario.
- Harmony Johnson, MHA, Coast Salish, born and raised in Tla'amin Nation, BC.
- Krista Stelkia, PhD candidate, Sylix/Tlingit from Osoyoos Indian Band, BC.
- Evan Adams, MD, MPH, born and raised in Tla'amin Nation, BC.
- Perry Kendall, MD, Canadian settler born in England.
- Bonnie Henry, MD, MPH, FRCPC, fifth-generation Canadian Settler, Scottish highland and Welsh ancestry. Born and raised in Mi'kmaq territory, PEI and NB.

REFERENCES

Czyzewski, Karina. "Colonialism as a Broader Social Determinant of Health." *International Indigenous Policy Journal*, 2, no. 1 (2011).

First Nations Health Authority. "Summary Service Plan: An Operational Plan for the Fiscal Year 2019/2020." FNHA (2019): 64.

First Nations Data Governance Initiative. "Data Governance Discussion Paper: Rolling Draft." First Nations Health Authority, First Nations Information Governance Centre, West Vancouver, BC, 2019.

First Nations Health Authority and the Office of the Provincial Health Officer of British Columbia. "The First Nations Population Health and Wellness Agenda Summary of Findings." FNHA (2020): 18.

Iwama, Marilyn, Murdena Marshall, Albert Marshall, and Cheryl Bartlett. "Two-Eyed Seeing and the Language of Healing in Community-Based Research." *Canadian Journal of Native Education*, 32, no. 2 (2009): 3–23.

Kimmerer, Robin Wall. "Braiding Sweetgrass: Indigenous Wisdom." *Scientific Knowledge and the Teachings of Plants* (Minneapolis: Milkweed Editions, 2013), 320.

Raibmon, Paige. "Provincializing Europe in Canadian History; Or, How to Talk About Relations Between Indigenous Peoples and Europeans." Active History, 2018. http://activehistory.ca/2018/10/provincializing-europe/

Ratima, Mihi, Debbie Martin, Heather Castleden, and Treena Delormier. "Indigenous Voices and Knowledge Systems: Promoting Planetary Health, Health Equity, and Sustainable Development Now and for Future Generations." *Global Health Promotion* 26, no.3 (2019): 3–5.

Stephen, Craig and Julie Wittrock. "First Nations Population Health & Wellness Agenda: Options for Finding a Path Forward to Incorporate Ecological Health Indicators." Report to the First Nations Health Authority (2017).

Yazzie, Melanie, and Cutcha Risling Baldy. "Introduction: Indigenous Peoples and the Politics of Water." *Decolonization: Indigeneity, Education & Society*, 7, no. 1 (2018): 1–18.

15 Conserving Nature for Health Protection and Climate Change Resilience

Colleen Duncan and Tricia L. Fry

CONTENTS

> "... the care of the earth is our most ancient and most worthy and, after all, our most pleasing responsibility. To cherish what remains of it, and to foster its renewal, is our only legitimate hope." (Wendell Berry, 2018, *The Art of the Commonplace: The Agrarian Essays of Wendell Berry*)

NATURE BUILDS CLIMATE RESILIENCE

In 1949, Aldo Leopold wrote that by "preserving the integrity, stability and beauty of the biotic community," we support the health of ecosystems, communities, populations, and individuals (Leopold, 1949). Now, as the World Health Organization (WHO) and others declare climate change the greatest threat to global health in the 21st century (Costello et al., 2009), the biotic community is needed more than ever to build climate resilience. Broadly defined as the ability to recover from or adjust easily to change, resilience is strengthened through the development and maintenance of healthy populations. The determinants of health model (see Chapter 2) includes abiotic, biotic, and social elements that interact to promote health (PHAoC, 2001; AFMC, 2017). These elements transcend species such that improvements of animal health, which includes the environmental systems that support them, confers an important public health benefit (Stephen and Duncan, 2017; Wittrock et al., 2019). Through this process, Nature contributes to the health of all species.

Health threats from climate change are here and are projected to increase. A wide range of direct and indirect health effects evolve from rising temperature, extreme weather, increased airborne pollution, access to food and clean water, and alterations in the distribution and abundance of pathogens. The health impacts are global, with disproportionate consequences for those less prepared to cope (USGCRP, 2016). Substantial health care costs have already been associated with climate change. In the United States, the direct and indirect health costs of six climate-related events exceeded $14 billion (Knowlton et al., 2011). While much of the health literature on climate change focuses on humans, the effects on animal health cannot be discounted. The pathogenesis of climate-associated disease is presumed to transcend species; however, the topic has received less attention within veterinary circles. Biodiversity loss, declining populations, and species extinctions are attributed to climate change and predicted to increase substantially under projected climate models (Thomas et al., 2004; Thuiller et al., 2011; Wiens, 2016). Given the interrelatedness of human and animal systems, health impacts from climate change in one sector will undoubtedly affect the other.

Interventions to improve health in the face of climate change include both mitigation, actions that help to avoid the unimaginable; and adaptation, actions that help us manage the unavoidable. Climate change mitigation refers to human-driven interventions to reduce emissions or enhance the sinks of greenhouse gases (IPCC, 2018). Mitigation efforts can be characterized as a "public good" whose benefits will be conferred upon many who are not directly involved in the action itself. Mitigation efforts that decrease pollution can have positive health consequences for individuals in other parts of the world or future generations (Landrigan et al., 2018). Climate change adaptation refers to processes that moderate harm or exploit beneficial opportunities within the changing climate (IPCC, 2018). Adaptation exists on a variety of scales, from local to global, and includes physical and social attributes. Examples of climate change adaptations include preparing health care systems to manage the increasing burden of climate-associated illness (Bell, 2011), changes in livestock production systems to maintain animal productivity (Hristov et al., 2018) and wild animals altering their distributions to meet their survival needs (Millar and Westfall, 2010). The protection and promotion of Nature and natural systems contributes to both mitigation and adaptation efforts and is therefore a critical component in the development of resilient systems.

WHY NATURE-BASED ACTION IS NEEDED

The term Nature refers to living and non-living things that occur naturally, ranging in scope from completely natural, wild, systems to elements of the natural world that exist within built environments (Hartig et al., 2014). The interconnectedness of Nature, health, and climate make it a logical area through which health promotions and protection can be maximized. This idea is not new. The idea that one's local built and natural environments influence health dates to Hippocrates

FIGURE 15.1 Schematic representation of pathways through which the natural environment can affect human health. (Figure by Ah Young Kim, based on Hartig et al., 2014.)

(Stephen and Ribble, 2001). There is evidence that open space that includes plants and animals has health benefits, and that higher species' richness increases those benefits even in an urban landscape (Fuller et al., 2007). The extensive human health benefits of Nature include a wide range of cardiovascular, pulmonary, neurological, endocrine, and emotional benefits (Twohig-Bennett and Jones, 2018). These benefits occur through direct and indirect pathways, including air quality, physical activity, social cohesion, and stress reduction (Figure 15.1, Hartig et al., 2014). While there remain gaps in knowledge regarding the type, frequency, and duration of exposure to Nature that confers health benefits, there is ample evidence of the positive effects on physical and mental health (Williams, 2017). Nature access as a health promotion strategy is well documented, and practices such as forest bathing, outdoor preschools, and urban planning projects that promote Nature are being implemented globally (Tan et al., 2013; Hartig and Khan, 2016; Williams, 2017; Xue et al., 2017; White et al., 2019).

Human health and well-being benefits from natural systems extend far beyond the individual. Services that humanity derives from Nature were classified by the Intergovernmental Science-Policy Platform on Biodiversity and Ecosystem Services (IPBES) as "nature's contributions to people" (Díaz et al., 2015). These include provisioning (e.g. production of food and water), regulating (e.g. control of climate and disease), supporting (e.g. nutrient cycles and oxygen production), and cultural (e.g. spiritual and recreational benefits) services (MEA, 2005). Biodiversity, the variety of species in an ecosystem, is the foundation for ecosystem functioning and one of the metrics we can use to measure these services.

Unfortunately, biodiversity is being lost at unprecedented rates (Dirzo et al., 2014). Loss and degradation of habitat is responsible for declining biodiversity and the dysregulation of natural system (Cardinale et al., 2012; Hooper et al., 2012; Rosenberg et al., 2019) and health (Dirzo et al., 2014; Sandifer et al., 2015; Stephens and Athias, 2015). The loss of biodiversity has been deemed as great a threat as the threats posed by climate change. The loss of species and its cascading effects have long been recognized; however, declines in abundance within populations are likely to have even more immediate impactful consequences (Ceballos and Ehrlich, 2002; Gaston and Fuller, 2008; Dirzo et al., 2014). Increasing global temperatures could lead to the extinction of more than one-eighth of Earth's species by the time global temperatures rise 4.3°C above pre-industrial averages (Díaz, 2019). The synergistic impacts of climate change and declines in biodiversity that cannot be mitigated have the potential to topple many species, communities, and ecosystems beyond their tipping point, resulting in potentially unmitigated changes to economies, agriculture, and health.

Climate change potentiates a wide variety of health outcomes and determinants, such as the built environment in which people reside. In their 2016 assessment of the global burden of disease, the WHO attributed 23% of 12.6 million deaths globally to modifiable environmental risks (Prüss-Ustün et al., 2016). While the WHO report did not specifically focus on the role of Nature, it highlighted the urgent need to address climate and ecosystem change as the most challenging and significant future environmental health risks. Several infectious disease epidemics have been attributed to the loss of Nature and biodiversity, particularly the loss of appropriate animal habitat (see *Ecohealth*, 2019, 16:4; Ahmed et al., 2019). For example, in North America, the encroachment of suburbia into forest edges increased the risk of human and companion animals contracting Lyme disease (Ostfeld and Keesing, 2012). The destruction of rainforests in Indonesia triggered the emigration of fruit bats into human-inhabited areas, resulting in the outbreak of the Nipah virus (Epstein et al., 2006). However, the scope and scale of the health impacts of healthy, or unhealthy, environments are considerably greater than these single-agent examples.

Biodiversity preservation is also important in agriculture. Worldwide there is a general decline in the varieties and breeds of plants and animals. The United Nations Food and Agricultural Organization (FAO) highlighted the reliance of food systems on biodiversity and the continual decline of biodiversity within the agricultural sector (FAO, 2019). This lack of diversity compromises agroecosystems resilience against climate change, pests, and pathogens (Díaz, 2019). The United Nations Sustainable Development Goals (SDGs) are a "blueprint to achieve a better and more sustainable future for all" (UN, 2020) (see Chapter 3 for more on the sustainable devcelopment goals). The interconnectedness of health with the other 16 SDGs highlights the strong interdependence of these global issues (Figure 15.2). Health and climate change mitigation is synergistic with other SDGs such as conserving ocean resources, clean energy and, responsible consumption and production (Díaz, 2019). Climate mitigation and adaptation pathways must ensure that other SDGs, such as poverty, water, and energy access, are not compromised in the process (IPCC, 2018).

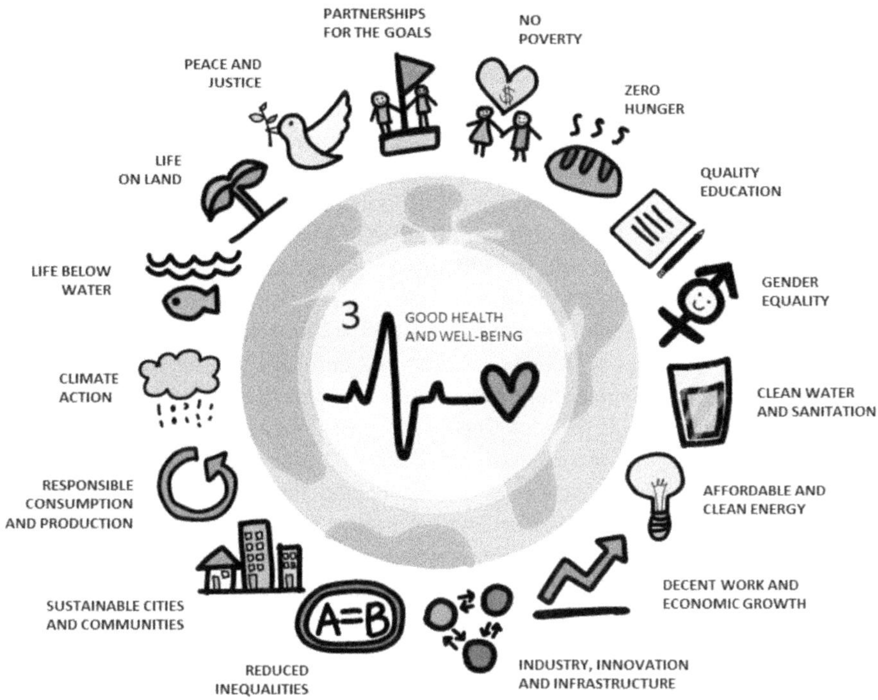

FIGURE 15.2 Relationship between the 17 UN Sustainable Development Goals and environment-health links. (Figure by Ah Young Kim, based on Prüss-Ustün et al., 2016.)

CAN NATURE-BASED ACTION BE DONE?

As our landscapes become more urban and removed from Nature, we need new reminders, novel approaches, and networks to assure the health-affirming benefits of Nature are accessible and sustainable. This is especially true when climate insecurities are pervasive. What constitutes Nature is debated by environmental philosophers and ethicist, and varies by the observer (Nash, 2001; Callicott and Nelson, 1998). What is common is that Nature confers health benefits, whether the gardens of Versailles, a community vegetable plot, or the vast wilderness that ties ecosystems together.

Preserving Nature can be done, even in the most urban environments. Singapore with a population density of 7,615 people per km^2 (Xue et al., 2017) succeeded in increasing herbaceous cover through urban greening (Tan et al., 2013). Singapore's Nature is vertical green space. It is manicured, horticultural, and tamed. Singapore's mission to be a "Garden City" uses native and non-native foliage and technology to blur the lines between city and Nature (Tan et al., 2013). These gardens climb the exterior walls of skyscrapers to provide habitats as well as mitigate the effects of urban heating exacerbated by climate warming.

These features ranked as one of the most important elements to Singapore residents (Roth, 2007; Tan et al., 2013). Singapore's urban development guidelines require not only mitigating green spaces removed by development but also expansion of green spaces within in the urban ecosystem.

The Yellowstone to Yukon (Y2Y) Initiative, a civil action supported by over 300 organizations, is another example of purposeful greening of landscapes (Locke and Heuer, 2015). Y2Y is a 1.2 million km^2 corridor across the Northern Rocky Mountains (Locke and Heuer, 2015; Y2K, 2020). The foundation for its creation was to protect biodiversity and the wilderness character of the land (Locke and Heuer, 2015) with the goal of protecting large landscapes as a solution in support of industry, climate change, and wildlife (Y2Y, 2020). Achieving conservation at such a massive scale was built on small actions, commitments that helped support clean air, water, and diverse plant and animal communities. A component of the Y2Y initiative's success is the idea that actions taken locally are exploited to provide large-scale habitat for a diversity of species (Pearce et al., 2008).

Between curated green space of Singapore and the wilds of the Yukon are many grassroots conservation efforts with missions founded on protecting Nature (e.g. Box 15.1). Wild areas help clean water and air, providing global climate change mitigation services. It takes large-scale efforts like those supported by Wildlife Conservation Society and World Wildlife Foundation as well as governments to protect true wilderness; also important is the support of individuals who advocate for Nature through their election of conservation-minded officials and regard of Nature as a resource for all.

PATH FORWARD

Incorporating the extrinsic and intrinsic benefits of Nature with the health sectors is a critical health protection and climate action strategy. While the hurdles are formidable (Rosenberg et al., 2019), conserving and sustainably using Nature can be achieved through "transformative changes across economic, social, political and technological factors" (Díaz et al., 2019). The health co-benefits of climate action are innumerable with simultaneous ecological benefits at the global scale: clean air, clean water, and the preservation of carbon sinks within habitats that protect biodiversity, populations, and species from the effects of climate change (Warren et al., 2013) and offer opportunities to build health and resilience.

Climate change health impacts may be gradual, like increasing temperatures, or sudden as with extreme weather events. The extreme events have the most potential to cause the most damage to humans (Streets and Glantz, 2000). To minimize the impact of catastrophic events, we must enhance coping capacity within human and animal systems (Stephen et al., 2015). This capacity can be expanded, in part, through conservation and preservation of biodiversity and Nature. Health benefits of Nature transcend species; and, therefore, foster reciprocal care/interdependence of health. Of importance is the need to link health professionals with environmental and climate advocacy groups (Barrett et al., 2015). Intersectoral

BOX 15.1 UMBRELLA SPECIES CONSERVATION FOR CLIMATE MITIGATION

Improving health through mitigation can begin with a single species, even one that is only a few inches long. The monarch butterfly protects Nature across the urban to rural gradient from paved to forested landscape. It is an umbrella species whose protection and management conserve habitat and other species well beyond the umbrella species itself. Natural and anthropogenic influences have resulted in 80% decline in the monarch population over the past 20 years (Rendon-Salinas, 2015). The interest in protectinghabitat for monarch butterflies, a species that completes its 3,200 km migration in up to five generations, burgeoned with action at local, state/provincial, federal, and transnational levels. Monarchs require milkweed as a source of nutrition and a substrate for laying eggs. Milkweed is easy to grow in prairies, along road edges, and in backyard gardens. The advent of butterfly gardens in urban landscapes results in open spaces with a diverse plant and animal community. Conservation efforts encouraged planting milkweed and native prairie species along field and road edges to support monarchs and other species. Mexico protected millions of acres of habitat and banned logging on monarch overwintering sites. In 2015, the United States committed to creating a "butterfly highway" traversing North America, a "safe path" for migration. These efforts help protect and conserve thousands of additional species, and their habitats, which in turn act as carbon sinks and aid in the mitigation of climate change, thus supporting Nature and health.

health teams need to value and employ greater input from the environmental sector (Stephen and Karesh, 2014; Barrett and Bouley, 2015; Destoumieux-Garzón et al., 2018). Framing climate change as a health issue can accelerate response and position health professionals as leaders in this battle (Watts et al., 2015). There is a need for more climate change mitigation and adaptation innovations as animal and human populations become increasingly vulnerable to climate change effects (Stephen and Wade, 2020). While high-functioning teams and diverse collaborations are critical components, climate mitigation and adaptations can take place while those are developed.

Mechanisms are in place to protect natural spaces through the hundreds of non-governmental organizations as well as governments that support biodiversity. The use of biodiversity-friendly practices is reported to be increasing (FAO, 2019), but loss of Nature continues to outpace protections. The ecosystems we share provide health protection and promotion for all species. Keeping each other healthy and resilient is one of the impactful ways we promote climate adaptation. The world can reverse this biodiversity crisis, but doing so will require proactive environmental policies, the sustainable production of food and other resources along with a concerted effort to reduce greenhouse gas emissions.

Aldo Leopold (1949) noted that "wild things were taken for granted until progress began to do away with them," resulting in more than the loss of landscapes. Conserving Nature in the urban landscape and in the wilderness, for pleasure and food, affords us the beauty, services, and resilience needed to adapt to a climate change. It is difficult to refute the importance that Nature plays in health outcomes and that degradation of Nature increases pathologic conditions. While the influence of individuals and civil society can result in big actions, the importance of government policy on the issue of climate change and Nature protection cannot be underestimated. The list of how to protect Nature is overwhelming, but actions need not be. Simple actions – such as planting a garden, riding your bike, appreciating Nature – have the power to promote the protection of Nature and the persistence of your health but must be accompanied by advocacy and voting to promote government change.

REFERENCES

AFMC. "Determinants of Health and Health Inequalities," in *AFMC Primer on Population Health*, 2nd Edition (Association of Faculties of Medicine of Canada, 2017).

Ahmed, Kamruddin, Mohammad Saffree Jeffree, Tom Hughes, and Peter Daszak. "Editorial: Can the Health Implications of Land-use Change Drive Sustainability?" *EcoHealth*, 16, no. 4 (2019): 585–586. doi: 10.1007/s10393-019-01462-y.

Barrett, Meredith A., and Timothy A. Bouley. "Need for Enhanced Environmental Representation in the Implementation of One Health." *EcoHealth*, 12, no. 2 (2015): 212–219. doi: 10.1007/s10393-014-0964-5.

Barrett, Bruce, Joel W. Charles, and Jonathan L. Temte. "Climate Change, Human Health, and Epidemiological Transition." *Preventive Medicine*, 70 (2015): 69–75. doi: https://doi.org/10.1016/j.ypmed.2014.11.013.

Bell, Erica. "Readying Health Services for Climate Change: A Policy Framework for Regional Development." *American Journal of Public Health*, 101, no. 5 (2011): 804–813. doi: 10.2105/AJPH.2010.202820.

Berry, Wendell. *The Art of the Commonplace: The Agrarian Essays of Wendell Berry* (Counterpoint, 2018).

Callicott, J. Baird, and Michael P. Nelson (eds.). *The Great New Wilderness Debate* (Athens: University of Georgia Press, 1998), 697.

Cardinale, Bradley J., J. Emmett Duffy, Andrew Gonzalez, David U. Hooper, Charles Perrings, Patrick Venail, Anita Narwani et al. "Biodiversity Loss and Its Impact on Humanity." *Nature*, 486, no. 7401 (2012): 59–67.

Ceballos, Gerardo, and Paul R. Ehrlich. "Mammal Population Losses and the Extinction Crisis." *Science*, 296, no. 5569 (2002): 904–907.

Costello, Anthony, Mustafa Abbas, Adriana Allen, Sarah Ball, Sarah Bell, Richard Bellamy, Sharon Friel et al. "Managing the Health Effects of Climate Change: Lancet and University College London Institute for Global Health Commission." *Lancet*, 373, no. 9676 (2009):1693–1733.

Destoumieux-Garzón, Delphine, Patrick Mavingui, Gilles Boetsch, Jérôme Boissier, Frédéric Darriet, Priscilla Duboz, Clémentine Fritsch et al. "The One Health Concept: 10 Years Old and a Long Road Ahead." *Frontiers in Veterinary Science*, 5 (2018): 14. doi: 10.3389/fvets.2018.00014.

Díaz, Sandra, Sebsebe Demissew, Julia Carabias, Carlos Joly, Mark Lonsdale, Neville Ash, Anne Larigauderie et al. "The IPBES Conceptual Framework: Connecting Nature and People." *Current Opinion in Environmental Sustainability*, 14 (2015): 1–16.

Díaz, Sandra, Josef Settele, Eduardo S. Brondízio, Hien T. Ngo, John Agard, Almut Arneth, Patricia Balvanera et al. "Pervasive Human-Driven Decline of Life on Earth Points to the Need for Transformative Change." *Science*, 366, no. 6471 (2019).

Díaz, S., J. Settele, E.S. Brondízio, H.T. Ngo, M. Guèze, J. Agard, A. Arneth et al. (eds.) *Summary for Policymakers of the Global Assessment Report on Biodiversity and Ecosystem Services of the Intergovernmental Science-Policy Platform on Biodiversity and Ecosystem Services* (Bonn: IPBES Secretariat, 2019), 56.

Dirzo, Rodolfo, Hillary S. Young, Mauro Galetti, Gerardo Ceballos, Nick J. B. Isaac, and Ben Collen. "Defaunation in the Anthropocene." *Science*, 345 (2014): 401–406. doi: 10.1126/science.1251817.

Epstein, Jonathan H., Hume E. Field, Stephen Luby, Juliet R.C. Pulliam, and Peter Daszak. "Nipah Virus: Impact, Origins, and Causes of Emergence." *Current Infectious Disease Reports*, 8, no. 1 (2006): 59–65.

FAO. *The State of the World's Biodiversity for Food and Agriculture*, eds. J. Bélanger and D. Pilling (Rome: Food and Agriculture Organization Commission on Genetic Resources for Food and Agriculture Assessments, 2019), 572.

Fuller, Richard A., Katherine N. Irvine, Patrick Devine-Wright, Philip H. Warren, and Kevin J. Gaston. "Psychological Benefits of Greenspace Increase with Biodiversity." *Biology Letters*, 3, no. 4 (2007): 390–394.

Gaston, Kevin J., and Richard A. Fuller. "Commonness, Population Depletion and Conservation Biology." *Trends in Ecology and Evolution*, 23, no. 1 (2008): 14–19.

Hartig, Terry, and Peter H. Kahn. "Living in Cities, Naturally." *Science*, 352, no. 6288 (2016): 938–940.

Hartig, Terry, Richard Mitchell, Sjerp de Vries, and Howard Frumkin. "Nature and Health." *Annual Review of Public Health*, 35, no. 1 (2014): 207–228.

Hooper, David U., E. Carol Adair, Bradley J. Cardinale, Jarrett E. K. Byrnes, Bruce A. Hungate, Kristin L. Matulich, Andrew Gonzalez et al. 2012. "A Global Synthesis Reveals Biodiversity Loss as a Major Driver of Ecosystem Change." *Nature*, 486, no.7401 (2012):105–108.

Hristov, A.N., A.T. Degaetano, C.A. Rotz, E. Hoberg, R.H. Skinner, T. Felix, H. Li et al. "Climate Change Effects on Livestock in the Northeast US and Strategies for Adaptation." *Climatic Change*, 146, no. 1 (2018): 33–45.

IPCC. "Special Report Global Warming of 1.5°C," eds. V. Masson-Delmotte, P. Zhai, H.-O. Pörtner, D. Roberts, J. Skea, P.R. Shukla, A. Pirani et al. (Intergovernmental Panel on Climate Change, 2018).

Knowlton, Kim, Miriam Rotkin-Ellman, Linda Geballe, Wendy Max, and Gina M. Solomon. "Six Climate Change–Related Events in the United States Accounted for About $14 Billion in Lost Lives and Health Costs." *Health Affairs*, 30, no. 11 (2011): 2167–2176.

Landrigan, Philip J., Richard Fuller, Nereus J.R. Acosta, Olusoji Adeyi, Robert Arnold, Abdoulaye Bibi Baldé, Roberto Bertollini et al. "Lancet Commission on Pollution and Health." *Lancet*, 391, no. 10119 (2018): 462–512.

Leopold, Aldous. *A Sand County Almanac* (New York: Oxford University Press, 1949), 222.

Locke, Harvey, and Karsten Heuer. "Yellowstone to Yukon: Global Conservation Innovations Through the Years," in *Protecting the Wild: Parks and Wilderness, the Foundation for Conservation*, eds. George Wuerthner, Eileen Crist, and Tom Butler (Washington: Island Press/Center for Resource Economics, 2015), 120–130.

Millar, Constance I., and Robert D. Westfall. "Distribution and Climatic Relationships of the American Pika (*Ochotona Princeps*) in the Sierra Nevada and Western Great Basin, USA; Periglacial Landforms as Refugia in Warming Climates." *Arctic, Antarctic, and Alpine Research*, 42, no. 1 (2010): 76–88.

MEA (Millennium Ecosystem Assessment). *Ecosystems and Human Well-Being: Synthesis* (Washington, DC: World Resources Institute, 2005), 36.

Nash, Roderick Frazier. *Wilderness and the American Mind*, 4th Edition (New Haven: Yale University Press, reprint 2001).

Ostfeld, Richard S., and Felicia Keesing. "Effects of Host Diversity on Infectious Disease." *Annual Review of Ecology, Evolution, and Systematics*, 43 (2012): 157–182.

Pearce, Jennie L., David Anthony Kirk, Cynthia P. Lane, Marguerite H. Mahr, John Walmsley, Daniel Casey, Judy E. Muir et al. "Prioritizing Avian Conservation Areas for the Yellowstone to Yukon Region of North America." *Biological Conservation*, 141, no. 4 (2008): 908–924.

PHAoC. "What Determines Health?" Public Health Agency of Canada, Government of Canada, 2001. http://www.phac-aspc.gc.ca/ph-sp/determinants/index-eng.php.

Prüss-Ustün, Annette, J. Wolf, Carlos F. Corvalán, R. Bos, and Maria Neira. *Preventing Disease Through Healthy Environments: A Global Assessment of the Burden of Disease from Environmental Risks* (Geneva: World Health Organization, 2016), 147.

Rendon-Salinas E., F. Martinez-Meza, and A. Fajardo-Arroy. *Forest Surface Occupied by Monarch Butterfly Hibernation Colonies in December 2014* (Mexico: World Wildlife Fund, 2015), 4.

Rosenberg, Kenneth V., Adriaan M. Dokter, Peter J. Blancher, John R. Sauer, Adam C. Smith, Paul A. Smith, Jessica C. Stanton et al. "Decline of the North American Avifauna." *Science*, 366, no. 6461 (2019): 120–124.

Roth, Matthias. "Review of Urban Climate Research in (Sub)Tropical Regions." *International Journal of Climatology: A Journal of the Royal Meteorological Society*, 27, no. 14 (2007): 1859–1873.

Sandifer, Paul A., Ariana E. Sutton-Grier, and Bethney P. Ward. "Exploring Connections Among Nature, Biodiversity, Ecosystem Services, and Human Health and Well-being: Opportunities to Enhance Health and Biodiversity Conservation." *Ecosystem Services*, 12 (2015): 1–15.

Stephen, Craig, John Berezowski, and Vikram Misra. "Surprise Is a Neglected Aspect of Emerging Infectious Disease." *EcoHealth*, 12, no. 2 (2015): 208–211.

Stephen, Craig, and Colleen Duncan. "Can Wildlife Surveillance Contribute to Public Health Preparedness for Climate Change? A Canadian Perspective." *Climatic Change*, 141, no. 2 (2017): 259–271.

Stephen, C., and W.B. Karesh. "Is One Health Delivering Results? Introduction." *Revue Scientifique et Technique (International Office of Epizootics)*, 33, no. 2 (2014): 375–392.

Stephen, Craig, and Carl Ribble. "Death, Disease and Deformity: Using Outbreaks in Animals as Sentinels for Emerging Environmental Health Risks." *Global Change and Human Health*, 2, no. 2 (2001): 108–117.

Stephen Craig, and Joy Wade. "Missing in Action: Sustainable Climate Change Adaptation Evidence for Animal Health." *The Canadian Veterinary Journal* 2020.

Stephens, Carolyn, and Renato Athias. "Focus on Biodiversity, Health and Wellbeing." *Environmental Research Letters*, 10, no. 12 (2015): 120401.

Streets, David G., and Michael H. Glantz. "Exploring the Concept of Climate Surprise." *Global Environmental Change*, 10, no. 2 (2000): 97–107.

Tan, Puay Yok, James Wang, and Angelia Sia. "Perspectives on Five Decades of the Urban Greening of Singapore." *Cities*, 32 (2013): 24–32.

Thomas, Chris D., Alison Cameron, Rhys E. Green, Michel Bakkenes, Linda J. Beaumont, Yvonne C. Collingham, Barend F.N. Erasmus et al. "Extinction Risk from Climate Change." *Nature*, 427, no. 6970 (2004): 145–148.

Thuiller, Wilfried, Sébastien Lavergne, Cristina Roquet, Isabelle Boulangeat, Bruno Lafourcade, and Miguel B. Araujo. "Consequences of Climate Change on the Tree of Life in Europe." *Nature*, 470, no. 7335 (2011): 531–534.

Twohig-Bennett, Caoimhe, and Andy Jones. "The Health Benefits of the Great Outdoors: A Systematic Review and Meta-Analysis of Greenspace Exposure and Health Outcomes." *Environmental Research*, 166 (2018): 628–637.

UN. "About the Sustainable Development Goals." United Nations, 2020. https://www.un.org/sustainabledevelopment/sustainable-development-goals/ (accessed June 2020).

USGCRP. "Populations of Concern" in *The Impacts of Climate Change on Human Health in the United States: A Scientific Assessment*, eds. J. Balbus A., J.L. Gamble, C.B. Beard, J.E. Bell, D. Dodgen, R.J. Eisen, N. Fann et al. (Washington, DC: U.S. Global Change Research Program, 2016), 312.

Warren, R., Jeremy VanDerWal, J. Price, Justin A. Welbergen, I. Atkinson, Julián Ramirez-Villegas, Timothy J. Osborn et al. "Quantifying the Benefit of Early Climate Change Mitigation in Avoiding Biodiversity Loss." *Nature Climate Change*, 3, no. 7 (2013): 678–682.

Watts, Nick, W. Neil Adger, Paolo Agnolucci, Jason Blackstock, Peter Byass, Wenjia Cai, Sarah Chaytor et al. "Health and Climate Change: Policy Responses to Protect Public Health." *Lancet*, 386, no. 10006 (2015): 1861–1914.

White, Mathew P., Ian Alcock, James Grellier, Benedict W. Wheeler, Terry Hartig, Sara L. Warber, Angie Bone et al. "Spending at Least 120 Minutes a Week in Nature Is Associated with Good Health and Wellbeing." *Scientific Reports,* 9, no. 1 (2019): 1–11.

Wiens, John J. "Climate-Related Local Extinctions Are Already Widespread Among Plant and Animal Species." *PLoS Biology*, 14, no. 12 (2016).

Williams, F. *"The nature fix: Why nature makes us happier, healthier, and more creative."* New York, NY: WW Norton & Company (2017).

Wittrock, Julie, Colleen Duncan, and Craig Stephen. "A Determinants of Health Conceptual Model for Fish and Wildlife Health." *Journal of Wildlife Diseases*, 55, no. 2 (2019): 285–297.

Xue, Fei, Zhonghua Gou, and Stephen Lau. "The Green Open Space Development Model and Associated Use Behaviors in Dense Urbn Settings: Lessons from Hong Kong and Singapore." *Urban Design International*, 22, no. 4 (2017): 287–302.

Y2Y. "Yellowstone to Yukon Conservation Initiative." Yellowstone to Yukon Conservation Initiative, 2020. https://y2y.net (accessed June 2020).

16 Managing Zoonotic Disease in Wildlife Populations: Priorities and Pitfalls of the Human Connection

Todd K. Shury, Ryan K. Brook, and Pushpakumara D.B. Nihal

CONTENTS

Livestock disease management and surveillance can be very challenging but are even more difficult in free-ranging wildlife, due to the disparate and often remote nature of wildlife populations (Miller et al., 2013; Silk et al., 2019). Wildlife are generally either highly valued components of society or considered pests which must be eradicated. They are often publicly owned, highly mobile, and do not respect regional or national borders. A One Health approach has been advocated as a viable tactic to deal with wildlife reservoirs of disease This approach argues for including human, animal, and environmental considerations in all social and ecological aspects of disease management, with the ultimate goal of improving or reducing threats through harm reduction (Miller and Olea-Popelka, 2013; Buttke et al., 2015). This is easier done in theory than in practice. There are few documented and evaluated examples where One Health has been successful in a wildlife context. An interdisciplinary One Health approach typically takes longer, is more costly, and more logistically challenging due to large numbers of stakeholders with conflicting values, which makes group decision-making more complex. We use two examples, one in Canada and the other in Sri Lanka, to highlight cross-cultural lessons learned and how critical it is to consider local socio-economic factors. We believe wildlife managers, conservationists, public and private landholders, and societies can apply these lessons for successful disease management.

BOVINE TUBERCULOSIS IN CANADA

Bovine tuberculosis (TB) is an insidious disease primarily of cattle, with the causative bacterium *Mycobacterium bovis*, having a substantial host range among many mammals, including humans. There are numerous wildlife reservoirs of TB worldwide which have complicated eradication or control efforts in South Africa, New Zealand, the United Kingdom, Spain, the United States, and Canada (Palmer, 2013). Canada has had a national TB eradication programme since 1923 and has been considered free of bovine TB in its cattle herd since 1985 (Wobeser, 2009). Following the first bovine TB-positive wild elk (*Cervus canadensis*) in the area around Riding Mountain National Park in 1992 in the province of Manitoba (Lees, 2004), a Manitoba Bovine Tuberculosis Task Force was created to help manage the disease in wildlife and cattle. The task force comprised four government agencies: two involving wildlife (Manitoba Sustainable Development, and the Parks Canada Agency), and two involving agriculture (Manitoba Agriculture and Resource Development, and the Canadian Food Inspection Agency); plus two non-governmental organizations, the Manitoba Beef Producers, and the Manitoba Wildlife Federation. The task force's primary goal was eradicating bovine TB from the Riding Mountain ecosystem. The primary reservoir species, undetermined at the beginning of the outbreak were elk, with white-tailed deer (*Odocoileus virginianus*), and cattle were the spillover species likely unable to maintain infection on their own without ongoing re-infection (Shury and Bergeson, 2011; Shury, 2014). A series of town hall meetings at the beginning of this process proved very confrontational and increased conflict (Brook and McLachlan, 2008; Brook, 2009). Vocal individuals advocated for extreme responses, including total park fencing, even though subsequent research showed that few people supported such responses (Brook, 2008). Research was initiated to learn more about cattle-wildlife interactions and to better understand the human dimensions behind some of the frustrations voiced by local cattle producers (Brook and McLachlan, 2008; Brook, 2009, 2010). These studies proved key in understanding both the wildlife-livestock interactions at a regional scale as well as the attitudes and beliefs of local ranchers, who proved key in helping eventually eradicate bovine TB from the National Park (Brook and McLachlan, 2006). Five integrated principles for collaborative disease management were learned from the successful eradication of bovine TB from the greater Riding Mountain Ecosystem (Table 16.1).

Adaptive management is a structured approach to intractable problems that involves instituting possible solutions, measuring or monitoring these actions, and adjusting these actions based on ongoing learning (Miller et al., 2013; Thirgood, 2009). Options for TB management came from locally generated research through partnerships with universities as well as from international conferences and symposia. Annual management plans prioritized surveillance and monitoring in both wildlife and livestock and a jointly developed, transparent budget. The Riding Mountain Eradication Area (RMEA) was created in 2000 as part of zoning efforts to compartmentalize the disease and focus surveillance efforts on

TABLE 16.1
Integrated Principles for Collaborative Socio-Ecological Management Learned From the Successful Eradication of Bovine TB at the Wildlife-Agriculture Interface in Manitoba, Canada

Use adaptive management
Have clear communication between decision-makers and communities
Incorporate multiple knowledge systems
Use long-term thinking
Prioritize innovation and flexibility during the planning, implementation, and evaluation phases

both livestock and wildlife (Lees et al., 2003). It also resulted in the creation of a wildlife laboratory to facilitate wildlife sample testing within the National Park.

A Stakeholder Advisory Committee and a Scientific Review Committee provided a forum to generate consensus for adaptive management, bringing together both wildlife and livestock interests with very different viewpoints and methods. The Stakeholder Advisory Committee used a professional facilitator, and the Science Advisory Committee was chaired by an independent scientist. These groups provided a mechanism for clear communication between decision-makers and communities. Mistrust between the National Park and local cattle producers was a serious obstacle at the beginning of the programme, with calls to eradicate the entire elk population within the park (Brook, 2009). This abated over time as more collaborative solutions, trust, and relationships were built within the structures of the task force.

Another effective method of communication was concurrent sociologic research, which resulted in numerous conversations between the primary researcher and local cattle producers over a cup of coffee at the farmer's kitchen table (Brook and McLachlan, 2006, 2009; Brook, 2013, 2015) (Figure 16.1). This created not only an effective way for cattle producers' voices to be heard, but also developed trust very quickly. Relationships were never perfect and other issues and conflicts that have occurred over decades over other issues, such as wolf control, remained (Stronen et al., 2007).

One of the most difficult aspects of the TB programme was the incorporation of multiple knowledge systems. Turning scientific knowledge as well as local ecological and traditional knowledge into effective policy decisions was very challenging (see Chapter 10 for more on healthy policy). Once it became clear, through epidemiological research, that bovine TB was restricted to one area in the western region of the park, management activities could focus on this area (Shury, 2015; Brook et al., 2013). Local ecological and scientific knowledge aligned very closely to identify this area (Brook and McLachlan, 2009). Subsequent actions included legislation to prohibit baiting of wild ungulates for hunting, construction of over 150 high fence hay storage areas to prevent

FIGURE 16.1 One-on-one discussion with rancher about elk and deer interactions with domestic animals on their farm. (Photo credit Ryan Brook.)

wildlife access to cattle feed, on-farm biosecurity measures, intensive livestock testing, and testing and removal of TB-infected wild ungulates through both liberal extended hunting seasons and helicopter capture of wild ungulates within the park. Indigenous hunters were recruited to assist with reducing ungulate population densities. Helicopter capture became less acceptable over time and eventually ceased. This adaptive management was enabled by incorporating different knowledge systems and stakeholder viewpoints.

Long-term thinking was another critical success factor. Communicating the persistent nature of bovine TB to policymakers, stakeholders, and communities ensured that expectations were realistic. "TB fatigue" affected government staff, local cattle producers, as well as hunters and indigenous communities, as the management programme spanned a period of over 20 years. It was vitally important that decision-makers understood that success may not be achieved within their four- or five-year mandate, but that progress was being made and eradication remained a valid long-term objective. Establishing trusting relationships between government agencies at different jurisdictional levels with very different mandates, and associated stakeholder groups, was important in creating a shared long-term vision at a landscape scale. Lack of trust is often the greatest barrier to successfully dealing with other wildlife reservoirs around the world.

The last lesson from TB management in Manitoba was the importance of innovation and flexibility in the planning, implementation, and evaluation phases of the cross-sectoral, disease management programme. Several novel diagnostic tests were adopted early for wildlife surveillance, some which were very effective in elk and white-tailed deer (Shury et al., 2014). Understanding local opinions and values through social science research was important to build bridges with local stakeholders and to understand what management activities would have

FIGURE 16.2 Hay storage fence to prevent wildlife from accessing cattle forage. (Photo by Todd Shury.)

a high likelihood of acceptance. This knowledge was very important for creating on-farm risk assessments which were used to tailor unique management approaches. Flexibility in how resources were spent resulted in shared management approaches, which resulted in over 150 hay storage fences being built on private ranches bordering the park (Gooding and Brook, 2014; Brook, 2015). Government agencies paid for the capital costs for these fences (Figure 16.2), while local landowners were responsible for operation and maintenance. This collaboration created a visible reminder to local cattle producers that they had a direct hand in managing bovine TB on their farms.

WILDLIFE SURVEILLANCE IN SRI LANKA

The need to be attentive to human dimensions of wildlife disease surveillance and control was also seen in Sri Lanka, where rabies is a zoonotic disease which occurs at the interface between humans, wildlife, and domestic animals (Sánchez-Soriano et al., 2019). Zoonotic diseases such as rabies, leptospirosis, and Japanese encephalitis remain significant public health threats in the country. A previous review of Sri Lanka's emerging infectious disease preparedness (Dissanayake et al., 2012) noted a national vulnerability due to lack of information on the status of wildlife diseases. South Asia has been identified as a hot spot for zoonotic infectious disease emergence, but limited capacity has kept wildlife departments from being active participants in preparedness and response efforts. Coordinating and integrating surveillance activities was a key challenge in controlling animal rabies in Sri Lanka, a country which has less developed wildlife health infrastructure and more limited funding than wealthier countries. Persistent challenges affecting collection of high-quality wildlife health data include lack of training, lack of access to facilities, and lack of communication amongst government departments with responsibilities for livestock, wildlife, and public health (Dissanayake et al., 2012). Logistic difficulties such as lack of instruments, materials, transport facilities, diagnostic facilities, advanced technologies, and trained personnel impede a cross-sectoral coordinated rabies surveillance system. Further wildlife disease surveillance constraints arise from reliance on convenience and opportunistic sampling. This prevents wildlife surveillance systems from meeting expectations for public health or domestic animal surveillance (Stephen et al., 2019) and leads to conflicting expectations between sectors on what a wildlife disease surveillance system should produce.

Sri Lanka's legislative mechanisms further complicated surveillance system development. The Fauna and Flora Protection Act No. 44 and its amendments, for example, require the collection and submission of wild animal samples to diagnostic laboratories to be done by staff of the Department of Wildlife Conservation (DWC). Transporting or keeping wild animals, their bodies, or parts by others are offenses under this Act. This discourages citizens or members of other government programmes from contributing samples for surveillance. Wildlife disease surveillance is a public good, but there is no encouragement of private citizens to submit wildlife surveillance samples. The DWC manages most wildlife habitats, including nature reserves, national parks, jungle corridors, and marine reserves, but other forests are managed by the Department of Forests. DWC wildlife veterinarians' main duties are to prevent and control wild animal and zoonotic diseases and mitigate human-wildlife conflict. DWC veterinarians are minimally involved in domestic animal–related cases and public outreach. Even though DWC wildlife veterinarians are directly involved in handling and treating wildlife and post-mortem procedures, they lack laboratory investigation capacity. This means disease investigation must go through the Department of Animal Production and Health (DAPH) veterinarians whose efforts are dedicated mostly to farm animals. Medical Officers of Health (MOH) and their laboratory infrastructure are also involved in diagnosing zoonotic wildlife disease, like rabies. Governance, accountability, and legal authorities in Sri Lanka lead to fragmented surveillance responsibilities and constrained options for collaborative rabies surveillance.

A university-government collaboration called the Sri Lankan Wildlife Health Centre (SLWHC) was developed to overcome some of these challenges (Nihal et al., 2020). The objective of the SLWHC is to track wildlife diseases and build capacity to inform conservation, agriculture, and public health actions (Kolla, 2020). The SLWHC helps centralize the flow of surveillance samples and information by engaging a diversity of stakeholders and programme implementers with interests in wildlife diseases. The SLWHC brings relevant government ministries responsible for domestic animals, human health, and wildlife management together for the first time to develop and govern a new wildlife diagnostics and surveillance programme. This allows for relationships to develop to facilitate better understanding of each other's needs and capabilities. Establishing or facilitating relationships between DWC and DAPH officers benefitted a collaborative surveillance system. The SLWHC initially provided training and capacity development to its partner agencies to develop a shared knowledge base across partner organizations as well as to overcome some rabies surveillance logistic challenges. By empowering key stakeholders and enhancing core capacities, it is possible to address some disease surveillance constraints (Halliday et al., 2012).

Through SLWHC-sponsored research, government agencies became aware of workplace or regulatory changes that reduced staff willingness to be involved in wildlife surveillance when wildlife disease management was not their primary responsibility. This was an issue particularly for field workers most likely to encounter sick or dead wild animals. Researchers met with local communities to understand their concerns about rabies and other wildlife diseases. Surveillance systems that are useful for communities and address the needs of regulatory agencies are often more sustainable and can enable the development of longer-term capacities (Malani and Laxminarayan, 2011). The SLWHC subsequently prioritized rabies as a targeted disease, because it was a priority shared across the DWC, DAPH, and MoH and local communities. Working on a shared priority can foster relationships needed to improve communications and trust across sectors dealing with wildlife health in the country. SLWHC partners explored innovative methods to use existing knowledge, personnel, and capacities. For example, a mobile phone-based surveillance system was shown to be effective for field-level disease investigation (Robertson et al., 2010). The Sri Lankan experience highlights the need to design surveillance systems with the constraints, capacities, priorities, and direct input of all stakeholders in mind.

SUMMARY

Experiences in Canada and Sri Lanka in engaging citizens and various levels of government for effective One Health interventions related to wildlife have demonstrated some degree of success, but more can be done. Honestly engaging key stakeholder communities and ensuring they are involved in decision-making and implementation regarding highly valued ecosystem components, such as wildlife, is critical to success. Adopting adaptive management as a strategy for managing disease issues in wildlife in a One Health context helps ensure successful and long-lasting positive outcomes for all sectors of society, including wildlife, domestic animal, and human populations. Managing disease in wildlife populations is often portrayed as a two-sided, zero sum battle where one sector must lose something to allow for positive outcomes in another sector. Experience in dealing with rabies surveillance and bovine TB control demonstrates that this does not have to be the case. Approaches that can benefit wildlife, domestic animals, and humans can be achieved if the practices outlined in this chapter are adopted and put in place for the long term. Strong leadership from government agencies, non-governmental organizations, and citizens will be necessary. This can happen with meaningful engagement at a local level where shared solutions are emphasized over drastic options which often only benefit one sector or group. Relationships are complicated and we should always expect a range of responses, but starting with those that are willing to engage and building trust through effective communication and knowledge sharing can result in transformative approaches that produce win-win scenarios.

REFERENCES

Brook, Ryan K. "Elk-Agriculture Conflicts in the Greater Riding Mountain Ecosystem: Building Bridges between the Natural and Social Sciences to Promote Sustainability." PhD thesis, Faculty of Graduate Studies, University of Manitoba (2008): 364.

Brook, Ryan K. "Historical Review of Elk-Agriculture Conflicts in and around Riding Mountain National Park, Manitoba, Canada." *Human-Wildlife Conflicts*, 3, no. 1 (2009): 72–87.

Brook, Ryan K. "Incorporating Farmer Observations in Efforts to Manage Bovine Tuberculosis Using Barrier Fencing at the Wildlife-Livestock Interface." *Preventive Veterinary Medicine*, 94, no. 3–4 (2010): 301–305.

Brook, Ryan K. "Farmer Perceptions of Hay Yard Barrier Fence Effectiveness in Mitigating Bovine Tuberculosis Transmission among Elk (*Cervus canadensis*), White-Tailed Deer (*Odocoileus virginianus*) and Cattle (*Bos tarus*)." *Canadian Wildlife Biology and Management*, 4, no. 1 (2015): 18–30.

Brook, Ryan K., and Stéphane M. McLachlan. "Factors Influencing Farmers' Concerns Regarding Bovine Tuberculosis in Wildlife and Livestock Around Riding Mountain National Park." *Journal of Environmental Management*, 80, no. 2 (2006): 156–166.

Brook, Ryan K., and Stéphane M. McLachlan. "Trends and Prospects for Local Knowledge in Ecological and Conservation Research and Monitoring." *Biodiversity and Conservation*, 17, no. 14 (2008): 3501–3512.

Brook, Ryan K., and Stéphane M. McLachlan. "Transdisciplinary Habitat Models for Elk and Cattle as a Proxy for Bovine Tuberculosis Transmission Risk." *Preventive Veterinary Medicine*, 91, no. 2–4 (2009): 197–208.

Brook, Ryan K., Eric Vander Wal, Floris M. van Beest, and Stéphane M. McLachlan. "Evaluating Use of Cattle Winter Feeding Areas by Elk and White-Tailed Deer: Implications for Managing Bovine Tuberculosis Transmission Risk from the Ground Up." *Preventive Veterinary Medicine*, 108, no. 2–3 (2013): 137–147.

Buttke, Danielle E., Daniel J. Decker, and Margaret A. Wild. "The Role of One Health in Wildlife Conservation: A Challenge and Opportunity." *Journal of Wildlife Diseases*, 51, no. 1 (2015): 1–8.

Dissanayake, D.M., B. Ravi, Craig Stephen, Samson Daniel, and Preeni Abeynayake. "Gap Assessment of Animal Health Legislation in Sri Lanka for Emerging Infectious Disease Preparedness." *Outlook on Agriculture*, 41, no. 3 (2012): 203–208.

Gooding, R.M., and R.K. Brook. "Modeling and Mitigating Winter Hay Bale Damage by Elk in a Low Prevalence Bovine Tuberculosis Endemic Zone." *Preventive Veterinary Medicine*, 114, no. 2 (2014): 123–131.

Halliday, Jo, Chris Daborn, Harriet Auty, Zacharia Mtema, Tiziana Lembo, Barend M. deC Bronsvoort, Ian Handel et al. "Bringing Together Emerging and Endemic Zoonoses Surveillance: Shared Challenges and a Common Solution." *Philosophical Transactions of the Royal Society B: Biological Sciences*, 367, no. 1604 (2012): 2872–2880.

Kolla, Rebecca. "Defining Opportunities and Challenges in One Health Agency Management in Sri Lanka." PhD dissertation, University of Saskatchewan, 2020.

Lees, V. Wayne. "Learning from Outbreaks of Bovine Tuberculosis Near Riding Mountain National Park: Applications to a Foreign Animal Disease Outbreak." *The Canadian Veterinary Journal*, 45, no. 1 (2004): 28–34.

Lees, V. Wayne, Shelagh Copeland, and Pat Rousseau. "Manitoba: Bovine Tuberculosis in Elk (*Cervus elaphus manitobensis*) near Riding Mountain National Park, Manitoba, from 1992 to 2002." *The Canadian Veterinary Journal*, 44, no. 10 (2003): 830–831.

Malani, Anup, and Ramanan Laxminarayan. "Incentives for Reporting Infectious Disease Outbreaks." *Journal of Human Resources*, 46, no. 1 (2011): 176–202.

Miller, Ryan S., Matthew L. Farnsworth, and Jennifer L. Malmberg. "Diseases at the Livestock-Wildlife Interface: Status, Challenges, and Opportunities in the United States." *Preventive Veterinary Medicine*, 110, no. 2 (2013): 119–132.

Miller, Michele, and Francisco Olea-Popelka. "One Health in the Shrinking World: Experiences with Tuberculosis at the Human-Livestock-Wildlife Interface." *Comparative Immunology, Microbiology and Infectious Diseases*, 36, no. 3 (2013): 263–268.

Nihal, Pushpakumara Don Bamunusinghage, Ashoka Dangolla, Ranjani Hettiarachchi, Preeni Abeynayake, and Craig Stephen. "Challenges and Opportunities for Wildlife Disease Surveillance in Sri Lanka." *Journal of Wildlife Diseases*, 56, no. 3 (2020): 538–546.

Palmer, M.V. "*Mycobacterium bovis*: Characteristics of Wildlife Reservoir Hosts." *Transboundary and Emerging Diseases*, 60, no. 1 (2013): 1–13.

Robertson, Colin, Kate Sawford, Samson L.A. Daniel, Trisalyn A. Nelson, and Craig Stephen. "Mobile Phone-Based Infectious Disease Surveillance System, Sri Lanka." *Emerging Infectious Diseases*, 16, no. 10 (2010): 1524–1531.

Sánchez-Soriano, Carlos, Andrew D. Gibson, Luke Gamble, Jordana L. Burdon Bailey, Samantha Green, Mark Green, M. Barend et al. "Development of a High Number, High Coverage Dog Rabies Vaccination Programme in Sri Lanka." *BMC Infectious Diseases*, 19, no. 1 (2019): 1–12.

Shury, Todd. "The Epidemiology of Bovine Tuberculosis (*Mycobacterium bovis*) in the Greater Riding Mountain Ecosystem." PhD dissertation, University of Saskatchewan, 2015.

Shury, Todd K., and Doug Bergeson. "Lesion Distribution and Epidemiology of *Mycobacterium bovis* in Elk and White-Tailed Deer in South-Western Manitoba, Canada." *Veterinary Medicine International*, 2011, (2011): 1–11.

Shury, Todd K., Doug Bergeson, Om Surujballi, Konstantin P. Lyashchenko, and Rena Greenwald. "Field Evaluation of Three Blood-Based Assays for Elk (*Cervus canadensis*) Naturally Infected with *Mycobacterium bovis*." *Preventive Veterinary Medicine*, 115, no. 3–4 (2014): 109–121.

Silk, Matthew J., David J. Hodgson, Carly Rozins, Darren P. Croft, Richard J. Delahay, Mike Boots, and Robbie A. McDonald. "Integrating Social Behaviour, Demography and Disease Dynamics in Network Models: Applications to Disease Management in Declining Wildlife Populations." *Philosophical Transactions of the Royal Society B*, 374, no. 1781 (2019): 20180211.

Stephen, Craig, Patrick Zimmer, and Michael Lee. "Is There a Due Diligence Standard for Wildlife Disease Surveillance? A Canadian Case Study." *The Canadian Veterinary Journal*, 60, no. 8 (2019): 841–847.

Stronen, Astrid V., Ryan K. Brook, Paul C. Paquet, and Stephane McLachlan. "Farmer Attitudes Toward Wolves: Implications for the Role of Predators in Managing Disease." *Biological Conservation*, 135, no. 1 (2007): 1–10.

Thirgood, Simon. "New Perspectives on Managing Wildlife Diseases." *Journal of Applied Ecology*, 46, no. 2 (2009): 454–456.

Wobeser, Gary. "Bovine Tuberculosis in Canadian Wildlife: An Updated History." *The Canadian Veterinary Journal*, 50, no. 11 (2009): 1169.

17 Navigating Social Norms and Animal Welfare in Hunted Animals

Pierre-Yves Daoust

CONTENTS

Humans hold a wide diversity of values and beliefs stemming from different paths taken over millennia by different cultures across the planet. Such cultural and social perspectives particular to a region or a community of people need to be considered and incorporated in any meaningful discussion of our long history of association, close or distant, with animals and Nature. Respect for animals, people, and the environment or community in which they live is, therefore, a tenet of this chapter. People are part of an ecosystem, and their different views must be considered if that ecosystem is to be preserved or improved in a sustainable manner. In many situations, not all components of the ecosystem can be fully accommodated, and compromise is required to achieve an optimal balance.

Concern by individuals or communities for the welfare of animals goes back centuries, but animal welfare as is understood in a Euro-American context has become a major topic of discussion and scientific study for only the past few decades, starting with emphasis on laboratory and domestic animals (Walker et al., 2014). Studies on animal welfare were eventually extended to wild animals but focused mainly on captive (zoo) animals and on techniques such as chemical immobilization, trap and release, and radio-tagging used for wildlife management (Walker et al., 2014; JWD Editorial Board, 2016). Relatively little work has addressed animal welfare as it relates to consumptive use of free-living wildlife, such as sport hunting, commercial hunting, trapping, and subsistence hunting by Indigenous people and others. Yet, the number of animals taken for these various purposes runs into the millions in North America alone. For example, it has been estimated that 10–11 million waterfowl annually in North America (Hicklin and Barrow, 2004) and 2.8 million white-tailed deer in 2016–2017 in the United States (Webb, 2018) were taken through sport hunting. What values are guiding, or should guide, these consumptive users to ensure respectful treatment of the harvested birds and mammals in the course of their activities?

Some studies have estimated the proportion of animals (mammals and birds) wounded but not retrieved during sport hunting. This proportion varied roughly between 10% and 48%, depending on the species hunted, the location where the animals were hunted, the tools used, and the hunters' skill (Nixon et al., 2001; Hicklin and Barrow, 2004; Gregory, 2005). Some of these studies proposed Codes of Practice for hunters involved in particular hunts, with the aim "to support sportsmanship and acceptable animal welfare practices" (Gregory, 2005). Whereas improvement in hunters' attitude towards animals through education remains the key to decreasing the suffering of hunted animals, the high proportion of "struck and lost" animals in sport hunting may be a sufficient argument for many to disapprove of such activity when it is done simply for recreation. However, where can a line be drawn between wealthy hunters from urban areas and those in remote communities who only have to step outside their home to hunt and who use their quarry to supplement their dining table? The Wild Harvest Initiative® is an example of an effort to assess the economic, social, and ecologic significance of hunting and angling to modern society. Moreover, the ethics of eating wild meat that has been "produced" locally should be weighed against that of eating meat from livestock that may have been transported for long distances to the slaughter house, with the associated severe stress imposed on the animals. To hunt successfully, experienced hunters need to have a thorough knowledge of the life history of the target animals and the environment in which they live. With such knowledge can come an appreciation of the animals and their environment, and most of these hunters often turn out to be strong custodians of Nature.

There are few instances of hunts of wild animals for commercial purposes (apart from fisheries), some of which have been tied to culling as a management tool for wild populations (Lewis et al., 1997; Urquhart and McKendrick, 2003). The Canadian commercial seal hunt, in the Gulf of St. Lawrence and off eastern Newfoundland and Labrador, is by far the most widely known commercial harvest of wild animals, based on the large number of seals that used to be taken annually during the industry's peak harvest years. Trapping of fur-bearing wildlife represents another important commercial activity involving wild animals. Hunting of several species of wild animals by Indigenous people for their own subsistence continues to be of fundamental importance. Wildlife thus remains a vital resource for some segments of the human population. Communities and their people who live far from urban centres, such as many First Nations and most Inuit in Canada, have depended for many generations, if not millennia, on resources from the land and the sea. This has shaped their approach to these resources and concurrently their social and cultural norms, and these are bound to differ from those of other demographic groups that have been removed for some generations from their food sources. Economically, some of these hunting activities undertaken for commercial purposes may represent one of the very few local sources of income in places that are often affected by a chronic lack of employment opportunities.

The Canadian commercial seal hunt exemplifies well the dichotomy between distant communities and urbanized regions. Its history goes back a few centuries, when Europeans settlers in Newfoundland and on the Magdalen Islands, Québec, in the Gulf of St. Lawrence, relied on abundant populations of harp seals that came down from the Arctic in late winter to early spring to give birth and mate. Originally hunted for their oil as fuel, the purpose of the hunt changed with time to include the pelt (primarily from young animals) and, currently, oil as a rich source of omega-3 fatty acids and meat for human consumption. In recent years, a hunt for grey seals off the Magdalen Islands has also been developed. At the outset, neither the seal hunt nor the trapping for fur considered animal welfare as an important element. In the seal hunt, animals needed to be killed quickly to be skinned promptly and correctly, but improper practices were apparently common prior to the mid-1960s, when government regulations to prevent such practices came into effect (Malouf, 1986). The welfare of trapped animals was also of little concern, based on the poor design of restraining traps that were then used. Pressure from animal rights groups, starting in the 1960s in relation to the seal hunt, challenged wildlife users to reconsider their harvesting methods. In the past 40 years, objective studies targeting elements of animal welfare have aimed to improve practices during commercial harvest of wildlife (AIHTS, 1997; Daoust and Caraguel, 2012; NAMMCO, 2009). For both the sealing and the trapping industries, the message became clear that if commercial users intend to develop a socially acceptable industry, they need to adopt and consistently adhere to methods meeting scientifically and legally appropriate standards of animal welfare, such as very short, preferably immediate, time to death and very low proportion of animals struck and lost. Ultimately, such practices also benefit the resource users since they often result in products of better quality from a commercial standpoint. Through mandatory training programs for sealers and trappers prior to the acquisition of their licence, the issue of animal welfare is now at least part of their conversation. Information to these user groups, based on objective research and delivered in a respectful manner, is again key to improving, if needed, their attitude towards animals, but this must be supported by appropriate legislation and enforcement.

The views of North American Indigenous people (such as Inuit, First Nations, and Métis within the territorial borders of Canada) regarding animal welfare have also been explored, not as an isolated concept, however, but rather as an integral part of an elaborate system of human-animal relationships (see, for example, Nadasdy, 2005; Watts, 2013; Robinson, 2014). It is not the intent of this author, a Euro-American white male, to pretend to fully understand, let alone explain, norms and belief systems that have evolved for millennia. It is instead incumbent on him to give voice as much as possible to Indigenous scholars, hunters, and elders who have reflected on human-animal relationships. Inuit and First Nations peoples have for millennia been primarily, if not totally, dependent on animal resources for their survival, and many, particularly those living in northern regions, continue to depend on these resources in one form or another. Because of this, they have developed a very strong relationship with their natural environment, which still exists today. Just as there is a diversity of perspectives on animal welfare among Euro-Americans, there also exist different views of animals among Indigenous peoples. Generalization is therefore not possible, but there appears to be much common ground among various Indigenous nations about their views of the natural world.

At the onset, North American Indigenous peoples were "neither ecologists nor conservationists but hunters" (Laugrand and Oosten, 2015; Nadasdy, 2005), and amongst many of them hunting continues to define their relationships with animals. The concept of animal welfare as is understood by Euro-Americans is not approached in the same way. The Euro-American perception of the natural world has been greatly influenced by Judeo-Christian teachings, in which humans are made to be distinct from Nature, supposedly having "dominion" over it (Freeman, 1999; Watts, 2013; Karetak and Tester, 2017). This distinction between people and Nature (and its resources) has followed into secular society and, interestingly, has resulted in two opposite attitudes: one utilitarian, in which "anything that is not human is defined as an object to be used for the benefit of human beings" (Karetak and Tester, 2017), and the other ecocentric, in which "the value of nature is inherent rather than contingent on its use by humans" (Nadasdy, 2005). Indigenous perspectives on Nature and animals, informed by their own, equally valid "complex set of beliefs about the proper relationship between humans and their spiritually powerful animal benefactors" (Nadasdy, 2005), do not subscribe to either of these views. Instead, "habitats and ecosystems are better understood as societies … [in which n]on-human beings are active members"; these societies do not involve interactions among human beings only, and humans are not put at the centre (Watts, 2013; Karetak and Tesser, 2017; Todd, 2018). "Not only animals, but also plants, rocks, water, and geographic locations can have an identity, personality, and spirit … everything on Earth is connected" (Robinson, 2014). Under the Nunavut Wildlife Act, one of the guiding principles of Inuit Qaujimajatuqangit ["the way of knowing that encompasses the past, present, and future of Inuit experiences and values, principles, skills, and beliefs that have evolved over time" (Sudlovenick, 2019)] affirms that "People are stewards of the environment and must treat all of nature holistically and with respect, because humans, wildlife and habitat are inter-connected" (Nunavut, 2012). In this context, so-called subsistence hunting becomes much more than simply collecting food from the environment. It is a means of sustaining and enhancing social and cultural relationships and promoting shared responsibilities for the well-being of a community of people (Freeman, 2018; Wenzel, 1995).

In many Indigenous views of the world, "all living things contain spirits" (Watts, 2013; Freeman, 1990; Karetak and Tester, 2017), yet humans' survival depends on killing animals. The attitude of Indigenous peoples towards animals can therefore be best described as a mixture of kinship, awe, and pragmatism (Robinson, 2014). The concept of respect and gratitude that derives from this dependence on animals "is far more complex and culturally dependent than most Euro-North American are aware" (Nadasdy, 2005). For example, it seems to have been a common belief among a number of Indigenous peoples that the animals offer themselves to the hunters (Freeman, 1999; Robinson, 2014). In a sense, one may view this as a form of humility and respect in contrast to the hunter boasting that they outsmarted their prey (Freeman, 1999). If the animal offers itself, this must come with obligations

FIGURE 17.1 James Simonee, from Pond Inlet, Nunavut, Canada, watches a ringed seal's breathing hole in Eclipse Sound, May 2016. (Photo credit: P-Y Daoust.)

on the hunter's part, and these take the form of respect for, and appropriate conduct towards, Nature and the animals; otherwise, animal spirits will be offended, and animals will not return. Respect can be shown in several ways. One that is of importance from the Euro-American perspective of animal welfare is to kill the animals quickly and thus minimize their suffering (Nunavut, 2012; Freeman, 1999). This, however, must consider the pragmatic nature of hunting in a harsh northern environment and the Indigenous hunters' own perspective. A ptarmigan may be killed instantly with one shot of a rifle or a shotgun. By comparison, one of the few methods available for Inuit to hunt ringed seals concealed under the ice in winter and spring requires the use of a gaff or a harpoon to catch the animal when it surfaces at a breathing hole (Figure 17.1), a technique which would likely be seen as inappropriate by many Euro-Americans. The bowhead whale hunt by Canadian Inuit and Alaskan Iñupiat may also be considered problematic in the Euro-American context. These whales cannot be killed with even the most powerful rifles. Instead, one or two grenades are needed, and possibly lances that target the unique blood supply to the brain of these animals, which derives from intercostal muscles (Marshall, 2002); the latter method had been used ancestrally for more than 1,000 years presumably because of its relative efficiency. Even so, immediate death is not guaranteed, and time to death may be prolonged (NAMMCO, 2015). Yet, the social and cultural importance of this hunt needs to be taken into account: "The reintroduction of whale hunts can not only be seen as an empowerment of Inuit but also a means of valorizing hunting and sharing practices as core elements of Inuit traditions" (Laugrand and Oosten, 2015). This does not diminish the value of promoting improvements in hunting methods that may decrease the time to death for the whales because, beyond the issue of animal welfare, this benefits the hunters' safety in a type of hunt that can be very dangerous. By comparison, it is interesting that, in some Indigenous cultures, fish are deeply important and respected animals (Todd, 2018), whereas they are the group of vertebrates least studied by Euro-American scholars from an animal welfare perspective (Walker et al., 2014).

Respect by Indigenous hunters for the animals as prey can manifest itself in several other ways besides aiming for a rapid death. These include minimizing disturbance to the animal populations that are hunted; restricting the number of animals that are killed to only those that are required for the needs of the hunters, their family, and their community; reducing to a minimum the number of animals struck and lost by developing hunting skills; limiting food wastage (although "waste" may have different meanings in different cultures); and using proper manners of disposing of parts of the carcasses (although the methods used may vary among Indigenous groups) (Freeman, 1999; Robinson, 2014). Humans are also responsible "to provide the conditions necessary for animals to thrive" (Robinson, 2014), which recognizes the importance of a healthy ecosystem for wild animals to live in. This very much aligns with the Euro-American concept of environmentalism, but it also has direct relevance to animal welfare (Paquet and Darimont, 2010). Overall, "[t]he rules governing the hunt, and indeed many of the rules governing interpersonal behaviour, are inherently conservationist by nature" (Freeman, 1990; Nadasdy, 2005).

It is not assumed that these various forms of respect for wild animals and Nature are followed universally by all Indigenous peoples, since there are various degrees of contemporary adherence to the ancestral culture, just as there are among Euro-Americans with regard to Judeo-Christian teachings. Ultimately, many Euro-Americans and many Indigenous scholars, hunters, and elders will likely agree that "[i]t is only in our relationship with the non-human [animals] that we become fully human" (Robinson, 2014).

ACKNOWLEDGEMENTS

The comments on Indigenous perspectives about animals and their well-being in this chapter benefited greatly from discussions and interactions with Solomon Awa, Gabrielle Daoust, Milton Freeman, Yoanis Menge, Noel Milliea, Megan Sheremata, James Simonee, and Greg Thompson.

REFERENCES

AIHTS (Agreement on International Humane Trapping Standards). "Agreement on International Humane Trapping Standards Between the European Community, Canada and the Russian Federation." *Official Journal of the European Communities*, M-BD-2/EN (1997). https://fur.ca/wp-content/uploads/2015/09/AIHTS-Copy-of-Agreement.pdf

Daoust, Pierre-Yves, and Charles Caraguel. "The Canadian Harp Seal Hunt: Observations on the Effectiveness of Procedures to Avoid Poor Animal Welfare Outcomes." *Animal Welfare*, 21, no. 4 (2012): 445.

Freeman, Milton M.R. "The Significance of Animals in the Life of Northern Foraging Peoples and Its Relevance Today," in *International Symposium on Human-Animal Relationships in the North* (1990): 5–12.

Freeman, Milton M.R. "Respect and Reciprocity as Key Elements in Arctic Sustainable Use Strategies." *SUI Technical Series*, 2 (1999): 89–100.

Freeman, Milton M.R. "Human Food Consists Entirely of Souls": Ethics, Adaptive (Co) Management and the Path to Sustainability in the Contemporary Arctic." Research Gate, 2018. https://www.researchgate.net/publication/327350603_Human_food_consists_entirely_of_souls_ethics_comanagement_and_the_path_to_sustainability_in_the_contemporary_Arctic

Gregory, N.G. "Bowhunting Deer." *Animal Welfare*, 14, no. 2 (2005): 111–116.

Hicklin, Peter W., and W.R. Barrow. "The Incidence of Embedded Shot in Waterfowl in Atlantic Canada and Hudson Strait." *Waterbirds*, 27, no. 1 (2004): 41–45.

JWD Wildlife Welfare Supplement Editorial Board. "Advances in Animal Welfare for Free-Living Animals." *Journal of Wildlife Diseases*, 52, no. 2s (2016): S4–S13.

Karetak, Joe, and Frank Tester. "Introduction: Inuit Qaujimajatuqangit, Truth and Reconciliation," in *Inuit Qaujimajatuqangit: What Inuit Have Always Known to be True*, eds. J. Katetak, F. Tester, and S. Tagalik (Black Point, Fernwood Publishing, 2017), 1–16.

Laugrand, Frédéric, and Jarich Oosten. *Hunters, Predators and Prey: Inuit Perceptions of Animals* (Oxford: Berghahn Books, 2015), 408.

Lewis, A.R., A.M. Pinchin, and S.C. Kestin. "Welfare Implications of the Night Shooting of Wild Impala (*Aepyceros melampus*)." *Animal Welfare*, 6, no. 2 (1997): 123–131.

Malouf, Albert H. "Seals and Sealing in Canada. Report of the Royal Commission on Seals and the Sealing Industry in Canada." *Canadian Government Publishing Centre, Supply and Services Canada*, 3 (1986): 679.

Marshall, C.D. "Morphology, Functional," in *Encyclopedia of Marine Mammals*, eds. W.F. Perrin, Bernd Würsig, and J.G.M. Thewissen (San Diego: Academic Press, 2002), 759–774.

Nadasdy, Paul. "Transcending the Debate over the Ecologically Noble Indian: Indigenous Peoples and Environmentalism." *Ethnohistory*, 52, no. 2 (2005): 291–331.

NAMMCO (North Atlantic Marine Mammal Commission). "Report of the NAMMCO Expert Group on Best Practices in the Hunting and Killing of Seals" (Copenhagen: North Atlantic Marine Mammal Commission, 2009), 30. http://nammco.wpengine.com/wp-content/uploads/2016/10/nammco-report-expert-group-on-best-practises-in-hunting-and-killing-of-seals.pdf

NAMMCO. "Report of the Expert Group Meeting on Assessing Time to Death Data from the Large Whale Hunts" (Copenhagen: North Atlantic Marine Mammal Commission, 2015), 46. http://nammco.wpengine.com/wp-content/uploads/2016/10/report-of-expert-group-meeting-on-ttd-data-for-large-whales.pdf

Nixon, Charles M., Lonnie P. Hansen, Paul A. Brewer, James E. Chelsvig, Terry L. Esker, Dwayne Etter, Joseph B. Sullivan, et al. "Survival of White-Tailed Deer in Intensively Farmed Areas of Illinois." *Canadian Journal of Zoology*, 79, no. 4 (2001): 581–588.

Nunavut. "Inuit Qaujimajatuqangit (Inuit Traditional Knowledge, Beliefs & Practices)," in *Consolidation of Wildlife Act* (Government of Nunavut, 2012): 11–12. https://www.canlii.org/en/nu/laws/stat/snu-2003-c-26/latest/snu-2003-c-26.pdf

Paquet, Paul C., and Chris T. Darimont. "Wildlife Conservation and Animal Welfare: Two Sides of the Same Coin." *Animal Welfare*, 19, no. 2 (2010): 177–190.

Robinson, Margaret. "Animal Personhood in Mi'kmaq Perspective." *Societies*, 4, no. 4 (2014): 672–688.

Sudlovenick, Enooyaq. "A Serological Survey and Inuit Qaujimajatuqangit of Ringed seals (Nattiit) in Frobisher Bay, Nunavut." *Master of Science, Faculty of Veterinary Medicine* (Canada: University of Prince Edward Island, 2019): 102.

Todd, Zoe. "Refracting the State Through Human-Fish Relations: Fishing, Indigenous Legal Orders and Colonialism in North/Western Canada." *Decolonization: Indigeneity, Education and Society*, 7 (2018): 60–75.

Urquhart, K. A., and I. J. McKendrick. "Survey of Permanent Wound Tracts in the Carcases of Culled Wild Red Deer in Scotland." *Veterinary Record*, 152, no. 16 (2003): 497–501.

Walker, Michael, Maria Diez-Leon, and Georgia Mason. "Animal Welfare Science: Recent Publication Trends and Future Research Priorities." *International Journal of Comparative Psychology*, 27, no. 1 (2014).

Watts, Vanessa. "Indigenous Place-Thought and Agency Amongst Humans and Non Humans (First Woman and Sky Woman Go on a European World Tour!)." *Decolonization: Indigeneity, Education and Society*, 2, no. 1 (2013): 20–34.

Webb, Stephen. "White-Tailed Deer Facts, Findings and Numbers." Nobel Research Institute, 2018. https://www.noble.org/news/publications/ag-news-and-views/2018/october/white-tailed-deer-facts-findings-and-numbers/.

Wenzel, George W. "Ningiqtuq: Resource Sharing and Generalized Reciprocity in Clyde River, Nunavut." *Arctic Anthropology* (1995): 43–60.

18 Healthy Animals for Sustainable Livelihoods and Poverty Alleviation

Craig Stephen

CONTENTS

Income is one of the most influential determinants for human health. It shapes living conditions, food access, health behaviours, and social circumstances. Poverty is multidimensional. Its definition varies with time, culture, and power, but its impacts are clear and include unemployment, social exclusion, and high vulnerability to disasters and other phenomena which prevent people from being productive (Anon, 2020). Low income is a significant risk factor for disease incidence and severity as well as life expectancy (Krisberg, 2016). Raising incomes improves the health of poor individuals, helps reduce health inequalities, and increases average population health (Lynch et al., 2004). For these reasons, many sustainable development plans, such as the United Nations Sustainable Development Goals, include poverty eradication.

People who rely on livestock, nature, or fishing for their food and income – over one-third of the world's population – are often the most vulnerable (Halwart et al., 2003). Poverty is predominantly rural in most countries. Between 80% and 90% of very poor rural households (approximately 900 million people) partake in some form of agriculture (Pradere, 2014). Agriculture has been the cornerstone of many rural development programmes and has been viewed by some as the most important provider of gainful employment in rural areas. Coastal communities have historically relied on wild fisheries for food and income, but with the decline of wild populations, aquaculture is increasingly complementing the rural poor's catches from traditional fisheries.

There are widely accepted links between biodiversity loss and poverty (Adams et al., 2004). Biodiversity offers a means of income to the poor and helps protect them from risks and shocks leading them deeper into poverty (CBD, 2010). Given the interdependencies between environmental conditions and human well-being, poverty eradication goes hand-in-hand with conservation and maintaining healthy ecosystems. Animals can be a direct source of income, can be a means to store wealth, can provide food and other resources that otherwise would need to be purchased, and can reduce gender income disparities (FAO, 2009). Livestock, wildlife, and fish are central to the livelihoods of many of the world's poorest and most vulnerable people. Agriculture and aquaculture development and conservation have, therefore, been used to try to bring people out of poverty.

Shocks, surprises, and creeping changes, such as natural disasters, diseases, or climate change, that affect animals can make it more difficult for people to reliably and sustainably access the natural, financial, or social capital derived from animals, thereby reducing opportunities for households (Stringer, 2014). For example, zoonotic infections have been described as a poverty trap. A poverty trap is something that forces people to remain poor. Poor livestock keepers are especially vulnerable to zoonoses due to their close contacts with animals, their consumption of animal products, and limited access to health care for themselves and their animals. Non-zoonotic animal diseases also contribute to the poverty trap. When poor farmers lack the resources to access animal health services, they inequitably suffer the effects of animal diseases. In less developed countries, livestock mortality can range from 7% to 50% depending on the species, animal age, and country (Pradere, 2014). The death of even a single animal can have dramatic consequences for a vulnerable family in low-income countries. As another example, seal hunting in Nunavut, Canada, in 2015 represented CAD 4–6 million in food value each year. Income from selling seal pelts could reach up to CAD 1 million annually, funds which in turn are used to buy the equipment and fuel necessary to access food from the land and the sea (Le Cercle, 2015). A EU ban on seal pelts caused an economic collapse of the sealing industry, which in turn reduced the buying capacity of Inuit communities and subsequently reduced community well-being (see Chapter 17 for further details of this hunt).

While there is a growing literature on the contributions of livestock, aquaculture, and conservation to reducing poverty, questions abound about whether the advantages provided are always a net gain. The promise of aquaculture, for example, has been a mixed experience, which in part reflects the varying circumstances under which aquaculture is used and the outcomes and time frames selected to gauge success. Conflicts over shared resources and damage to common environments linked to some form of aquaculture (such as declines in access to wild aquatic food sources or pollution of fresh water) necessitates balancing income improvements for some with negative impacts on common resources used by others (Burns et al., 2014). As a conservation example, inequality between the local rural residents and other stakeholders in accessing benefits from the Wolong Nature Reserve for Giant Pandas has been noted in China (He et al., 2008). Rural residents bore much of the cost of conservation, but most economic benefits went into the ecotourism sector. Rural households that benefited from ecotourism were located near main roads rather than households far from the road and closer to panda habitats. Another investigation concluded that the commodification of a rural village in a nature sanctuary that is part of gorilla conservation efforts in Uganda promoted the external control of conservation spaces by international organizations which catered to foreign tourists but ultimately contributed to, rather than alleviated, poverty and dependency in local communities (Laudati, 2010).

How strategies to improve animal-based income programmes are implemented can determine their success or failure. For example, improved animal production has contributed to reduced poverty in many countries, but this effect stagnates in countries where growth in animal production has been mainly through extensive agriculture (Pradere, 2014). One report from Kenya noted the failure of top-down oriented livestock improvement programmes, seeing instead the need for participatory and community-based development activities taking farmers' perspectives into account (Wollny et al., 2002).

The benefits of animals to income are highly context specific. Efforts to promote rural incomes through animal production or services need to reflect on some of the core principles of health promotion seen throughout this book, such as being oriented to the local context, being attentive to the effects on other determinants of health, paying attention to the equitable sharing of risks and benefits, and recognizing the complexity of human-animal-environment interactions to reduce the risk of unexpected effects or surprises.

CASE STUDY – AQUACULTURE AND SUSTAINABLE LIVELIHOODS IN SRI LANKA

In the early 2000s, the government of Sri Lanka was looking to aquaculture to foster rural poverty reduction and income stabilization. Government policies and farm practices needed to avoid pitfalls that had been affecting the short-to-medium-term sustainability of this food production sector in order to gain the advantages of aquaculture development. Fish and shrimp diseases had been the prevailing sources of instability in production and farm failure in Sri Lanka, hampering realization of the desired economic gains for rural communities. Animal disease control was, thus, a critical component of this poverty-reduction strategy. Supportive and effective on-farm management policies and national policies were needed to create an enabling environment that promoted sustainable farming and the resultant benefits beyond individual farms. The Sri Lankan government recognized the need to redesign and strengthen their aquaculture policy and extension services to increase the likelihood of effective sector-wide disease control that would lead to sustainable human development benefits. This case study focuses on some insights and experiences working with the shrimp farming sector.

Shrimp health was a sustainability pillar for the industry but there were social disincentives to implement disease management plans despite the recognition of this pervasive threat to farms (Westers et al., 2017). At the outset of this project, the shrimp farming sector was constrained by lack of access to health management expertise and services, limiting the ability of farmers to effectively implement practices necessary to avoid disease and survive a production cycle. A collaboration between local farm societies, academia, and government set out to

FIGURE 18.1 A small-scale shrimp farm in Sri Lanka. (Photo by C. Stephen.)

examine and enhance the social learning networks between farmers and between farmers, government, community, and academics to remedy some of the obstacles impeding the application of shrimp disease control management practices. Social-learning-networks are connections between people who change an individual's knowledge or motivation, resulting in behaviour change and alteration of practices. Such networks are important for smallholder farmer learning (Conley and Christopher, 2001) (Figure 18.1). A survey of 225 farmers in the eastern and northwest provinces of Sri Lanka found significant gaps and breakdowns in the flow of information between farmers, academics, and government (Wu et al., 2016). As they are the ultimate knowledge end users, knowledge is most critical for farmers, but farmers tended to have the weakest links to knowledge sources needed to make management decisions. These gaps created inefficiencies in the use of resources and capacities. The transitory nature of key knowledge brokers in government, lack of trust, and lack of history of farm level aquaculture extension services resulted in many farmers being isolated and not benefiting from others' expertise and experience.

Farmers who implemented a greater number of disease control best management practices were significantly less likely to report disease than farmers who implemented fewer, and farms with disease were less likely to be sustainable (Westers et al., 2017). The project team set out to assess the impact of low-cost ways to strengthen farmers' social learning networks. Cell phone access was widespread in Sri Lanka. Sixty farmers were recruited to regularly receive short, culturally appropriate SMS messages on their cell phones on disease best management practices as well as alerts on changing risk factors, all of which would enable them to act quickly with farm interventions. Farmers communicated about information received by SMS most readily with other farmers, which encouraged farmers to act in greater cooperation with each other, attend meetings, seek technical support, and share information. Geographic and demographic factors influenced the impact of education and policy on farm sustainability outcomes, emphasizing the need to tailor policy and education programmes to specific areas.

The project team also reviewed aquaculture development policy. Sustainable aquaculture development for rural poverty reduction was an oft-stated goal, but explicit links between policies, plans, decisions, and actions were few. Decision-making had been centralized, leading to some practices and requirements that were insensitive to local social and ecological realities. The lack of a coordinated, locally informed approach resulted in (i) inconsistencies and gaps in combating disease, (ii) industry expansion happening almost *de novo* in new areas as opposed being built from past lessons, and (iii) lack of trust and acceptance of required management activities. Co-management, with key stakeholders working towards the same goals and each playing their appropriate role, was seen as a reasonable and necessary step in Sri Lanka's aquaculture development to ensure a strategic approach to developing locally acceptable and effective plans for rural development (Stephen et al., 2012).

Follow-up activities applied these lessons more widely by (i) enhancing the social knowledge-sharing networks to boost adoption of better practices, risk reduction, and value chain strengthening; (ii) adopting structures and processes for "co-management," facilitated by knowledge sharing; and (iii) building capacity for knowledge and technology adoption in resource-poor coastal communities (specifically targeting women as key participants) and interacting with private sector actors developing markets and with state agencies building natural resource management plans (DeJager et al., 2014). These were complemented with critical capacity enhancements for disease management. Participating small-scale shrimp farmers (those with fewer than five ponds) increased production yields per acre by 26% over five years. Farmers increased individual average shrimp weight at harvest by 14%. These improvements resulted in net incomes averaging 382,400 LKR per acre at the end of the project compared to 111,375 LKR at the start. For a small farm with a single pond of 1.12 acres, this would translate into 35,690 LKR in monthly income. The official poverty line at the time was 3,028 LKR real total expenditure per person per month. Small-scale shrimp farming could now feasibly lift households out of poverty.

SUMMARY

Much attention in One Health has focussed on protecting human health by reducing the risk of people acquiring zoonotic infectious diseases. Endemic, non-zoonotic, production-limiting diseases are continually present. These diseases can keep poor farmers in poverty. This chapter illustrates how keeping animals healthy can underpin access to critical determinants of health for people, such as employment, income, and food security. Creating and enabling an environment in which captive animals can realize healthy lives can at the same time help reduce poverty, support sustainable food production, and promote household well-being.

ACKNOWLEDGEMENT

The work in Sri Lanka included in this chapter was graciously supported by the International Development Research Centre.

REFERENCES

Adams, William M., Ros Aveling, Dan Brockington, Barney Dickson, Jo Elliott, Jon Hutton, Dilys Roe et al. "Biodiversity Conservation and the Eradication of Poverty." *Science* 306, no. 5699 (2004): 1146–1149.

Anon. "UN Sustainable Development Goals. Goal 1: End Poverty in All Its Forms Everywhere." United Nations, 2020. https://www.un.org/sustainabledevelopment/poverty/.

Burns, Theresa E., Joy Wade, Craig Stephen, and Lorraine Toews. "A Scoping Analysis of Peer-Reviewed Literature About Linkages Between Aquaculture and Determinants of Human Health." *EcoHealth* 11, no. 2 (2014): 227–240.

CBD (Convention on Biological Diversity). "Linking Biodiversity Conservation and Poverty Alleviation: A State of Knowledge Review." CBD Technical Series No. 55 (2010): 71.

Conley, Timothy, and Udry Christopher. "Social Learning Through Networks: The Adoption of New Agricultural Technologies in Ghana." *American Journal of Agricultural Economics* 83, no. 3 (2001): 668–673.

DeJager, Tim, Sam Daniel, and Sunil Jayakody. "Scaling-up Sustainable Aquaculture Development in Sri Lanka. Final Technical Report. Project #107519." Government of Canada. International Development Research Centre, 2014. https://www.idrc.ca/en/project/scaling-sustainable-aquaculture-development-sri-lanka.

FAO (Food and Agriculture Organization). "Livestock, Food Security and Poverty Reduction" in *The State of Food and Agriculture*. Food and Agriculture Organization (2009): 168. http://www.fao.org/3/a-i0680e.pdf.

Halwart, Matthias, Simon Funge-Smith, and John Moehl. "The Role of Aquaculture in Rural Development." *Review of the State of World Aquaculture, FAO, Rome* (2003): 47–58.

He, Guangming, Xiaodong Chen, Wei Liu, Scott Bearer, Shiqiang Zhou, Lily Yeqing Cheng, Hemin Zhang et al. "Distribution of Economic Benefits from Ecotourism: A Case Study of Wolong Nature Reserve for Giant Pandas in China." *Environmental Management* 42, no. 6 (2008): 1017.

Krisberg, Kim. "Income Inequality: When Wealth Determines Health: Earnings Influential as Lifelong Social Determinant of Health." The Nation's Heath, American Public Health Association (2016): 1–17. http://thenationshealth.aphapublications.org/content/46/8/1.1.

Laudati, Ann. "Ecotourism: The Modern Predator? Implications of Gorilla Tourism on Local Livelihoods in Bwindi Impenetrable National Park, Uganda." *Environment and Planning D: Society and Space*, 28, no. 4 (2010): 726–743.

Le Cercle, Polaire. "D5.51-Scientific and Ethical Evaluation of the Impact of Indigenous Seal Hunting. Project 265863." Report to ACCESS Arctic Climate Change, Economy and Society (2015): 50.

Lynch, John, George Davey Smith, Sam A.M. Harper, Marianne Hillemeier, Nancy Ross, George A. Kaplan, and Michael Wolfson. "Is Income Inequality a Determinant of Population Health? Part 1: A Systematic Review." *The Milbank Quarterly* 82, no. 1 (2004): 5–99.

Pradere, J. P. "Improving Animal Health and Livestock Productivity to Reduce Poverty." *Revue Scientifique et Technique-Office International des Epizooties* 33, no. 3 (2014): 735–44.

Stephen, Craig, Tim Dejager, Sam Daniel, Sunil Jayakody, and Jenny Dawson. "Promoting Rural Income from Sustainable Aquaculture Through Social Learning in Sri Lanka." Final Technical Report. Grant # 106342. Government of Canada. International Development Research Centre, 2012. https://www.idrc.ca/en/project/promoting-rural-income-sustainable-aquaculture-through-social-learning-sri-lanka-cifsrf

Stringer, Andy. "Improving Animal Health for Poverty Alleviation and Sustainable Livelihoods." *Veterinary Record* 175, no. 21 (2014): 526–529.

Wollny, C.B.A., J.W. Banda, T.F.T. Mlewah, and R.K.D. Phoya. "The Lessons of Livestock Improvement Failure: Revising Breeding Strategies for Indigenous Malawi Sheep." *Proceedings of the Seventh World Congress on Genetics Applied to Livestock Production*, 33 (2002): 19–23.

Westers, Trisha, Carl Ribble, Sam Daniel, Sylvia Checkley, Jessica P. Wu, and Craig Stephen. "Assessing and Comparing Relative Farm-Level Sustainability of Smallholder Shrimp Farms in Two Sri Lankan Provinces Using Indices Developed from Two Methodological Frameworks." *Ecological Indicators*, 83 (2017): 346–355.

Wu, J., Thersa Burns, P.K. Kawadugama, T. DeJager, Trishia Westers, S. Scheckley, Carl Ribble et al. "Linkages Between Social-Learning Networks and Farm Sustainability for Smallholder Shrimp Farmers in Sri Lanka." *Asian Fisheries Society*, 29 (2016): 96–111.

19 Application of Harm Reduction Thinking to the Conservation of Uncharismatic Species

Joy Wade

CONTENTS

The conversations around planetary health are, for the most part, defeatist. There are individual stories of actions and successes, but these seem few and far between in popular and scientific publications. Nevertheless, we crave them. People, the party responsible for negative impacts on the planet, have growing feelings of powerlessness towards the likelihood of positive outcomes. But there are solutions if we should so choose to see them and work towards them. The first step is to define achievable, incremental improvements that people, whether individuals, governments, or conservation agencies, are willing to work towards together. It is important to set realistic expectations for desired outcomes. Too often we strive for a lofty goal that cannot be achieved, and this results in disappointment. Realism is not a failure; we must accept that our actions (or inactions) have consequences and sometimes those consequences are irreversible, and we must place our efforts elsewhere. Working towards an ideal of past generations is in most circumstances unrealistic. Not only has the situation changed, the societal expectations may also have changed.

With a growing list of species at risk, it is important to be realistic. In Canada, for example, there are over 1,000 species of mosses, lichens, plants, and animals on the public registry of species at risk. These species are afforded certain protections and there are legal requirements for certain documents, including recovery strategies and action plans. Given this increasingly enormous and diverse list of species at risk, their distribution over a country as large as Canada and multiple legislative jurisdictions affecting their conservation, it is unreasonable to expect equal attention to all species. The highly charismatic species, such as the North Atlantic right whale (*Eubalaena glacialis*), beluga whale (*Delphinapterus*

FIGURE 19.1 Photographs of the threatened Cowichan Lake lamprey (top), and endangered Morrison Creek lamprey (bottom).

leucas), or polar bear (*Ursus maritimus*), receive much more investment by individuals, governments, and conservation groups than, for instance, boreal felt lichen (*Erioderma pedicellatum*), dusky dune moth (*Copablepharon longipenne*), or the Banff longnose dace (*Rhinichthys cataractae smithi*), of which most people have never heard. We must accept that as humans, the value of animals and plants are different, and we have favourites. Although the inherent or ecological value of the beluga whale may be no greater than that of the dusky dune moth, they have a greater social value. A particular challenge for those trying to protect less charismatic species is finding motivations and methods to ensure their ability to be resilient in their current situation and to be able to adapt to future scenarios in the absence of social motivations. One way in which this can be achieved is through a harm reduction approach (Stephen et al., 2018). This chapter illustrates the application of this method using a case study for two fish species listed under Canada's Species at Risk Act (SARA) in British Columbia (BC): the Cowichan Lake lamprey (*Entosphenus macrostomus*) and the Morrison Creek lamprey (*Lampetra richardsoni marifuga*) (Figure 19.1).

CONSERVATION HARM REDUCTION: CONTEXT FOR THIS CASE

Cowichan Lake lamprey is a small [maximum length approximately 273 mm (Beamish 1982)], parasitic lamprey found in one watershed on Vancouver Island, BC, and nowhere else in the world (MacConnachie and Wade, 2016). It was first described in Beamish (1982) and was listed as Threatened on Schedule 1 of the Species at Risk Act in 2003. Its listing is largely a result of it being an extreme endemic, a population of fish found only in one location. Based on what is known of other similar species, most of its life is spent as a filter feeding ammocoete (larvae) burrowed in the sediment (Beamish and Wade, 2008). After many years (five or more based on other species), ammocoetes undergo metamorphosis into an adult, which feeds parasitically. It is estimated that it feeds for a year or year and a half before it spawns and dies. The precise details, including timing and habitat requirements, are largely unknown or estimated based on other species of lamprey. Cowichan Lake and the adjoining Mesachie Lake are part of a water retention system serving the Cowichan valley. There is a regulated weir which was installed in 1957 (Stephen and Wade, 2018) to maintain water levels in Cowichan Lake to meet downstream needs in the summer and early fall. The water licence is currently held by a local wood mill which regulates the amount of water flowing into Cowichan River to meet the economic and ecological needs downstream of the Lake.

This species exists in a complicated multi-user watershed. At its most basic level, the watershed supports many tangible human priorities, including an active logging industry, residential and recreational use, tourism, First Nations rights to access anadromous salmon, small-scale businesses relying on water in both the Lake and the River, sewage treatment, drinking water, agriculture, and a pulp mill (Wade and MacConnachie, 2016). Ecologically, both Cowichan Lake and River are habitat for resident fishes, crustaceans, amphibians, plants, and birds. The Cowichan River has been traditionally and currently is an important Chinook salmon (*Oncorhynchus tshawytscha*) river. The Lakes and surrounding terrestrial environment are highly valued both for supporting livelihoods as well as recreation. When making decisions regarding the watershed, and in this case the management of water resources, efforts are made to weigh all these needs. If those needs cannot be defined, they cannot be considered. This has been the case for Cowichan Lake lamprey. Although they are a protected species, they are poorly understood, largely understudied, and have low social value. They are a parasitic fish, which prey on highly valued recreational species and are seen to many as a pest, if not of no value at all. The "vampire" reputation of lampreys in general does the conservation of this species a great disservice. It is naïve to think the regulation of water in Cowichan Lake would be focused on the habitat needs of these lamprey rather than jobs or other highly valued species such as Chinook salmon. Their low social value and the general lack of knowledge of the needs of the species results in a situation where evidence-based arguments cannot be made to specify effective and acceptable conservation actions.

In recent years, applications have been approved to allow pumping of water from Cowichan Lake into the River when insufficient flow was available to sustain ecological and economic functions downstream. Depending on the time of year, this type of drawdown could decrease or eliminate the available spawning and rearing habitat for the Cowichan Lake lamprey. This was a compromise that the permitting agency felt appropriate under the circumstances. As an isolated event, it is reasonable to assume that the current spawning generation of lamprey would be mostly impacted from a drawdown as other year classes are present as ammocoetes that spend multiple years burrowed in the near-shore area and could move if conditions became unfavourable. Although a recent study with Pacific lamprey (*E. tridentatus*) ammocoetes has shown that they do not readily move under de-watering conditions. With a rapidly changing climate in the region, the likelihood of drawdowns is no longer a rarity. With what appears to be a new norm of low winter precipitation and extended summer droughts, available spawning and early rearing habitat is being compromised by low Lake water levels even without the added stress of an emergency water drawdown. Various committees, groups, and levels of government concluded that an increase in the height of the weir to accommodate more water storage in the rainier winter season and greater flexibility in regulation throughout the year is the best way to address ecological and social demands for water. Only a few lakeside landowners object to this plan because they would lose some beachfront in the winter storage period. Although plans are moving forward to replace the current weir, to date, no group has stepped forward to accept the sole cost, responsibility, and liability of the operation and maintenance of the weir. As this weir is mutually beneficial, it would be unreasonable to presume that the current single operator would continue to hold the water licence.

Morrison Creek lamprey is another example of an extreme endemic. It is found only in Morrison Creek and a few small tributaries flowing into it, in Courtenay, BC, on Vancouver Island. Morrison Creek is less than 24 km long and extends from rural headwaters through an urban town to the confluence with a larger river (Wade et al., 2015). Morrison Creek lamprey are unique in that they are thought to be an intermediate form between a parasitic and non-parasitic lamprey, and, thus, an example of an evolving species. The Morrison Creek lamprey population produces two distinct life history types: a freshwater non-parasitic lamprey, the Western Brook lamprey (*L. richardsoni*) and a parasitic (silver) variety *(L. richardsoni* marifuga) (Beamish, 1987). There are no other known living examples of two life history types of the same species of lamprey existing at the same time.

The watershed is in the coastal Douglas fir (*Pseudotsuga menziesii*) biogeoclimatic zone made up of interlinking wetlands with meadows, thick brush, beaver dams, and ponds (Wade et al., 2015). The area was heavily logged and is now comprised mostly of second growth trees in areas where urban development has not occurred. A great diversity of species exists in the watershed, including large mammals, such as black bear (*Ursus americanus*) and Roosevelt elk (*Cervus elaphus roosevelti*), and less obvious residents like amphibians, signal crayfish (*Pacifastacus lenisculus*), three-spined stickleback (*Gasterostreus aculeatus*), as well as resident and anadromous salmon (Wade et al., 2015). On the surface, Morrison Creek is a typical, quasi-urban Canadian creek. However, something unique in this watershed gave rise to a most unique animal, the Morrison Creek lamprey. What these conditions or situations may have been or may still be are unknown and likely to remain unknown. Morrison Creek lamprey as a population of both parasitic and non-parasitic forms are currently listed as endangered under SARA.

Apart from the required documents which must be produced as a result of listing a species under the Species at Risk Act, there have been only a few studies of this lamprey conducted since the 1980s when it was first described (Beamish et al., 2016). Most studies have attempted to be non-invasive and non-lethal as little is known of the potential impacts disturbance may have on the population. As a result, compounded in part by its cryptic nature, small size, and nocturnal activities, little is known of the basic biology of the animal. Morrison Creek lamprey reach a maximum size of approximately 142 mm (Beamish et al., 2016) and have not been observed feeding parasitically in the wild. Unlike the Cowichan Lake lamprey, which leaves substantial scars on highly valued recreational fish, there have been no reports of scars on fish in Morrison Creek. While this makes the Morrison Creek lamprey less of a threat to recreational fisheries values than the Cowichan Lake lamprey, it does not help to increase awareness for the animal or the need for habitat protection. Land-based activities can impact the aquatic habitat by altering water quality with sedimentation or pollutants, impacting riparian habitat, or altering run-off rates (National Recovery Team for Morrison Creek Lamprey, 2007). Urbanization is a primary driver of these threats. Urban development pressures are increasing around the creek and water courses have been disrupted for road construction and land development.

HARM REDUCTION AS A PATH TO MUTUALLY BENEFICIAL CONSERVATION

It is possible in these complex situations to take actions to reduce harm for these lampreys to provide the best conditions possible for their continued survival. The harm reduction approach is a viable method to facilitate action in the presence of conflicting values and scientific unknowns, a feature of utmost importance in these rapidly changing environments. There are six principles of harm reduction (modified from Stephen et al., 2018) which are applied in both these situations to minimize the total amount and impact of a harm (Table 19.1).

For Cowichan Lake lamprey, harm reduction principles (Table 19.1) are guiding actions to build a willing coalition of people by increasing the positive reputation of the species through education actions and ensuring that Cowichan Lake lamprey and its habitat are considered in water management decision-making from the individual citizen to the federal government. A multi-pronged approach has been used to achieve this, including (i) targeted scientific studies to provide advice on the relationship between weir height and spawning area coverage and identification of areas used for spawning and early rearing and (ii) public dissemination of information through various methods, including scientific publications, public talks, signage, and a lamprey poetry book for children. Community members, various levels of government, and scientists are now engaged and involved in these activities. Although the work is ongoing, attitudes towards the species are changing and the needs of the species are now being considered by the watershed board in plans to increase lake water storage.

For Morrison Creek lamprey, ensuring an ongoing secure water supply is essential as they do not have the buffering capacity of a lake or even a large river. Similar to the Cowichan Lake lamprey, a multi-pronged approach by community

TABLE 19.1
Principles of Harm Reduction for Lamprey Conservation (Based on Stephen et al., 2018)

Principle	Element
Focus on harms rather than eliminating hazards	Climate change and urbanization are complex and ongoing What can be done today to help the animals cope with the negative consequences of these pressures?
Pragmatism	What changes are likely to happen given existing knowledge, partnerships, and regulations?
Solution oriented	Rather than working to discover more about the lamprey, what can we learn about the strengths, possibilities, and opportunities to reduce negative consequences now?
Use a multi-pronged approach	Are there synergistic and re-enforcing opportunities to act throughout the lamprey's socio-ecological system?
Prioritize achievable goals	Are there incremental gains we can achieve that build trust, enthusiasm, and willingness to act within the current circumstances and state of knowledge?
Partnerships	How to engage a diversity of players in finding solutions throughout the chain of causation from hazard to harms

members, governments, and scientists has been used to reach this goal, including (i) public dissemination of information primarily through signage and engagement with local school children and town planning staff and (ii) targeted scientific studies to monitor habitat features and functions deemed important for lamprey. Because of decreases in catch rates of the silver form of Morrison Creek lamprey from the 1980s to the 2000s, Fisheries and Oceans Canada, the regulatory agency, has decided to conduct less research on the population to reduce the potential harm from capture and handling as a result of scientific activities. Efforts have instead focused on ensuring habitat is adequate for their continued survival rather than addressing knowledge gaps. As such, two structural barriers to lamprey movement were identified (Wade and Beamish, 2014), which fragmented the Creek and prohibited upstream movement. Funding was procured and major restoration efforts were undertaken to eliminate these barriers in 2015. In a recent success, community driven efforts protected 22.6 hectares of private land in the headwaters of Morrison Creek. The protection of this habitat is a direct benefit for not only the Morrison Creek lamprey but also for ten other species at risk that live in this watershed. The fate of the lamprey played a significant part in outreach to donors to protect these lands.

For both the Morrison Creek lamprey and Cowichan Lake lamprey, it is not too late to act to reduce harms and build systems resilience, even in the face of low social value and large deficits in knowledge. They may be uncharismatic, underfunded, and woefully underappreciated, but it is possible to change people's attitudes to promote the conservation of both the species and their habitat.

REFERENCES

Beamish, Richard J. "*Lampetra macrostoma*, a New Species of Freshwater Parasitic Lamprey from the West Coast of Canada." *Canadian Journal of Fisheries and Aquatic Sciences*, 39, no. 5 (1982): 736–747.

Beamish, Richard J. "Evidence That Parasitic and Non-Parasitic Life History Types Are Produced by One Population of Lamprey." *Canadian Journal of Fisheries and Aquatic Sciences*, 44 (1987): 1779–1782.

Beamish, Richard, J., and Joy Wade. "Critical Habitat and the Conservation Ecology of the Freshwater Parasitic Lamprey, *Lampetra macrostoma*." *Canadian Field-Naturalist*, 122, no. 4 (2008): 327–337.

Beamish, Richard, Ruth Withler, Joy Wade, and Terry Beacham. "A Nonparasitic Lamprey Produces a Parasitic Life History Type: The Morrison Creek Lamprey Enigma," in *Jawless Fishes of the World, Vol. 1*, eds. Alexei Orlov and Richard Beamish (Cambridge: Cambridge Scholars Publishing, 2016), 191–230.

MacConnachie, Sean E.M., and Joy Wade. "Information in Support of the Identification of Critical Habitat for the Cowichan (Vancouver) Lamprey (*Entosphenus macrostomus*)." *DFO Canadian Science Advisory Secretariat Research Document*, 109 (2016): 17.

National Recovery Team for Morrison Creek Lamprey. "Recovery Strategy for the Morrison Creek Lamprey (*Lampetra richardsoni var. marifuga*) in Canada." *Species at Risk Act Recovery Strategy Series, Fisheries and Oceans Canada*, Ottawa (2007): 24.

Stephen, Craig, and Joy Wade. "Wildlife Population Welfare as Coherence Between Adapted Capacities and Environmental Realities: A Case Study of Threatened Lamprey on Vancouver Island." *Frontiers in Veterinary Science*, 5 (2018): 227.

Stephen, Craig, Julie Wittrock, and Joy Wade. "Using a Harm Reduction Approach in an Environmental Case Study of Fish and Wildlife Health." *EcoHealth*, 15, no. 1 (2018): 4–7.

Wade, Joy, and Richard Beamish. "Identification of Barriers Affecting the Movement of Morrison Creek Lamprey (*Lampetra richardsoni marifuga*) Within Its Natural Range." *Canadian Manuscript Report of Fisheries and Aquatic Sciences*, 3043 (2014): 16.

Wade, Joy, and Sean MacConnachie. "Shared Values, Shared Success: Remediating Endangered Lamprey Habitat in British Columbia." *International Journal of UNESCO Biosphere Reserves*, no. 1 (2016) https://biospherejournal.org/vol1-1/remediating-endangered-lamprey-habitat-in-british-columbia-wade-macconnachie/

Wade, Joy, Nadine Pinnell, Gabrielle Kosminder, and Sean MacConnachie. "Information to Support the Identification of Critical Habitat for the Morrison Creek Lamprey (*Lampetra richardsoni var. marifuga*)." *DFO Canadian Science Advisory Secretariat Research Document*, 31 (2015): 30.

20 A Caribbean Call to Action: Behaviour Change Strategies to Reduce Local Plastic Waste

Luis Cruz-Martinez, Luis-Pablo Hervé-Claude, and Craig Stephen

CONTENT

INTRODUCTION

The continuous accumulation of plastic debris in marine environments worldwide is a global threat (Van Sebille et al., 2015). An impactful example is the Great Pacific Garbage Patch, where approximately 80,000 tons of floating plastic have accumulated in an area of 1.6 million km^2 (Lebreton et al., 2018). Sea turtles, sea birds, and marine mammals are experiencing morbidity and mortality mainly from entanglement and ingestion of this persistent plastic debris (Gall and Thompson, 2015).

Urgent action is needed to prevent plastic reaching the oceans (Law, 2017) and to promote local actions that combat this global threat (MacDonald et al., 2015), especially in small island nations where recycling and reuse options are limited. A call to action informed by principles of behaviour change is needed to prompt people to think and act on this marine pollution issue. Here we share a series of interventions to encourage people to reduce the amount of waste from disposable food items (plastic and Styrofoam, primarily), at an institution of higher medical education in the Caribbean, specifically Ross University School of Veterinary Medicine, showing how behaviour change theory can lead to success.

BEHAVIOUR CHANGE FRAMEWORK

A small team of faculty, staff, and students was formed to undertake a series of steps to promote reduced plastic pollution. We used the health belief model, theory of planned behaviour, and the transtheoretical model (Glanz et al., 2008) as our theoretical foundations (see Chapter 9 for a description of these concepts). We complemented them with elements from human behaviour disciplines and concepts, including Maslow's hierarchical human needs framework, behavioural economics, conservation psychology, and positive psychology. We used guiding questions to (i) identify people's attitudes and beliefs (their mindset) about the problem, solutions, and capacity to create change; and (ii) to identify the social norm that guided their mindset (Table 20.1). This information is crucial for

TABLE 20.1
Guiding Questions to Explore a Person's Mindset on Problems, Solutions, and Capacity and Willingness to Act

Drivers of Mindset	Guiding Questions
Attitudes (positive, neutral, or negative personal evaluations) Beliefs (thoughts, opinions, and perceptions that are considered as truth) (Including cognitive and emotional responses)	Is it severe? Will I be affected? Is it urgent? Is it affecting me/us now that I need to take immediate action? What are the costs and barriers? Are the benefits greater or less than the costs? Do I have enough resources to act? Am I in control of my actions? Will my actions matter towards the solution or are they irrelevant? Can I make a difference with available resources and conditions? Will I be capable of complying?
Norms and conditions (expectations from others, what other people are doing, what I'm supposed to comply with)	What are most people doing about the problem? What should I do when little is being done by others? Do I want to comply with society/industry/organization's expectations? What are influential people doing? Why should I comply with my superiors when they are not complying and are contributing to the problem? What am I expected to do in this group situation?

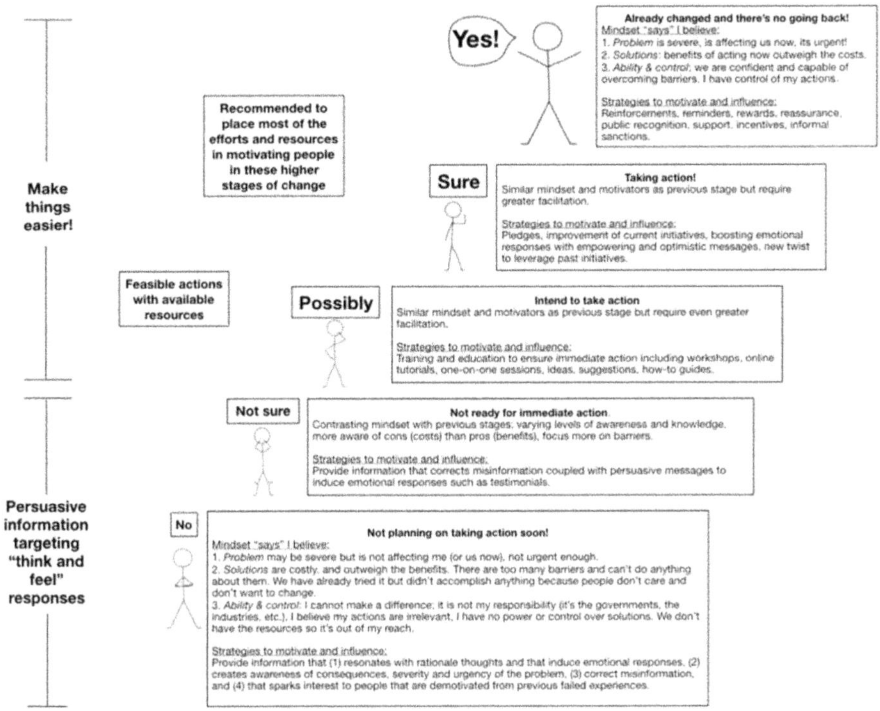

FIGURE 20.1 Strategies for matching behaviour change interventions with the mindset (attitudes and beliefs) of people in the different stages of change. This diagram was created based on the transtheoretical model.

creating and applying specific strategies that resonate with people based on their readiness for change (Figure 20.1).

OUR CHALLENGES

How to transmit an effective call to action on campus that (i) inspires people to take immediate action, (ii) convinces people to believe their actions (large or small) are impactful, and (iii) encourages people to disseminate the message further? We identified the driving mindsets and social norms through surveys, one-on-one conversations, and by participating in various committees and activities.

STEPS TO PROMOTING CHANGE

Our team found that most people were aware of the problem and generally agreed on the need to address it but were at an early stage of change. Their mindset and the social norms were generally not supportive of change. There was limited sense of urgency: "the problem is not affecting us right now," there was a general belief that the problem was not their responsibility, and their actions were irrelevant because they did not have power or control over the solutions. The beliefs about potential solutions emphasized that the costs were much greater than the immediate benefits and the barriers were too great to overcome. A minority of people who were contemplating action or had already tried to do something were discouraged to resume action because of their perceived low ability to influence change.

Our targeted higher education institute is multicultural: composed of residents of St. Kitts and Nevis, people from other islands in the Caribbean, and expats from North America and Europe (including students, faculty, and staff). The prevailing social norm, how most people act when they are in a group, appeared to drive people's behaviour regardless of their nationality. For example, students native to the United States and Canada readily engaged in unsustainable behaviours (i.e. using disposable Styrofoam containers and disposable plastic water bottles) even though such behaviours were increasingly discouraged in their home country. A complicating factor was that most people on the island did not have an intimate connection to the ocean. It became evident that promoting change in attitudes and behaviours about marine conservation and ocean pollution would be challenging due to the lack of relatability of islanders to the ocean.

Leading by Example to Promote Sustainability Values in the Prep School

A Prep School was available for children of students, faculty, and staff of this higher education institute. The norm and mindset described above were strongly held by the school's administrators, with an emphasis in beliefs of little to no control over the solutions, along with a plethora of perceived barriers. Most of the school activities produced a substantial amount of waste, particularly from disposable plastic and Styrofoam items. The school's administrators were open to listening to suggestions and ideas from us; however, the implementation of these alternatives was not deemed possible.

We focused on leading by example, by disseminating the positive changes resulting from the local event we organized for the 2016 One Health Day, specifically on how faculty, staff, and veterinary students responded to our call for action for making simple changes in the planning of events that would reduce the amount of waste. We highlighted the support from a local business located on the campus that agreed to start selling compostable/biodegradable food ware, an action that facilitated this change. During school events, we refused plastic and Styrofoam items, explaining our congruency with the initiatives for reducing waste in the institution. This strategy resonated with parents and teachers who were not ready for immediate action but intended to take action and prompted them to start talking about the problem and adopt some of the new behaviours being promoted. As we noticed increased responses to our call to action, we praised and commended people's efforts and used the momentum to urge them to encourage other parents to act similarly. Once we had motivated people within the administrative structure, we used targeted questions that further motivated them to develop new ideas and lead their own initiatives. A different mindset that encouraged Prep School activities to be environmentally responsible grew. Balloon use was discouraged, disposable plastic water bottles were no longer provided, and disposable plastic or Styrofoam items were no longer used at the school. Most of the decorative items in the school were reused in subsequent events, and parents and children were encouraged to bring their own containers, plates, and cups. To facilitate this process, in any event where food was served, the school provided a small number of reusable plates and cups along with readily available water fountains.

Linking Empowerment of Women with Environmental Sustainability

The institution's parent company created a programme for the advancement of women in the workplace across their higher education institutions. The programme intended to raise awareness about gender equality and promotion of women into leadership positions. By participating in this group, our team saw an opportunity to contribute to their cause while promoting our call to action for environmental sustainability based on the premise that gender equality is a common denominator for achieving sustainable environmental goals (UNDP, 2012).

Most of the participants from this group were local women. As a few of our team were expats, we first ensured that our participation was accepted and welcomed into this group by being active, committed, and responsible contributors to the group's activities. We focused first on understanding how our goal of environmental sustainability was compatible with this group's goals by focusing our initial involvement on their goals. This strategy gained their respect and allowed us to meet and interact with many participants, from which we identified or categorized the members of the group based on their readiness to change (stages 1–5 of change are shown in Figure 20.1). More importantly, we identified community leaders who embraced our call to action and became strong supporters.

The next step was to suggest the possibility for the group's following event to generate the least amount of waste as possible, especially from Styrofoam containers. This call to action was conveyed through a blast email to all the members of the group with the following message:

> I would like to propose we have a potluck as environmentally sustainable as possible!
>
> At its core, environmental sustainability is a way to live our daily lives making sure we create the conditions that support our lives – clean and available water, safe food to eat, a healthy environment – and the lives of our kids and grandkids, even for those that are not born yet. Let's show the power and strength of women for protecting the environment. See this link from United Nations for an example of the power of women for environmental sustainability.
>
> Here are a couple of easy and simple ideas for this:
>
> - Let's bring our own plate or container along with our own silverware and cups/glasses. If bringing these from home is uncomfortable, let's borrow them from the kitchen.

- Let's avoid plastic and Styrofoam items as much as possible because these items do not biodegrade and have the potential to end up forming part of the islands of garbage floating in the oceans.
- Please feel free to add more suggestions.

This message was well received and as an encouraging result, no Styrofoam containers were used. People found ways of reducing waste such as using napkins instead of plates and using silverware instead of disposable food ware. To maintain momentum, we engaged the leaders of this group and wrote a formal proposal to the institution's administration requesting funds for a "green starter kit" that included high-quality reusable mug and glass food containers that would remind members of the link between simple life choices (refuse plastic and Styrofoam) and a sustainable future. This green starter kit was a great motivation for the current members as well as an incentive for new members to join. The local leaders of the women's empowerment group became more inclined towards the "green" mindset when planning events.

Using a Multicultural Party to Reinforce Behaviour Change – The Melting Pot

The Melting Pot is a campus-wide initiative that brings the multicultural community to celebrate diversity and inclusion with a common and effective catalyzer, namely food. For the inaugural event in 2017, we encouraged the organizer (who is also an active member of the women empowerment group) to include a "green" calling in the event's invitation with the goal of reducing waste. We provided support by showing how reusable items could replace disposable ones and identified locally available alternatives and options.

The following year, the event organizer independently included the following sentence in the invitation: "We continue to show by example by refusing disposable items and encourage you to do your part for making this a green event." We conducted an exchange programme for trading disposable containers (Styrofoam plates and cups) for high-quality reusable food and beverage containers. We also gave out similar products (reusable bags, metal straws, cloth grocery bags, etc.) that served as triggers to remind people of the alternative available items. During this trading, our call to action started with the phrase: "many of us are guests in this island and we have to make every possible effort for making this beautiful island healthier." This was briefly followed by messages on the power we all have for making small changes in our daily lives, which over time lead to a positive impact for environmental conservation that ultimately makes us healthier.

This message resonated very well with the local community and helped inspire local leaders to actively advocate and promote the message further. Two years after this event, we saw many people still using the "green" gifts from that event. The following year, we delivered a stronger call to action targeting influential colleagues (Dean, Campus Administrator, Research Center Directors, etc.) to embrace and promote reusable food ware or environmentally responsible alternatives (such as compostable food ware). We have evidence that this environmental call to action became an essential component of campus event planning, serving as an example for other campus-wide activities.

CONCLUSIONS

These three stories show positive outcomes and gradual changes in people's attitudes and behaviours over the course of four years. We believe this success was due to the application of principles of behaviour change for tailoring the interventions. There were three essential guidelines that were applied across the interventions. Firstly, we motivated people on the premise that change is a process and targeted most time and resources to the people who were contemplating or taking action. These individuals quickly became supporters and promoters of the call to action. Secondly, we got involved with the target audience and met them where they were in the change process. This meant having a clear and inspiring message (or story) that could be easily adapted to resonate with people at different stages of change. To reach this goal, it was imperative to listen actively without intervening or judging to identify a person's mindset (cognitive and emotional factors) and social norms that shape their thoughts, feelings, and actions towards the problem, the solution, and their ability (real or perceived) to make a difference. Once such factors are identified, one can briefly tell the call to action in relatable and passionate ways for inducing positive emotional and rational responses.

We encountered two main challenges to meeting people where they were. The first was the risk of misclassifying members of the target audience within the stages of change (readiness for action; Figure 20.1). When misclassification occurred, it resulted in frustration, resistance, and lack of interest because the strategies we used did not resonate with their mindset. The second challenge was that potential leaders – individuals who were advocates for environmental conservation and sustainability – tended to promote change forcefully (telling people what to do) instead of matching the recommended strategies based on the target audience's mindset. This caused tension when these potential leaders felt unappreciated or restricted when their suggestions were not approved because of the mismatch between their proposed strategies and target audience's mindset and cultural norms.

The focus of these interventions was to promote changes in people's mindset (attitudes and beliefs) rather targeting individual behaviours. Focusing on positive, influential, and persuasive messages showing our rationale (our why) proved to be an effective primer for change. A small number of attitudes and beliefs determine a greater number of individual behaviours (Vaske and Manfredo, 2012); therefore, this approach prevented a patronizing and ineffective approach of "telling people what to do." Our current focus is in maintaining momentum from these interventions to continue creating the conditions for these changes to persist and to be further promoted by local leaders, who will potentially promote the call to action in their household and community.

REFERENCES

Gall, Sarah C., and Richard C. Thompson. "The Impact of Debris on Marine Life." *Marine Pollution Bulletin*, 92, no. 1–2 (2015): 170–179.

Glanz, Karen, Barbara K. Rimer, and Kasisomayajula Viswanath. "The Scope of Health Behaviour," in *Health Behaviour and Health Education: Theory, Research, and Practice*, eds. Karen Glanz, Barbara K. Rimer, and Kasisomayajula Viswanath (San Francisco: Jossey-Bass, 2008), 552.

Law, Kara Lavender. "Plastics in the Marine Environment." *Annual Review of Marine Science*, 9 (2017): 205–229.

Lebreton, L., Boyan Slat, Francesco Ferrari, Bruno Sainte-Rose, Jen Aitken, R. Marthouse, Sara Hajbane et al. "Evidence That the Great Pacific Garbage Patch Is Rapidly Accumulating Plastic." *Scientific Reports*, 8, no. 1 (2018): 1–15.

MacDonald, Edith, Taciano Milfont, and Michael Gavin. "Thinking Globally But Not Acting Locally?: Expert and Public Perceptions of Environmental Threats and Conservation Actions." *Human Dimensions of Wildlife*, 20, no. 2 (2015): 123–132.

UNDP. "Powerful Synergies: Gender Equality, Economic Development and Environmental Sustainability." United Nations Development Program (2012):192. https://www.undp.org/content/undp/en/home/librarypage/womens-empowerment/powerful-synergies.html.

Van Sebille, Erik, Chris Wilcox, Laurent Lebreton, Nikolai Maximenko, Britta Denise Hardesty, Jan A. Van Franeker, Marcus Eriksen et al. "A Global Inventory of Small Floating Plastic Debris." *Environmental Research Letters*, 10, no. 12 (2015): 124006.

Vaske, Jerry J., and Michael J. Manfredo. "Social Psychological Considerations in Wildlife Management." *Human Dimensions of Wildlife Management* (2012): 43–57.

21 Living with Rats: Could an Ecosystem Lens Provide New Insights into Urban Rat Control?

Chelsea Gardner Himsworth

CONTENTS

THE RAT PROBLEM

Brown rats (*Rattus norvegicus*, henceforth referred to as rats) have coexisted alongside people for millennia. They are one of the most ubiquitous animal species on the planet. Having spread around the world in association with human transport, rats are now found throughout every continent except Antarctica (Feng and Himsworth, 2014). Rats are unique among free-ranging wildlife in that they not only tolerate human presence, they thrive on it. Indeed, rats are so well adapted to utilizing human resources that they are seldom found outside of human habitations (Feng and Himsworth, 2014). Cities, in particular, provide optimal rat habitat. High human population densities provide abundant food and harborage, sustaining large numbers of rats within small geographic areas (Davis, 1953; Feng and Himsworth, 2014). Unfortunately, rats make for unwelcome bedfellows, as they have been associated with several negative consequences for humans. Rats can damage infrastructure and contaminate and consume food stuff; they are a source of a number of infectious diseases (Wundram and Ruback, 1986; Himsworth et al., 2013; Himsworth and Feng, 2014). Given global trends towards urbanization and densification, urban rat-related issues are more than likely to increase in the future (Parsons et al., 2018).

When one considers the fact that human communities have been trying to rid themselves of rats since almost the dawn of civilization, it is surprizing that the "rat problem" persists. Indeed, rats have proven themselves virtually impossible to extirpate. This may be partially due to the characteristics of the rats themselves. They are highly exploratory, opportunistic, and adaptable, which allows them to exploit a broad range of resources (Barnett, 1976; Colvin and Jackson, 1999). However, they are also neophobic and can learn from negative experiences, which decreases the efficacy of traps and poisons (Barnett, 1976; Clapperton, 2006). Finally, they reproduce at a staggering rate, therfore infestations are quick to become established and, after a control attempt, the infestation can quickly rebound to pre-control levels (Davis, 1953; Colvin and Jackson, 1999). So, is it any wonder that many cities seem to be plagued by rats? Or do the cities themselves bear some responsibility for their rat problems?

COMMON ISSUES WITH URBAN RAT CONTROL AT A MUNICIPAL LEVEL

There are three major areas in which cities tend to err in their approach to rats. The first is a lack of good governance or strategic planning (Colvin and Jackson, 1999). Rat-related issues are often addressed through a hodgepodge of unrelated policy and programming. Thus, municipal governments may respond only to infestations that occur on public properties or in scenarios requiring permitting (i.e. demolition), while local health authorities may respond only to infestations in food-producing establishments or where there is a demonstrated health risk. At best, municipal leadership is highly fragmented, at worst it is absent altogether. For most of the urban landscape, people are left to fend for themselves.

The second problem is the dearth of urban rat surveillance tools and techniques, in combination with the fact that decision-makers frequently focus on action even in the absence of knowledge. Data collection is often neglected in the design and execution of municipal rat control interventions, so much so that there is usually not enough information to answer even the most basic questions like: "How many rats are there? Where do they live? Why are they there? Is the problem getting worse?" Without this information, it is virtually impossible to efficiently allocate resources, to design informed and effective policies and programmes, or to assess the return on investments. After all, without knowing what the rat problem looked like beforehand, there is no way of knowing whether an intervention made the problem any better.

This leads to the third problem: the "ambulance approach" to rats. Without data or strategy for guidance, the most common response to rat infestations is to wait until they are causing a significant problem and then swoop in and try to trap or poison as many animals as possible. This reactionary approach is probably the least effective way to deal with rats and is rarely successful at controlling or eliminating infestations in the long term (Clinton, 1969; Colvin and Jackson, 1999; Himsworth et al., 2013). Additionally, waiting for an infestation to reach a critical threshold before there is an intervention is neither the most efficient nor the most effective point at which to intervene, and this approach completely precludes any possibility of preventing infestations before they occur. We need only to look to preventative medicine, our current standard of health care, to see the value of the idea that "an ounce of prevention is worth a pound of cure," and that it is never a good idea to wait until someone needs an ambulance before dealing with their health problems.

Some might argue that ecologically based rodent management (i.e. pest control techniques that incorporate an understanding of the biology and ecology of rodents and the environment in which they reside) is the panacea to the urban rat problem, and that the reason it has not been widely adopted in urban centres is a lack of political will (Singleton et al., 1999; Colvin and Jackson, 1999). However, given that no city on any axis of space, time, or politics has yet to emerge with an efficient or effective rat management strategy, the problem may lie deeper. The problem may be fundamentally one of perspective.

SOLUTIONS ACHIEVED USING AN ECOSYSTEM LENS

Rather than viewing the city as an artificial landscape completely under human control and the rat as a foreign invader of our territory, we need to see the rat for what it is – a wild animal that shares the urban ecosystem with us. Although cities are not often thought of as ecosystems, adoption of an ecosystem-based perspective may provide new insights and opportunities for dealing with rats and rat-related issues.

Urban ecosystems are often conceptualized as complex adaptive systems (CAS) (Desouza and Flanery, 2013; Meerow et al., 2016). A CAS is a set of component parts that interact to form a system that has emergent properties unique from the components in isolation, and that has the capacity to change over time (see Chapter 13 on complexity and One Health). Each component may, itself, be complex (Levin, 1998; Meerow et al., 2016). Indeed, the urban ecosystem has been described as CAS composed of intertwined socio-economic, infrastructural, material and energy, and governmental networks (Meerow et al., 2016). It would be beneficial to supplement this characterization with an additional level that includes biological networks (Figure 21.1) inclusive of all life forms existing in a city – human, wildlife, microbiota, etc.

Once we recognize that we are managing a system, it becomes clear that leadership and strategic planning are critical (Ernston et al., 2010). The very concept of a system is that the whole is more than the sum of its parts, which is antithetical to our current reductionist approach to rats. It also follows that if we are going to manage a system, then we must understand it. This, in turn, provides justification for substantive, long-term investments in collecting data on rat infestations, the specific conditions that support them, and the issues that they may cause (Parsons et al., 2018).

The CAS approach can be used to provide solutions that better address those issues. Rat control interventions are often undertaken with the goal of reducing rat-associated public health risks – particularly the risk of pathogen transmission from rats to people. These interventions usually focus on trapping or poisoning based on the perspective that, regardless of the nature of the problem, reducing

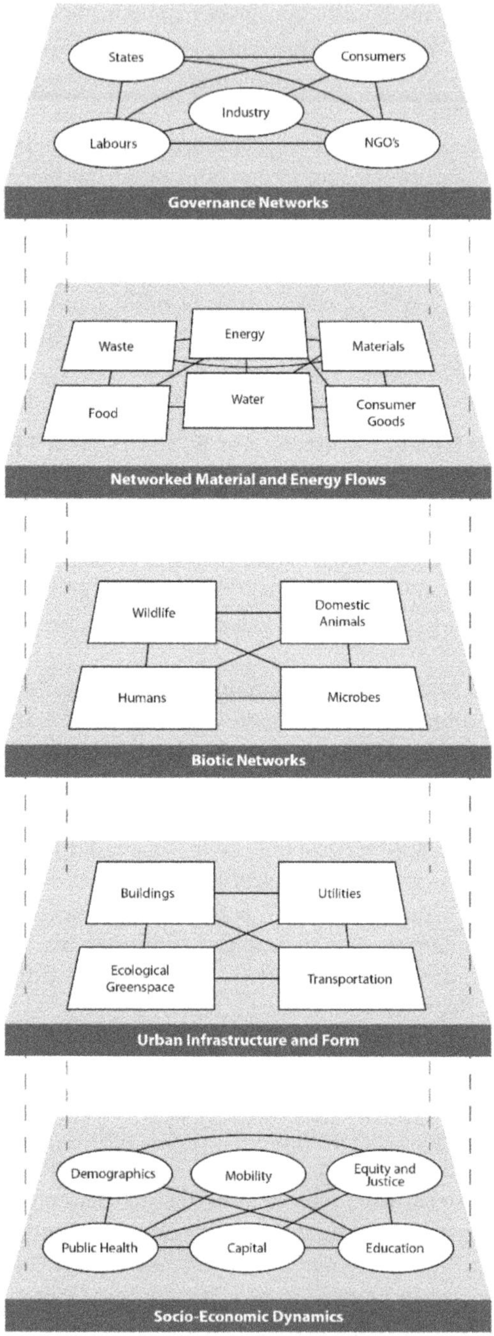

FIGURE 21.1 A conceptual model of an urban ecosystem based on the framework proposed by Meerow et al., 2016.

the number of rats will somehow make it better. However, it has been shown that, for certain rat-related diseases, rats are more likely to carry zoonotic pathogens after the intervention compared to before (Lee et al., 2018). This surprising finding is predictable once rats are considered in the broader context of their own ecology, the ecology of the pathogens they carry, and their interactions with the urban environment. Specifically, urban rats have evolved to form tight family units with limited home ranges owing to a resource-rich environment and barriers to movement (e.g. rats avoid crossing roads) (Davis, 1953). For pathogens that are transmitted among rats through specific social interactions within the colony, removal of key family members through trapping and poisoning can result in social strife and increased opportunities for disease transmission as rats fight to restore social order (Lee et al., 2018).

Revisiting the rat control programme described earlier with a CAS approach would require us to understand how rats interact with each other, how they interact with the microbiota they carry, and how both interact with urban infrastructure and resources. Using this perspective, rather than killing rats, one might seek to alter the environment in a manner that decreases the number of rats it can support (e.g. through altering waste management, repairing aging infrastructure), so there is a gradual decline in total rat population while maintaining social order (Colvin and Jackson, 1999).

The rats, however, are not the only complex beings in the urban ecosystem. Just as the impact of humans on rats can be multifaceted and unpredictable, so too can the impact of rats on humans. Rats are usually considered to be a health risk to people only if they are found to be a source of infectious disease. However, it is increasingly being recognized that chronic exposure to rats can have significant negative mental health impacts, particularly among impoverished, inner city residents (Wundram and Ruback, 1986; Lam et al., 2018; Byers et al., 2019). It has been shown that rats can elicit fear, anger, stress, worry, and exhaustion, as well as sleep disturbances and avoidance of activities (e.g., urban gardening) that could lead to rat contact (Byers et al., 2019). A CAS lens further enriches this understanding by identifying the fact that rat infestations are just one component of a constellation of negative environmental factors afflicting impoverished city dwellers; rats, in other words, are inextricably intertwined with a variety of social justice issues (Wundram and Ruback, 1986; Lam et al., 2018; Byers et al., 2019). For example, being forced to live with rats can be seen as a symptom or symbol of an individual's lack of control over their lives and living conditions, which is compounded by the perceived neglect of those who may have the power to improve these conditions (e.g. municipal governments and slum landlords) (Byers et al., 2019). A more sophisticated understanding of the potential consequences of rat-human contact not only brings to light a significant and widespread public health risk that has hitherto been completely ignored, but it also highlights the importance of understanding the interface between rats and people.

Further to the concept of interfaces, a CAS lens would also find fault with our current approach to rats, which is largely focused on "things" – the rat, the human, the environment, etc. By contrast, a key aspect of systems thinking is the understanding that the relationships between the components of a system often contribute more to the function of the system than the components themselves (Levin, 1998; Meerow et al., 2016). Thus, a CAS lens would have us direct the majority of our time and energy to identifying, understanding, and ultimately modifying these relationships or interfaces. For example, a safe, healthy, and controlled physical home environment is a multidimensional component of an individual's mental and physical health (Cornwell, 2014; Pruss-Ustun et al., 2017). Rats have a propensity for invading this environment and can be extremely difficult to get rid of. Thus, the home environment is the place where the relationship between rats and people is least tenable (Wundram and Ruback, 1986; Himsworth et al., 2013, Lam et al., 2018). In our current rat management paradigm, private residences are often the areas most ignored by municipal powers; however, a CAS lens would show us that they are an area of particular vulnerability and should be prioritized for municipal intervention and assistance. Similarly, since factors associated with urban poverty promote rat infestations and rat-human conflict (Clinton, 1969; Wundram and Ruback, 1986; Himsworth et al., 2013), impoverished inner city areas should also receive special attention.

Finally, the CAS approach puts some completely novel tools in our toolbox for dealing with urban rats. Instead of focusing on responding to problems as they occur, the CAS lens introduces the concept of resilience (Desouza and Flanery, 2013; Ernston et al., 2010; Meerow et al., 2016). Resilience, or the ability of a system and the members within in it, to function in the face of disturbances (Desouza and Flanery, 2013; Meerow et al., 2016), is a helpful concept, because it shifts the perspective from one that is reactionary and myopic (i.e. focused on a single issue) to one in which we can predict, prepare for, or even prevent a multitude of challenges. One example of resilience-based rat control would be a comprehensive waste management programme that prevents rats from accessing garbage (a common food source), thereby reducing the capacity of the urban environment to sustain rat infestations (Colvin and Jackson, 1999). Another would be the development and enforcement of more stringent bylaws that enshrine the right of low-income tenants to live in a healthy and rat-free environment. This latter strategy is particularly pertinent as a disordered living environment is a central contributor to a number of physical and psychological ailments (Cornwell, 2014; Pruss-Ustun, 2017), so this one intervention could increase resilience to diverse array of insults. All-in-all, then, a focus on resilience introduces many new opportunities and approaches to managing the urban rat problem, very few of which have anything to do with killing rats or even with the rats themselves.

It is of note that the concept of resilience has many parallels with the public health concept of harm reduction (see Chapter 6 for details on harm reduction and Chapter 19 for an example of its use in practice). Harm reduction is mainly used to deal with the impact of drug use, and it refers to policies, programmes, and practices that aim to minimize negative health, social, and legal impacts. Similar to interventions rooted in resilience theory, those based on harm reduction principles may have a broad range of predicted and unforeseen benefits. For example, in Vancouver, Canada, supervised injection sites are an effective harm reduction tool that produce a range of health and community benefits, including reductions in overdose deaths, public drug use, and syringe sharing, as well as increased uptake of detoxification services and other community and medical resources (Wood et al., 2006; Marshall et al., 2011). Given that a major risk factor for rat exposure among impoverished inner city residents is engaging in outdoor injection drug use (McVea et al., 2018), these injection sites may also help to reduce rat exposure and associated harms.

OPERATIONALIZING A SYSTEMS APPROACH TO RATS

One of the challenges of a CAS approach is that it is so complex that it may be difficult for municipal leaders and decision-makers to operationalize. There are, however, several different perspectives that could be employed to overcome this obstacle.

For example, an understanding of the urban ecosystem and its interconnected networks can be used as a scaffolding on which to develop more comprehensive, efficient, and effective rat monitoring and mitigation programmes (Figure 21.1). Governance networks must be understood in order to identify stakeholder concerns and objectives (i.e. why do stakeholders care about rats and what do they want done about them), as well as jurisdictions, resources, and tools for intervention (i.e. the area or scenario being governed, availability of funding, expertise, and other supports; and methods to deploy, enforce, monitor, and maintain the intervention). Socio-cultural and socio-economic networks will need to be understood in order to establish priorities (i.e. who is most at risk and/or most likely to benefit from an intervention) and to develop programmes with the greatest chance for success (i.e. what are potential barriers and opportunities for acceptance of and compliance with the intervention). Urban infrastructure and form networks must be understood to identify factors that sustain infestations and to ensure that rodent management is considered during urban planning and management of the built environment and infrastructure.

Ultimately, though, the benefit of a systems approach lies in its ability to change our perspective on rats and rat-related issues. Therefore, it may be more helpful to shift the conversation away from developing a single set of policies or procedures that best embody the CAS approach, to using CAS as a way of thinking, or a lens through which a problem can be viewed (Meerow et al., 2016). This will allow us to reap the benefits of this thinking regardless of the scale, scope, or nature of the programme in question.

TURNING THE LENS BACK ON OURSELVES

Seeing through the lens of a CAS requires a paradigm shift in our way of viewing our world and ourselves in it. It requires that we focus on the forest rather than the trees; that we become aware of relationships rather than entities. Our goal ought to be that we work to minimize friction that is potentially imbedded in the infinite interaction zones within a CAS. Such a viewpoint shifts the emphasis away from winners and losers, or oppressors versus the oppressed, to one of creating *resilience*, so that members of an ecosystem may peacefully exist alongside one another.

Ultimately, adopting this lens may help humans acquire the kind of humility that is necessary to truly understand that we cannot control the world – even one as seemingly under our control as a city. However, the “adaptive” component of the CAS approach teaches us that through innovation, leadership, thoughtful action, and strategic planning, we can “trigger a transition of the system to a more preferable regime” (Ernston et al., 2010). We are all, and always will be, members of a CAS, and as such, human well-being depends on our stewardship of that system. With regard to rats, it will likely never be the case that rats are welcome into our homes; however, it is possible for us to live with them in a mutually healthy arrangement.

REFERENCES

Barnett, S.A. *The Rat: A Study in Behaviour* (Canberra: Australian National University Press, 1976): 318.

Byers, Kaylee A., Susan M. Cox, Raymond Lam, and Chelsea G. Himsworth. "“They're Always There”: Resident Experiences of Living with Rats in a Disadvantaged Urban Neighborhood." *BMC Public Health*, 19 (2019): 853.

Clapperton, B. Kay. “A Review of the Current Knowledge of Rodent Behavior in Relation to Control Devices.” *Science for Conservation*, 263 (2006).

Clinton, James M. “Rats in Urban America.” *Public Health Reports*, 84 (1969): 1–7.

Colvin, Bruce A., and William B. Jackson, “Urban Rodent Control Programs for the 21st Century,” in *Ecologically Based Management of Rodent Pests*, eds. G.R. Singleton, L.A. Hinds, H. Leirs, and Z. Zhang (Canberra: Australian Centre for International Agricultural Research, 1999), 243–257.

Cornwell, Erin York. “Social Resources and Disordered Living Conditions: Evidence from a National Sample of Community-Residing Older Adults.” *Research in Aging*, 26 (2014): 399–430.

Davis, David E. “The Characteristics of Rat Populations.” *The Quarterly Review of Biology*, 28 (1953): 373–401.

Desouza, Kevin C., and Trevor H. Flanery. “Designing, Planning, and Managing Resilient Cities: A Conceptual Framework.” *Cities*, 35 (2013): 89–99.

Ernston, Henrik, Sander E. van der Leeuw, Charles L. Redman, Douglas J. Meffert, George Davis, Christine Alfsen, and Thomas Elmqvist. “Urban Transitions: On Urban Resilience and Human-Dominated Ecosystems.” *Ambio*, 39 (2010): 531–545.

Feng, Alice, and Chelsea G. Himsworth. “The Secret Life of the City Rat: A Review of the Ecology of Urban Norway and Black Rats (*Rattus norvegicus* and *Rattus rattus*).” *Urban Ecosystems*, 17, no. 1 (2014): 149–162.

Himsworth, Chelsea G., Alice Y.T. Feng, Kirbee Parsons, Thomas Kerr, and David M. Patrick. “Using Experiential Knowledge to Understand Urban Rat Ecology: A Survey of Canadian Pest Control Professionals.” *Urban Ecosystems*, 16 (2014): 341–350.

Himsworth, Chelsea G., Kirbee L. Parsons, Claire Jardine, and David M. Patrick. “Rats, Cities, People and Pathogens: A Systematic Review and Narrative Synthesis of Literature Regarding the Ecology of Rat-Associated Zoonoses in Urban Centres.” *Vector Borne and Zoonotic Diseases*, 13, no. 6 (2013): 349–359.

Lam, Raymond, Kaylee Byers, and Chelsea G. Himsworth. "Beyond Zoonoses: The Mental Health Impacts of Rat Exposure on Inner City Residents." *Journal of Environmental Health*, 81, no. 4 (2018): 8–12.

Lee, Michael J., Kaylee A. Byers, Christina M. Donovan, Julie J. Bidulka, Craig Stephen, David M. Patrick, and Chelsea G. Himsworth. "Effects on Culling on *Leptospira interrogans* Carriage by Rats." *Emerging Infectious Diseases*, 24, no. 2 (2018): 356–360.

Levin, Simon A. "Ecosystems and the Biosphere as Complex Adaptive Systems." *Ecosystems*, 1 (1998): 431–436.

Marshall, Brandon D.L., M.-J. Milloy, Evan Wood, Julio Montaner, and Thomas Kerr. "Reduction in Overdose Mortality After the Opening of North America's First Medically Supervised Safer Injecting Facility: A Retrospective Population-Based Study." *Lancet*, 377 (2011): 1429–1437.

McVea, David, A., Chelsea G. Himsworth, David M. Patrick, L. Robbin Lindsay, Michael Kosoy, and Thomas Kerr. "Exposure to Rats and Rat-Associated *Leptospira* and *Bartonella* Species Among People Who Use Drugs in an Impoverished, Inner-City Neighborhood of Vancouver, Canada." *Vector Borne and Zoonotic Diseases*, 18, no. 2 (2018): 82–88.

Meerow, Sara., Joshua P. Newell, and Melissa Stults. "Defining Urban Resilience: A Review." *Landscape and Urban Planning*, 146 (2016): 38–49.

Parsons, Michael, H., Peter B. Banks, Michael Deutsch, Robert M. Corrigan, and Jason Munshi-South. "Trends in Urban Rat Ecology: A Framework to Define the Prevailing Knowledge Gaps and Incentives for Academia, Pest Management Professionals and Public Health Agencies to Participate." *Journal of Urban Ecology*, 3, no. 1 (2018): 1–8.

Pruss-Ustun, A., J. Wolf, C. Corvalan, T. Neville, R. Bos, and M. Neira. "Diseases Due to Unhealthy Environments: An Updated Estimate of the Global Burden of Disease Attributable to Environmental Determinants of Health." *Journal of Public Health*, 39, no. 3 (2017): 464–475.

Singleton, Grant R., Herwig Leirs, Lyn A. Hinds, and Zhibin Zhang. "Ecologically-based Management of Rodent Pests – Re-Evaluating our Approach to an Old Problem," in *Ecologically Based Management of Rodent Pests*, eds. Grant R. Singleton, Lyn A. Hinds, Herwig Leirs, and Zhibin Zhang (Canberra: Australian Centre for International Agricultural Research, 1999), 17–29.

Wood, Evan, Mark W. Tyndall, Julio S. Montaner, and Thomas Kerr. "Summary of Findings from the Evaluation of a Pilot Medically Supervised Safer Injecting Facility." *Canadian Medical Association Journal*, 175, no. 11 (2006): 1399–1404.

Wundram, Ina Jane, and R. Barry Ruback. "Urban Rats: Symbol, Symptom, and Symbiosis." *Human Organization*, 45, no. 3 (1986): 212–219.

22 Making a Case for Harm Reduction in Invasive Species Management: The St. Kitts "Monkey Problem"

Christa Gallagher

CONTENTS

THE CONTEXT

African green monkeys (*Chlorocebus sabaeus*) were transported to St. Kitts and Nevis from West Africa through the colonial slave trade in the 1600s, and have been part of the islands' landscape ever since (Denham, 1987). The green monkeys flourished because they were highly adaptive to many habitats, had no natural predators, and were generally free of high-consequence diseases unlike their West African relatives. What likely started as 50–250 individuals from multiple incursions (McGuire, 1974) from Senegal or Gambia (Denham, 1987), has grown to an estimated 40–50,000 in 2020, a number rivaling St. Kitts' human population. From their arrival until modern day, these monkeys have been viewed as pests and agricultural crop raiders. As their population grew and range expanded, their impacts intensified.

St. Kitts is a small island (176 km^2) within the twin-island Federation of St. Kitts and Nevis, located in the Lesser Antilles of the Caribbean. St. Kitts has a rich history of sugar cane production, a monoculture which sustained the island for hundreds of years until 2005, when it ceased sugar production in favour of tourism as the primary economy. When the sugar industry prospered, large busy sugar cane plantations occupied much of the arable land and were well protected by rangers, who by gunfire noise and shooting kept the foraging monkeys primarily confined to the island's interior (Dore et al., 2018). The 2005 closure of the sugar cane industry changed the agricultural landscape as well as the monkeys' territory. Today, hundreds of small-scale farmers grow crops and raise livestock and poultry where sugar plantations once existed. No longer scared off or shot by rangers, the monkeys have encroached on human-inhabited areas on the island, resulting in increasing human-monkey conflicts.

THE "MONKEY PROBLEM"

The "monkey problem" is a widely used local term given to the negative interactions of people and green monkeys. It is commonly spoken of in casual conversation between community members and appears in media and scientific publications. It describes the spectrum of social, economic, and ecological harms caused by the monkeys' insidious encroachment and their notorious invasive and destructive behaviours. To complicate matters, these harms have not been equitably experienced by all within the islands' communities. The most prominent and greatest documented problem caused by the monkeys is crop raiding and the subsequent losses sustained by the islands' farmers. Due to monkeys' assaults on their farms, these already impoverished farmers are often unable to yield crops for their own use or sale, and therefore face economic losses, food insecurity, and threats to their livelihood. They often feel marginalized as they have been mostly left to their own resources to deal with this continuous and unrelenting threat. About 50–75% (Dore, 2018) of St. Kitts farms face damage from monkeys. This agricultural loss adds to the islands' food insecurity, and monkey contact with agricultural products risks the transmission of zoonotic disease.

There have been limited and irregular efforts by the local government to deal with the "monkey problem" as they grapple with deeply embedded conflicting ideas, values, and scientific uncertainties. There has yet to be consensus among government, researchers, farmers, or other citizens on how to minimize the negative impacts of the monkeys that would be practical, affordable, and socially accepted. Processes to empower people to work together towards mutually beneficial results have been lacking.

ENTER HARM REDUCTION

Non-human primates are menaces in many Asian and African countries, yet no single solution for their management has emerged. It is highly unlikely that the St. Kitts "monkey problem" can be eradicated in the near future by socially acceptable, economically feasible, and scientifically proven means. A novel approach is needed to advance collective action towards this invasive species issue. Harm reduction, a subset of health promotion, may provide a process for progress on this complex problem where there are multiple stakeholders, including the animals themselves, with competing values and priorities.

Harm reduction is designed to minimize the effects of persistent and high-consequence harms, without eliminating the hazard at the root of the harms. It has its conception and evolution in the public health realm, where it is mostly focused on harms resulting from high-risk behaviour of addictions in marginalized or underserved populations (Jalloh et al., 2017). Harm reduction aims to improve health-related quality of life using a dignified, humanistic, and non-judgemental approach to decrease harms at the individual, community, and societal level (Anon, 2005; Hawk et al., 2017). It supports grassroots participatory efforts to engage with public and private experts and authorities. It includes all individuals and sectors involved with a problem to participate in working towards a solution. Like health promotion, it is multi-level, multidisciplinary and empowers people to take control of their health and lives (Inciardi and Harrison, 2000).

Harm reduction is a pragmatic approach to decreasing harm through practical solutions to health challenges. It meets people "where they are" with an existing problem and takes incremental steps to minimize harms like injury, disease or death. It works in a hierarchical manner, achieving the most immediate and realistic goals first (Anon, 2005). Harm reduction is known to be dynamic, allowing flexibility as people and problems fluctuate. Lastly, harm reduction recognizes that behaviours, their associated harms, and proposed solutions are highly dependent on belief systems and culture within a setting, so harm reduction is greatly contextual in nature and must be culturally sensitive. (For more information on harm reduction, see Chapter 6.)

While many programmes serve to reduce harm at the human-animal interface, there are few examples of the application of the harm reduction approach outside of public health. The St. Kitts "monkey problem" has many features of a problem suited to harm reduction. The lack of social will and proven methods make monkey eradication an unlikely and socially undesirable solution. Harm reduction presents a novel way to collaborate towards solutions that are community based and community driven, but at the same time can be supported by authorities when mutually agreed upon and developed.

CAN A HARM REDUCTION STARTING PLACE BE FOUND IN ST. KITTS?

To start the harm reduction process, it is necessary to find an entry point, the point at which to "meet the people at the problem." Historically, the St. Kitts "monkey problem" has been framed as an agricultural problem. However, the perspective of all stakeholders has not been taken into account when characterizing the problem, precluding opportunities to see entry points other than agriculture around which coalitions can be built to implement actions to incrementally reduce social and ecological harms. One has to first know the system and the unique and intricate relationships and interconnections contained within that system in order to identify critical intervention points to initiate systems change. Therefore, systems mapping was the initial necessary step to gain working knowledge of the St. Kitts human-monkey system. There is no better way to describe a system than to ask the people who are part of it.

Qualitative and participatory methodologies were used to glean information surrounding two main themes: (i) the nature of the "monkey problem", and (ii) possible solutions to the "monkey problem". To get to the nature of the "monkey problem", participants, drawn from across a spectrum of stakeholders, attended focus groups or participated in a semi-structured interview. Participants represented agriculture, government, local business, tourism, biomedical research, academia, and community members. Within the focus groups, participants answered questions about who were the stakeholders in the human-monkey system and what their relationships were to each other. They were encouraged to tell their own versions of the monkey story to gain as much contextual information as possible. Next, they were asked to categorize and prioritize individual harms caused by the monkeys. They were asked two questions: "Are the monkeys a problem for you?" and "Why, when, and where are they a problem?"

The participants listed the harms, ranked them, and performed proportional piling of the harms, a Participatory Rural Appraisal methodology, to visually show a graduated emphasis of a ranked item (Mariner and Paskin, 2000). Despite the emphasis being placed on harms caused by monkey-human interactions, benefits of the monkeys were also revealed. Drawing on topographical maps was another method used to identify locations of stakeholder involvement and high-risk areas of human-animal interactions. The maps were used by the participants to point out areas where the monkey problem fluctuated due to seasonality or within a given timeline. The participants were asked about shared values, things they cared about that may be at risk by the harms caused by the monkeys. This question enabled some projection of fears participants had if the monkey problem was not adequately dealt with.

Participants were next asked about possible solutions to the "monkey problem". They were asked what they have personally done, if anything, to thwart the monkeys' activities. What tools are available to help alleviate the problem? Who may be in a position to help? What would inspire you to take action on this problem? What is the order of harms you would like to see addressed? Very importantly, they were asked: What are the obstacles for action taken against harms caused by the monkeys? When trying to mitigate a problem with interventions, it is possible to create new problems, or unintended consequences. The participants were, therefore, asked if they could imagine any unintended consequences to the posed solutions. Additional probing questions were asked to gather more detailed information on a response or to verify information in a sort of data quality control (Mariner and Paskin, 2000).

Through this process, a more fulsome view of the "monkey problem" was emerging. This monkey population was seen as a populous of non-human primates on a well-travelled island that could bridge primate associated diseases between South or Central America and the Caribbean and North America. With its shift towards tourism, St. Kitts and Nevis has a tremendous influx of global tourists who can introduce disease to this insular species or acquire and spread diseases from monkeys. There were, therefore, growing concerns about the role of the monkeys in amplifying or maintaining emerging infectious diseases. This became more acute when Zika virus emerged and spread in the Americas. The monkeys were described as being destructive to backyard fruit trees, gardens, and other property, and although infrequent, they also were seen as a physical threat to people and their pets. Emerging information was pointing to the monkeys' deleterious effect on the islands' ecosystems and biodiversity. Although the attitudes towards the monkeys seemed to favour the negative among many groups, there were some stakeholders that viewed this species as beneficial. Government and tourism operators capitalized on the iconic value of the monkeys to attract and entertain tourists (see Figure 22.1). Local biomedical professionals and university academics valued the monkeys as contributors to research. Some residents consumed monkey meat and viewed them as a significant local protein source, and still other groups considered the existent ecological niche of this well-established species and advocated for their welfare and protection.

FIGURE 22.1 Local St. Kitts tourism operators known as "monkey men" charge tourists for a photo with the captivating juvenile African green monkeys. (Photo by C. Gallagher.)

SUMMARY

The "monkey problem" has transcended scales of time and place, underscoring that the interconnections of people and animals in a closely shared environment are dynamic. There were multiple stakeholders with strong opposing values and ideas on how and what to do with the islands' green monkey population, and it was causing social conflict between groups that wanted to promote and protect them versus groups that wanted to control or cull them. Further complicating the situation is the Federation's inextricable reliance on the tourism industry, so the worldview of this problem from an animal welfare point was becoming profoundly influential.

Both the people affected by the monkeys and the local government seemed to be in search of a single intervention to solve this highly variable and contentious problem. Like many problems at the interface of animals, societies, and environments, one intervention is unlikely to remedy all negative harms while protecting all benefits. The "monkey problem" is in actuality a human-monkey system problem that calls for systems-based solutions.

TO WHERE FROM HERE?

The systems mapping of the socio-ecological harms from the St. Kitts "monkey problem" provided a representation that was accessible and understandable to all stakeholders. It is allowing for a wider conversation on needs and opportunities for action, helping people see how their preferred options for action might impact other citizens or species. It is providing a common language for managers and researchers to assess what is scientifically and economically feasible with a lens of what people might accept or feel empowered to do. It is helping establish what people perceive to be important and who they believe are able to act. All of these understandings can help select options for feasible actions that can make incremental improvements by encouraging people to change attitudes and behaviours. This is consistent with theories of change introduced in Chapter 9 and illustrated in Chapter 20. Perhaps as important, the harm reduction process we have initiated is helping to find common values and shared concerns about smaller problems around which collaborative actions can be proposed, to help build trust between stakeholders and ready them to work together on the more challenging aspects of the problem.

The St. Kitts "monkey problem" is a good example of challenging One Health problems that exist all over the globe. Problems that traverse human, animal, and ecosystem dominions are exceedingly complex and involve multiple stakeholders with diverging perspectives and values. There is often no clear path forward for which groups seeking answers and action can reach consensus to work cooperatively towards solutions. To aggravate conditions, these types of problems are fraught with high-powered dynamics, surprises, and scientific uncertainties. Harm reduction is showing the potential to advance collective action in stagnant One Health problems like the St. Kitts "monkey problem." Harm reduction has been tested and has achieved success in the public health realm due to its ability to be inclusive, participatory, and pragmatic. Further research and experience continues to test the harm reduction model as an approach to address problems at the human-animal interface.

REFERENCES

Anon. "Harm Reduction: A British Columbia Community Guide." British Columbia Ministry of Health, 2005. https://www.health.gov.bc.ca/library/publications/year/2005/hrcommunityguide.pdf

Denham, Woodrow W. "West Indian Green Monkeys: Problems in Historical Biogeography." *Contributions to Primatology*, 24 (1987).

Dore, Kerry M. "Ethnoprimatology Without Conservation: The Political Ecology of Farmer-Green Monkey (*Chlorocebus Sabaeus*) Relations in St. Kitts, West Indies." *International Journal of Primatology*, 39, no. 5 (2018): 918–944.

Dore, Kerry M., Andrea R. Eller, and Jackie L. Eller. "Identity Construction and Symbolic Association in Farmer-Vervet Monkey (*Chlorocebus aethiops sabaeus*) Interconnections in St. Kitts." *Folia Primatologica*, 89, no. 1 (2018): 63–80.

Hawk, Mary, Robert W.S. Coulter, James E. Egan, Stuart Fisk, M. Reuel Friedman, Monique Tula, and Suzanne Kinsky. "Harm Reduction Principles for Healthcare Settings." *Harm Reduction Journal*, 14, no. 1 (2017): 70.

Inciardi, James A., and Lana D. Harrison (eds.). *Harm Reduction: National and International Perspectives* (Thousand Oaks: Sage Publications, 2000), 243.

Jalloh, Chelsea, Shohan Illsley, John Wylie, Paula Migliardi, Ethan West, Debbie Stewart, and Javier Mignone. "What Goes Around: The Process of Building a Community-Based Harm Reduction Research Project." *Harm Reduction Journal*, 14, no. 1 (2017): 73.

Mariner, Jeffrey C., Roger Paskin, Food and Agriculture Organization of the United Nations (FAO), and EMPRES. "Manual on Participatory Epidemiology: Methods for The Collection of Action-Oriented Epidemiological Intelligence." *Animal Health*, 10 (2000).

McGuire, M.T. "The St. Kitts Vervet (*Cercopithecus aethiops*)." *Journal of Medical Primatology*, 3, no. 5 (1974): 285–297.

23 Socio-Economic and Behavioural Dimensions of Antimicrobial Use and Resistance in Animals

E. Jane Parmley, Irene Lambraki, Shannon E. Majowicz, and Carolee Carson

CONTENTS

WHAT IS ANTIMICROBIAL RESISTANCE AND WHY IS IT IMPORTANT?

Antimicrobial resistance (AMR) (Box 23.1) is a complex and growing health issue that is threatening people and animals around the world through increased morbidity and mortality. Canada (Council of Canadian Academies, 2019), the United States (CDC, 2019), and the United Kingdom (O'Neill, 2016) all report on and predict large direct and indirect economic and social costs associated with resistance, in addition to the health costs. Antimicrobial use (AMU) in humans (O'Neill, 2016), animals (Van Boeckel et al., 2015), crops (Finley et al., 2013), and elsewhere is widely recognized as the main driver of resistance (Pinto Ferreira, 2017).

AMR occurs when bacteria and other microorganisms can replicate in the presence of antimicrobials at levels that normally suppresses their growth or kills them. Resistance can occur due to naturally occurring characteristics of bacteria or can be acquired through genetic mutations or via transfer of genes (Aarestrup

BOX 23.1 ANTIMICROBIALS VERSUS ANTIBIOTICS AND ANTIMICROBIAL RESISTANCE VERSUS ANTIBIOTIC RESISTANCE

Antimicrobials refer to all compounds that kill or inhibit the growth of microorganisms. These microbes include viruses, bacteria, protozoa, fungi, and parasites. The use of any of these compounds can be selected for resistance.

Antibiotics are compounds that kill or inhibit the growth of bacteria.

In general, throughout this chapter, we have used the terms antimicrobial and antimicrobial resistance. However, unless otherwise stated, we are speaking specifically about antibiotics and antibiotic resistance.

et al., 2008; Boerlin and Reid-Smith, 2008). Not all resistant bacteria are harmful, but non-pathogenic resistant bacteria can serve as reservoirs of resistance genes (Boerlin and Reid-Smith, 2008). While any use of antimicrobials can select for resistance, inappropriate use, such as using antibiotics to treat viral infections or selecting the wrong antimicrobial for the specific pathogen causing the illness, is considered the major driver (Shallcross and Davies, 2014; Castro-Sánchez et al., 2016; Dar et al., 2016; Holmes et al., 2016).

Antimicrobials are one of the most successful medical advances of the 20th century. Diseases that once annually killed thousands of people and animals and made even more sick were dramatically reduced with the introduction of antimicrobials. By reducing death and preventing and controlling bacterial infections, antimicrobials allowed for development and routine implementation of many life-saving medical advances (e.g. cancer chemotherapy) and procedures to improve quality of life (e.g. joint replacements) (Shallcross and Davies, 2014; Laxminarayan et al., 2013). AMU in agriculture improved feed efficiency in animals and enabled food to be produced more efficiently and profitably (Durso and Cook, 2014; Grace, 2015).

The benefits of widespread AMU come with a cost. As our use of and dependence on antimicrobials increased, bacteria and other microbes evolved to become resistant, and available drugs became ineffective. AMU practices have prioritized short-term individual human and animal health improvements over long-term population, community, and ecosystem health.

HOW DOES AMR AFFECT HUMAN, ANIMAL, AND ENVIRONMENTAL HEALTH?

AMR has been recognized since the first antimicrobials were identified and used in medicine in the early 1900s. The predicted impacts of unrestricted use on human and animal morbidity and mortality have been described for decades (see

Fleming, 1945), but the full breadth of health and non-health consequences to humans, animals, nature, and society is only now starting to be characterized.

AMR is a global health crisis (Toner et al., 2015; WHO, 2015a). Its impacts on health and well-being are far-reaching and hard to predict. The impacts already identified include increased duration of disease, increased mortality associated with infections, and increased cost of health care delivery because of the greater severity and duration of illness (Cassini et al., 2019; Council of Canadian Academies, 2019; CDC, 2019). According to WHO (2015b), over 400,000 people die annually of foodborne diseases and hundreds of millions get sick. If these common bacteria and other microorganisms become resistant, many more will get sick and potentially die (WHO, 2017). O'Neill (2016) predicted that 10 million people will die annually from AMR by 2050. Additional indirect impacts of AMR on humans may include reduced access to life-saving treatments and surgeries which depend on antimicrobials to ward off secondary infections (Shallcross and Davies, 2014).

Farm and companion animal health will be similarly compromised. Although animals are affected in a comparable manner, estimates of the impacts of AMR (e.g. morbidity and mortality) in animal populations are rarely available (Robinson et al., 2016).

The impacts of rising levels of resistance on food production systems and interventions to reduce AMU threatens global food security (Grace, 2015; FAO, 2016). In Canada, Europe, and the United States, most antimicrobials (by weight) are used in animal agriculture (PHAC, 2018; EFSA, 2017; O'Neill, 2014). It is likely that agricultural AMU exceeds human use globally. The total amount (kg) of antimicrobials used in agriculture is expected to increase over the coming decade, especially in low- and middle-income countries (LMIC) (Schar et al., 2018; van Boeckel et al., 2015). Without access to effective antimicrobials to prevent, control, and treat infectious disease, livestock production may no longer be feasible using current large-scale, conventional production methods. Changing agricultural production systems and methods could be costly and could potentially destabilize food prices and agricultural communities. Inability of some producers to continue in business without the same level of access and use of antimicrobials may negatively affect the productivity and profitability of the agricultural sector and have a particularly inequitable effect on rural communities.

Although antimicrobials are rarely used in natural environments, these areas are also affected by AMR. Antimicrobial residues and resistant microbes have been detected in rural and remote natural environments – in soils, water, and wild animals (Aga et al., 2018; FAO, 2018; Greig et al., 2015). These residues and resistant microbes may come from run-off/spillage from farms, septic systems, hospital and lab waste discharge sites, and more. Aquaculture also presents a threat to antimicrobial distribution and selection of resistant organisms in the aquatic environments. International trade in livestock and animal products as well as global tourism is also increasing resistant organisms' geographic distribution.

The health challenges presented by growing levels of AMR will not be distributed equally around the world. In LMIC, the issue of AMR has more to do with access to appropriate diagnostics and drugs. In these regions of the world, physicians and veterinarians have fewer available options, and animal owners have less disposable income to afford effective and appropriate treatments.

WHY IS AMR SUCH A MESSY AND COMPLEX ISSUE?

The decision to use antimicrobials is affected by many different factors. AMU in humans and companion animals is driven by the clinical context, such as treating versus preventing infectious disease. It is also influenced by socio-economic factors, including culture, behaviour and expectations of patients/clients, lack of social infrastructure (e.g. more precarious employment status with no sick days), lack of education about the risks associated with AMR, and how AMR risks are perceived compared to other health threats.

Similar factors influence farmer and veterinarian decisions to use or not to use antimicrobials. Farmer demand for antimicrobials is likely also affected by the type(s) of animals they are raising, the perceived disease threats to those animals, past experience with infectious disease in the herd or the flock, season, cost of production, and market value of the animal product. Current North American agricultural practices rely on antimicrobials to limit disease and maximize growth targets in the animals being farmed. Removal of antimicrobials will require a transformation of how we produce and raise livestock and/or the development of new antimicrobial alternatives (Lhermie et al., 2019).

The multitude of drivers that influence and are influenced by AMU and AMR makes AMR a very complex health challenge with multiple dependencies throughout the system. This makes it very hard to predict the short-, medium-, and long-term effects of new policies and regulations. Several teams have worked to describe the breadth and complexity of the AMR problem (Majowicz et al., 2018; Department of Health, 2014), including non-disease factors such as environmental sources, genetics, economics, food security, trade, agriculture sustainability, or other indirect drivers and consequences. Comprehensive views of AMR as an emergent property of a complex adaptive system (Jayasinghe, 2011) can enable researchers to integrate social, biological, and ecological perspectives and provide a framework for development and implementation of effective and sustainable interventions that could reduce the human, animal, and environmental health burden of AMR.

Because of the multitude of drivers of AMU and AMR, the issue has been framed as a wicked problem (Xiang, 2013; Hutchinson, 2017) and even as a super-wicked problem (Littmann, 2014). Wicked problems are those that cannot be fully characterized nor eliminated or solved. Intervention effects are difficult to predict; all interventions will have unexpected consequences across the system and no intervention can be reliably transferred to a new setting. A problem becomes super-wicked when time is running out and where there are multiple dependencies across the system. All these characteristics apply to the growing challenge of AMR.

THE NEED FOR NEW PERSPECTIVES: APPLYING HEALTH PROMOTION AND HARM REDUCTION LESSONS TO AMU AND AMR

The World Health Organization's (WHO) Ottawa Charter for Health Promotion (1986) defines health promotion as "the process of enabling people to increase control over, and to improve, their health" (for more information on health promotion see Chapter 2). It recognizes that health is affected by a wide array of determinants and the complex interactions between individual, social, economic, and environmental factors that shape them. Health promotion engages and empowers individuals, groups, communities, and institutions to increase control over the determinants of health and make healthier choices easier choices (WHO, 1997; WHO, 1986). Health promotion has tackled a wide range of issues, such as child and maternal health, mental health, obesity, and tobacco control. In addressing these complex issues, a wide range of activities have been employed. The five key health promotion action areas (Box 23.2) played an important role in driving down population-level tobacco use rates in many countries. Specific examples include preventing tobacco use, protecting people from second-hand smoke exposure, assisting people who smoke to quit, and countering tobacco industry tactics (U.S. Department of Health and Human Services, 2014; Ahluwalia et al., 2019).

When implemented together at multiple levels, these five action areas represent a comprehensive approach that, over time, alters the environment, changes social norms, and helps build individual and community capacity to empower and improve knowledge and skills to make healthier choices and reduce illness (U.S. Department of Health and Human Services, 2014). These changes are made

BOX 23.2 FIVE KEY HEALTH PROMOTION ACTION AREAS (WHO, 1986)

1. *Building healthy public policies*
   Legislations, regulations, and tax changes coordinated across policy departments, not just health
2. *Creating environments that support health*
   Environments encompass all the places where people (and animals) live, work, and play
3. *Strengthening community action to improve health*
   Forming multisector inter organizational arrangements (e.g. coalitions) to coordinate and drive change
4. *Developing personal skills*
   Equipping people with the knowledge and skills to reduce health risks
5. *Reorienting health care services*
   Moving beyond treating illness and disease towards imbedding health promotion activities into care

possible through multisector coordinated actions from local to broader levels (Potvin and Jones, 2011). Coordination is enabled via bringing sectors together to learn from surveillance and monitoring, research, and practice to better understand the complexities of a problem and what works for whom and under what conditions (WHO, 1997). This coordination requires shared vision, leadership, and investment across the human, animal, and environmental health sectors at all levels to support the cause and build trust between individuals and institutions (Seaton et al., 2018).

Strategic priority areas for AMR action focus on surveillance, stewardship, infection prevention and control, and research and innovation (Council of Canadian Academies, 2019; HM Government, 2019; U.S. White House Office, 2015; Government of Canada, 2017; WHO, 2015a). Further health promotion-based efforts need to focus on developing multi-pronged and collaborative approaches that enable all individuals, including physicians, veterinarians, farmers, patients, and others, to make informed decisions about AMU, advocate for health through associated changes in policy, and mediate dialogue between stakeholders with different perspectives (WHO, 1986).

Regardless of the success of current and future interventions, antimicrobials will continue to be used. In recognition of this use, a harm reduction approach offers alternative actions that can be applied to reduce AMR impacts. Historically, harm reduction has been used to minimize the negative health, social and legal impacts associated with illegal drug use, and other chronic health conditions (Hawk et al., 2017) (see Chapter 6 for more on harm reduction), but it could be applied to guide AMU with the fewest negative impacts. This will require research to determine the most appropriate route, dose, and duration of AMU to minimize AMR emergence and spread; exploration of antimicrobial alternatives that will prevent and control infection without contributing to resistance; as well as new innovations to support rapid and reliable diagnostic tests to ensure that the most appropriate antimicrobials are prescribed. This knowledge and availability of tools will help physicians, veterinarians, farmers, and patients make more informed decisions about AMU.

Beyond human and animal patient harm reduction, systemic and economic support is needed to change current health care practices, mainstream food production practices, and on-farm management. Investments in farm infrastructure have been made over decades while antimicrobials have been used, and these will be hard and potentially costly systems to replace. Education to increase awareness of prescribers, users, producers, and consumers about the threats posed by AMR and their contribution to the problem is needed but will not be enough. Further exploration of antimicrobial alternatives and alternative production and management practices are needed to support those industries and sectors that may take longer than others to shift to lower AMU. Like illegal substance use, dependence on AMU is not uniform across the agricultural sector. Lessons from harm reduction can help support industries and individuals that are struggling to change their AMU practices while providing incentives to help shift the industry to a new reality.

HOW TO WORK TOGETHER TO ADDRESS THIS GROWING HEALTH CHALLENGE

Growing international attention and priority placed on AMR now articulate the need for collaborative action between multiple sectors (McEwen and Collignon, 2018; Robinson et al., 2016). Intersectoral collaboration can be difficult but is possible when political will and public pressure exist, when places to meet and discuss are available, and when there is a shared understanding of the problem (Wernli et al., 2017).

With complex health issues such as AMR, there is a need to balance critical outcomes needed today with solutions that are more challenging to initiate but are more likely to have a greater influence on the long-term outcomes. By engaging multiple stakeholders, we can start new conversations about roles and responsibilities, with a key goal of fostering trust among the players. Transparent discussions will provide a launching point to begin to work together on sustainable solutions with shared responsibility. Traditional scientific evidence about how AMU affects AMR is mounting and what is now needed is a process to share knowledge, recognize and include different perspectives and motivations, develop and support relationships, and appreciate how the different types of evidence, generated at smaller points within the system, relate and drive each other across the system as a whole.

CONCLUSIONS

AMR is a wicked health emergency that requires all our knowledge, experience, and resources to address sustainably and effectively. Recommendations in national action plans and reports for surveillance, infection prevention and control, stewardship, and research and innovation can all contribute to a better understanding of AMR, more effective interventions, better infectious disease prevention, changes in farm and animal management, and control of resistant organism spread. However, we need to go beyond these recommendations to incorporate health promotion and harm reduction principles that have proven effective for other formidable public health challenges (e.g. tobacco control and substance use). The AMR emergency is an exemplar of the value to One Health in action of the perspectives, methods, and knowledge presented in this book.

More and better implementation science will enable us to improve our understanding of how values and beliefs determine success of AMR interventions across the system (Wernli et al., 2020; World Bank, 2019). We will need to create spaces for ongoing dialogue at all levels of government as well as in non-government circles to successfully respond to AMR, to build healthier and more resilient populations and communities. We need to build on our shared experiences and advocate for release of emergency funds to bring people and organizations together for ongoing dialogue. Resources should create structured spaces for exchange of ideas and discussions between individuals, communities, organizations, and countries.

REFERENCES

Aarestrup, Frank M., Henrik C. Wegener, and Peter Collignon. "Resistance in Bacteria of the Food Chain: Epidemiology and Control Strategies." *Expert Review of Anti-Infective Therapy* 6, no. 5 (2008): 733–750.

Aga, Diana, Julian Davies, Sumanth Gandra, Barbara Kasprzyk-Hordern, Joakim Larsson, Jean McLain, Andrew Singer et al. "Initiatives for Addressing Antimicrobial Resistance in the Environment: Current Situation and Challenges." *Prepared for Wellcome, U.S. Centers for Disease Control and Prevention, and UK Science and Innovation Network* (2018): 94.

Ahluwalia, Indu B., René A. Arrazola, Luhua Zhao, Jing Shi, Anna Dean, Edward Rainey, Krishna Palipudi et al. "Tobacco Use and Tobacco-Related Behaviors: 11 Countries, 2008–2017." *Morbidity and Mortality Weekly Report*, 68, no. 41 (2019): 928.

Boerlin, Patrick, and Richard J. Reid-Smith. "Antimicrobial Resistance: Its Emergence and Transmission." *Animal Health Research Reviews*, 9, no. 2 (2008): 115–126.

Cassini, Alessandro, Liselotte Diaz Högberg, Diamantis Plachouras, Annalisa Quattrocchi, Ana Hoxha, Gunnar Skov Simonsen, Mélanie Colomb-Cotinat et al. "Attributable Deaths and Disability-Adjusted Life-Years Caused by Infections with Antibiotic-Resistant Bacteria in the EU and the European Economic Area in 2015: A Population-Level Modelling Analysis." *Lancet Infectious Diseases*, 19, no. 1 (2019): 56–66.

Castro-Sánchez, Enrique, Luke S.P. Moore, Fran Husson, and Alison H. Holmes. "What Are the Factors Driving Antimicrobial Resistance? Perspectives from a Public Event in London, England." *BMC Infectious Diseases*, 16, no. 1 (2016): 465.

Centers for Disease Control and Prevention. "Antibiotic Resistance Threats in the United States, 2019." U.S. Department of Health and Human Services, CDC (2019): 140.

Council of Canadian Academies. "When Antibiotics Fail. The Expert Panel on the Potential Socio-Economic Impacts of Antimicrobial Resistance in Canada." *Council of Canadian Academies* (2019): 237.

Dar, Osman A., Rumina Hasan, Jørgen Schlundt, Stephan Harbarth, Grazia Caleo, Fazal K. Dar, Jasper Littmann et al. "Exploring the Evidence Base for National and Regional Policy Interventions to Combat Resistance." *Lancet*, 387, no. 10015 (2016): 285–295.

Department of Health. "Antimicrobial Resistance (AMR) Systems Map." *United Kingdom Department of Health* (2014): 10.

Durso, Lisa M., and Kimberly L. Cook. "Impacts of Antibiotic Use in Agriculture: What Are the Benefits and Risks?" *Current Opinion in Microbiology*, 19 (2014): 37–44.

EFSA (European Food Safety Authority). "ECDC/EFSA/EMA Second Joint Report on the Integrated Analysis of the Consumption of Antimicrobial Agents and Occurrence of Antimicrobial Resistance in Bacteria from Humans and Food-Producing Animals." *EFSA Journal* 15 (2017): 4872.

FAO (Food and Agriculture Organization of the United Nations). "Drivers, Dynamics and Epidemiology of Antimicrobial Resistance in Animal Production." *FAO* (2016): 59.

FAO (Food and Agriculture Organization of the United Nations). "Antimicrobial Resistance in the Environment Summary Report of an FAO Meeting of Experts." *FAO Antimicrobial Resistance Working Group* (2018): 3.

Finley, Rita L., Peter Collignon, D.G. Joakim Larsson, Scott A. McEwen, Xian-Zhi Li, William H. Gaze, Richard Reid-Smith et al. "The Scourge of Antibiotic Resistance: The Important Role of the Environment." *Clinical Infectious Diseases* 57, no. 5 (2013): 704–710.

Fleming, Sir Alexander. "Penicillin" *Nobel Lecture*, 1945. https://www.nobelprize.org/prizes/medicine/1945/fleming/lecture/

Government of Canada. "Tackling Antimicrobial Resistance and Antimicrobial Use: A Pan-Canadian Framework for Action." *Government of Canada, Minister of Health* (2017): 33.

Grace, Delia. "Review of Evidence on Antimicrobial Resistance and Animal Agriculture in Developing Countries." *United Kingdom (UK): Evidence on Demand* (2015).

Greig, J., A. Rajić, I. Young, M. Mascarenhas, L. Waddell, and J. LeJeune. "A Scoping Review of the Role of Wildlife in the Transmission of Bacterial Pathogens and Antimicrobial Resistance to the Food Chain." *Zoonoses and Public Health* 62, no. 4 (2015): 269–284.

Hawk, Mary, Robert W.S. Coulter, James E. Egan, Stuart Fisk, M. Reuel Friedman, Monique Tula, and Suzanne Kinsky. "Harm Reduction Principles for Healthcare Settings." *Harm Reduction Journal*, 14, no. 1 (2017): 70.

HM Government. "Tackling Antimicrobial Resistance 2019–2024. The UK's Five-Year National Action Plan." *Government of the United Kingdom* (2019): 98.

Holmes, Alison H., Luke S.P. Moore, Arnfinn Sundsfjord, Martin Steinbakk, Sadie Regmi, Abhilasha Karkey, Philippe J. Guerin et al. "Understanding the Mechanisms and Drivers of Antimicrobial Resistance." *Lancet* 387, no. 10014 (2016): 176–187.

Hutchinson, Eleanor. "Governing Antimicrobial Resistance: Wickedness, Competing Interpretations and the Quest for Global Norms." *Global Health Centre Working Paper*, 14 (2017): 28.

Jayasinghe, Saroj. "Conceptualising Population Health: From Mechanistic Thinking to Complexity Science." *Emerging Themes in Epidemiology*, 8, no. 1 (2011): 2.

Laxminarayan, Ramanan, Adriano Duse, Chand Wattal, Anita K.M. Zaidi, Heiman F.L. Wertheim, Nithima Sumpradit, Erika Vlieghe et al. "Antibiotic Resistance: The Need for Global Solutions." *Lancet Infectious Diseases*, 13, no. 12 (2013): 1057–1098.

Lhermie, Guillaume, Didier Wernlii, Peter Søgaard Jørgensen, Donald Kenkel, Loren William Tauer, and Yrjo Tapio Gröhn. "Global Resistance to Antimicrobials and Their Sustainable Use in Agriculture." *Lancet Planetary Health,* 3, no. 3 (2019): e109–e110.

Littmann, Jasper R. "Antimicrobial Resistance and Distributive Justice." PhD dissertation, University College London, 2014.

Majowicz, Shannon E., E. Jane Parmley, Carolee Carson, and Katarina Pintar. "Identifying Non-Traditional Stakeholders with Whom to Engage, When Mitigating Antimicrobial Resistance in Foodborne Pathogens (Canada)." *BMC Research Notes* 11, no. 1 (2018): 170.

McEwen, Scott A., and Peter J. Collignon. "Antimicrobial Resistance: A One Health Perspective." *Antimicrobial Resistance in Bacteria from Livestock and Companion Animals* (2018): 521–547.

O'Neill, Jim. "Antimicrobial Resistance: Tackling a Crisis for the Health and Wealth of Nations." *The Review on Antimicrobial Resistance*, 20 (2014): 1–16.

O'Neill, Jim. "Tackling Drug-Resistant Infections Globally: Final Report and Recommendations. The Review on Antimicrobial Resistance." *Report Commissioned by the Government of The United Kingdom* (2016): 81.

PHAC (Public Health Agency of Canada). "Canadian Antimicrobial Resistance Surveillance System Update 2018." *Public Health Agency of Canada* (2018): 60.

Pinto Ferreira, Jorge. "Why Antibiotic Use Data in Animals Needs to Be Collected and How This Can Be Facilitated." *Frontiers in Veterinary Science*, 4 (2017): 213.

Potvin, Louise, and Catherine M. Jones. "Twenty-Five Years After the Ottawa Charter: The Critical Role of Health Promotion for Public Health." *Canadian Journal of Public Health*, 102, no. 4 (2011): 244–248.

Robinson, Timothy P., D.P. Bu, Juan Carrique-Mas, Eric M. Fèvre, Marius Gilbert, Delia Grace, Simon I. Hay et al. "Antibiotic Resistance Is the Quintessential One Health Issue." *Transactions of the Royal Society of Tropical Medicine and Hygiene*, 110, no. 7 (2016): 377–380.

Schar, Daniel, Angkana Sommanustweechai, Ramanan Laxminarayan, and Viroj Tangcharoensathien. "Surveillance of Antimicrobial Consumption in Animal Production Sectors of Low-and Middle-Income Countries: Optimizing Use and Addressing Antimicrobial Resistance." *PLoS Medicine*, 15, no. 3 (2018).

Seaton, Cherisse L., Nikolai Holm, Joan L. Bottorff, Margaret Jones-Bricker, Sally Errey, Cristina M. Caperchione, Sonia Lamont et al. "Factors that Impact the Success of Interorganizational Health Promotion Collaborations: A Scoping Review." *American Journal of Health Promotion*, 32, no. 4 (2018): 1095–1109.

Shallcross, Laura J., and Dame Sally C. Davies. "Antibiotic Overuse: A Key Driver of Antimicrobial Resistance." *British Journal of General Practice* (2014): 604–605.

Toner, Eric, Amesh Adalja, Gigi Kwik Gronvall, Anita Cicero, and Thomas V. Inglesby. "Antimicrobial Resistance Is a Global Health Emergency." *Health Security*, 13, no. 3 (2015): 153–155.

U.S. Department of Health and Human Services. "The Health Consequences of Smoking: 50 Years of Progress. A Report of the Surgeon General." U.S. Department of Health and Human Services, Centers for Disease Control and Prevention, National Center for Chronic Disease Prevention and Health Promotion, Office on Smoking and Health, 2014.

U.S. White House Office. "National Action Plan for Combating Antibiotic-Resistant Bacteria." United States Government (2015): 62. https://www.cdc.gov/drugresistance/pdf/national_action_plan_for_combating_antibotic-resistant_bacteria.pdf

Van Boeckel, Thomas P., Charles Brower, Marius Gilbert, Bryan T. Grenfell, Simon A. Levin, Timothy P. Robinson, Aude Teillant et al. "Global Trends in Antimicrobial Use in Food Animals." *Proceedings of the National Academy of Sciences of the United States of America*, 112, no. 18 (2015): 5649–5654.

Wernli, Didier, Peter S Jørgensen, E Jane Parmley, Max Troell, Shannon Majowicz, Stephan Harbarth, Anaïs Léger, et al. "Evidence for action: a One Health learning platform on interventions to tackle antimicrobial resistance" Lancet Infectious Diseases. Published online August 24, 2020 https://doi.org/10.1016/S1473-3099(20)30392-3.

Wernli, Didier, Peter S. Jørgensen, Chantal M. Morel, Scott Carroll, Stephan Harbarth, Nicolas Levrat, and Didier Pittet. "Mapping Global Policy Discourse on Antimicrobial Resistance." *BMJ Global Health*, 2, no. 2 (2017): e000378.

WHO. "The Ottawa Charter for Health Promotion." World Health Organization, 1986. https://www.who.int/healthpromotion/conferences/previous/ottawa/en/ (accessed March 2020).

WHO. "The Jakarta Declaration: On Leading Health Promotion into the 21st Century." World Health Organization, 1997. https://apps.who.int/iris/handle/10665/63698 (accessed June 2020).

WHO. "Global Action Plan on Antimicrobial Resistance." World Health Organization, 2015a. https://apps.who.int/iris/handle/10665/193736 (accessed June 2020).

WHO. "WHO Estimates of the Global Burden of Foodborne Diseases: Foodborne Disease Burden Epidemiology Reference Group 2007–2015." *World Health Organization* (2015b): 255.

WHO. "Antimicrobial Resistance in the Food Chain." World Health Organization, 2017. https://www.who.int/foodsafety/areas_work/antimicrobial-resistance/amrfoodchain/en/ (accessed June 2020).

World Bank. "Pulling Together to Beat Superbugs: Knowledge and Implementation Gaps in Addressing Antimicrobial Resistance." *The World Bank Understanding Policy* (World Bank, 2019). https://www.worldbank.org/en/topic/agriculture/publication/pulling-together-to-beat-superbugs-knowledge-and-implementation-gaps-in-addressing-antimicrobial-resistance

Xiang, Wei-Ning. "Working with Wicked Problems in Socio-Ecological Systems: Awareness, Acceptance, and Adaptation." *Landscape and Urban Planning*, 110 (2013): 1–4.

Index

Note: Locators in *italics* represent figures and **bold** indicate tables in the text.

D

E

F

J

K

L

Q

R

S

T

Y

Z

For Product Safety Concerns and Information please contact our EU
representative GPSR@taylorandfrancis.com
Taylor & Francis Verlag GmbH, Kaufingerstraße 24, 80331 München, Germany

www.ingramcontent.com/pod-product-compliance
Ingram Content Group UK Ltd.
Pitfield, Milton Keynes, MK11 3LW, UK
UKHW022115190926
475702UK00013B/1459
9780367336226